Mathematics
for
New Technologies

Don Hutchison
Clackamas Community College

Mark Yannotta
Clackamas Community College

Boston San Francisco New York
London Toronto Sydney Tokyo Singapore Madrid
Mexico City Munich Paris Cape Town Hong Kong Montreal

Publisher: Greg Tobin
Editor in Chief: Maureen O'Connor
Acquisitions Editor: Jennifer Crum
Project Editor: Suzanne Alley
Assistant Editor: Jolene Lehr
Managing Editor: Ron Hampton
Production Supervisor: Sheila Spinney
Marketing Manager: Dona Kenly
Marketing Coordinator: Lindsay Skay
Software Development: TestGen, Gail Light
Prepress Buyer: Caroline Fell
Manufacturing Buyer: Hugh Crawford
Text and Cover Design: Dennis Schaefer
Cover Illustration: PhotoDisc, DigitalVision

Library of Congress Cataloging-in-Publication Data

Hutchison, Donald, 1948–
Mathematics for new technologies / Don Hutchison, Mark Yannotta.—1st ed.
p. cm.
Includes index.
ISBN 0-201-77137-3
1. Mathematics. I. Yannotta, Mark. II. Title
QA39.3.H88 2004
510—dc21 2002038342

3 4 5 6 7 8 9–CRW–07 06 05 04

Dedication

To Adriana, who makes the future look so much brighter.

Don Hutchison

To my wife Amy, who is the most important constant in my life, and to Cayley, who makes me smile every day.

Mark Yannotta

Contents

Preface

In the fall of 1999, I was asked to attend a meeting of the curriculum development committee for the computer technician and support programs at Clackamas Community College. As I had originally been hired to teach both mathematics and computer science, I was considered the most likely candidate to answer the question of the day: Why are our computer tech students so ill-prepared to handle the modest mathematical components of our program? The committee hoped that I could point to an existing course that would meet the needs of their students and their program. When I asked them to list the topics with which students struggled, the following items emerged:

Word Problems and Units Analysis: Although every mathematics course covers word problems, they needed a course for which this was a primary emphasis. If these word problems included units relevant to the program, that would be a bonus.

Properties of Exponents: The group pointed to a common type of problem that few of their students were prepared to answer. Given a computer that has 2^{32} bytes of memory, how much memory remains if x of it is lost? The group also talked about the students' lack of ability to read simple scientific notation from a calculator or computer display.

Elementary Logic: The texts used in the computer science courses required in the computer tech program emphasized the application of logic to programming and circuitry. The assumption was that students had picked up the fundamentals of logic in an earlier class.

Conversion of Number Bases: Unlike the other topics mentioned, this one was assimilated by the students rather quickly in their computer science classes. However, it seemed to the faculty that students should have learned this in a mathematics class rather than a CS class.

We discussed technical math, intermediate algebra, and even college algebra as possible prerequisite classes for the programs. Unfortunately, the topics in those continuous-mathematics courses did not come close to meeting their needs. The only class that offered these topics was discrete math. But discrete math was woefully inadequate in its coverage of simple word problems. It also required college algebra as a prerequisite. This would have added a year to the computer tech program and disqualified many students who would have otherwise been successful in the program.

By the end of the meeting, I was convinced that a completely new course was in order. The next challenge was finding an adequate textbook.

Although several of us followed dozens of leads given to us by publishing representatives, none of the texts we looked at came close to meeting the program needs. By the spring, we had decided to create materials for a new course that would be offered the next fall. With the help of an internal college grant, we were able to work on the project. Over the next four terms, the examples, lecture notes, and worksheets became a preliminary textbook. That project eventually evolved into this text.

As we developed the material and taught the course, it became increasingly apparent that we had both an opportunity and an obligation to teach mathematics in context. Each term we added more applications but still felt a certain discontinuity (no pun intended) with the class. This perception was resolved when we added the chapters on circuitry and color representation. With this addition, the course evolved in a way in which few mathematics courses can. Almost all of the disparate parts introduced throughout the term were revisited in the final two weeks of the class.

Related to the applicability of the material is the joy of eliminating the bane of the mathematics teacher. We cannot overstate the satisfaction of having a good answer any time the question, "Why do we have to learn this stuff?" arises. Even better is the fact that the question doesn't reappear after the first week of class.

Since that initial meeting, we have worked with representatives of other programs, notably electronics and machine technology, to create a text that is relevant to programs in the realm of new technologies.

We have also developed other curriculum projects and assignments that are helpful in teaching this course. The actual assignment guidelines, along with suggested topics, can be found there. We also have provided PowerPoint slides of student projects that have been produced to meet some of the course requirements. We encourage you to look at the MyMathLab Web site for this text.

Features

- **Practice Problems** Computer math students require a large number of exercises in order to practice the material that they have just learned. These exercises actively engage students in the learning process and reinforce newly learned concepts and skills. Answers are provided just before each section's exercise set.

- **Bits of History** Positioned at the beginning of each section, the Bits of History feature lets students in on the background and the origins of the concepts they are currently learning. This feature helps students better understand the theories behind their actions.

- **Definition Boxes** Important definitions are boxed and highlighted throughout each chapter to emphasize their importance to students, and they are easy to find for review purposes.
- **Ample Opportunity for Review** At the end of each chapter is a comprehensive review section. Starting with a chapter Summary, students are given the main concepts of the chapter. The Glossary is a comprehensive listing of key terms from throughout the chapter. Review Exercises require students to solve problems without the help of section references. Additionally challenging problems are highlighted by a triangle symbol around the exercise number. A Self-Test follows, allowing students to test themselves on the materials they just learned in the chapter. Cumulative Review Exercises gather various types of exercises from the preceding chapters to help students remember and retain what they are learning throughout the course.

Available Supplements

For the Student

Student's Solutions Manual (ISBN 0-321-17333-3)

The *Student's Solutions Manual* provides detailed solutions to the odd-numbered section exercises and to all Review Exercises, Self-Tests, and Cumulative Review Exercises.

Addison-Wesley Math Tutor Center

The Addison-Wesley Math Tutor Center is staffed by qualified college mathematics instructors who tutor students on examples and exercises from the textbook. Tutoring is provided via toll-free telephone, toll-free fax, e-mail, and the Internet. Interactive Web-based technology allows students and tutors to view and listen to live instruction in real-time over the Internet. The Math Tutor Center is accessed through a registration number that can be packaged with a new textbook or purchased separately. (Note: MyMathLab students obtain access to the Math Tutor Center with their MyMathLab access code.)

MyMathLab

MyMathLab is a complete on-line course for Addison-Wesley mathematics textbooks that provides multimedia instruction correlated to the textbook content. MyMathLab is easily customizable to suit the needs of students and instructors and provides a comprehensive and efficient on-line course-management system. Instructors can create, copy, edit, assign, and track all tests for their course as well as track student's results. The print supplements are available on-line, side-by-side with the textbook. For more information, visit our Web site at www.mymathlab.com or contact your Addison-Wesley sales representative for a live demonstration.

For the Instructor

Instructor's Solutions Manual (ISBN 0-321-17331-7)

The *Instructor's Solutions Manual* provides complete solutions to all text exercises.

Printed Test Bank/Instructor's Resource Guide (ISBN 0-321-17330-9)

The Printed Test Bank of this manual contains one multiple-choice test, two free-response forms per chapter, and two final exams. The Instructor's Resource Guide portion contains a "starter kit" that will include a sample syllabus, course outline, description of a term project, and an example of a term project.

TestGen with QuizMaster (ISBN 0-321-16939-5)

TestGen enables instructors to build, edit, print, and administer tests using a computerized bank of questions developed to cover all the objectives of the text. Instructors can modify test bank questions or add new questions by using the built-in question editor, which allows users to create graphs, import graphics, insert math notation, and insert variable numbers or text. Tests can be printed or administered on-line via the Web or other network. TestGen comes packaged with QuizMaster, which allows students to take tests on a local area network. The software is available on a dual-platform Windows/Macintosh CD-ROM.

MathXL

Available on-line with an ID and password, this testing-and-tracking system allows an instructor to administer tests on-line and track students' grades. Instructors can create their own tests and assessments using Addison-Wesley's TestGen software or assign existing tests that correspond with the Chapter Tests in the book.

MyMathLab

MyMathLab is a complete on-line course for Addison-Wesley mathematics textbooks that provides multimedia instruction correlated to the textbook content. MyMathLab is easily customizable to suit the needs of students and instructors and provides a comprehensive and efficient on-line course-management system. Instructors can create, copy, edit, assign, and track all tests for their course as well as track student's results. The print supplements are available on-line, side-by-side with the textbook. For more information, visit our Web site at www.mymathlab.com or contact your Addison-Wesley sales representative for a live demonstration.

Acknowledgments

This project owes much thanks to the reviewers who had the insight to see the importance of creating a new course and this new text. Several of these reviewers have been developing similar courses at their institutions. We thank them for their many contributions.

Robin Albert, *North Dakota State University*
Teri Anderson, *Sheridan College, Gillette Campus*
John Arena, *Augusta Technical College*
Eddie Cheng, *Oakland University*
Richard Christenson, *North Hennepin Community College*
Anthony Duben, *Southeast Missouri State University*
Thomas English, *College of the Mainland*
Kenny Fister, *Murray State University*
William Fox, *Francis Marion University*
Michael L. Gargano, *Pace University*
Jun Ji, *Valdosta State University*
Nickolas Jovanovic, *University of Arkansas at Little Rock*
Maryann Justinger, *Erie Community College, South Campus*
Jim McCleery, *Skagit Valley College*
Yves Nievergelt, *Eastern Washington University*
Leela Rakesh, *Central Michigan University*
E. Rodney Scaggs, *Southern West Virginia Community and Technical College*
Donna Shumate, *Johnston Community College*
Steve Stevenson, *Clemson University*
Jeffrey Stuart, *Pacific Lutheran University*
William Thacker, *Winthrop University*
Srini Vasan, *Albuquerque Technical Vocational Institute*
Tseng-Yuan Tim Woo, *Durham Technical College*
Kevin Yokoyama, *College of the Redwoods*
Woodford Zachary, *Howard University*

Thanks are due to Emily Hui and Sue Schroeder for accuracy checking manuscript and to Patrick Riley and Vincent Koehler for accuracy checking page proofs.

Also deserving our thanks are the people at Addison-Wesley who saw the value of this project before the market existed. Jennifer Crum, Ron Hampton, Suzanne Alley, Sheila Spinney, Jolene Lehr, Dona Kenly, Lindsay Skay, Scott Silva, and Jim McLaughlin deserve particular thanks. They demonstrated insight and enthusiasm throughout the process.

About the Authors

Don Hutchison received his MST in mathematics with an emphasis on probability and statistics from Portland State University in 1982. Two years later, he was part of the first AMATYC summer institute focused on discrete mathematics. As a result of this background, Don has spent much of his teaching career integrating discrete topics into the community college curriculum. For five years he chaired the TiME (Technology in Mathematics Education) committee of AMATYC. He was also a member of the ACM committee that undertook the writing of computer curriculum for the two-year college. This text is a result of both the work described here and Don's belief that the mathematics topics we teach students in professional-technical programs must begin to evolve with the programs

Mark Yannotta received his MA in mathematics from the University of Missouri—Columbia in 1997 and began teaching at Clackamas Community College in 1998. Mark has copresented at both AMATYC and ICTCM and is currently a doctoral student in mathematics education at Portland State University. His research interests include teaching with technology, mathematics history, and mathematical representation theory. This text project represents the confluence of Mark's academic pursuits and his teaching experience.

To the Student

No class we have ever taught has brought us more enjoyment than the class for which this text was written. Our enjoyment did not stem from the particular topics of the course, although we admit to finding these topics among the most interesting in all of mathematics. It came from the enthusiasm of the students taking the class.

In the teaching of mathematics, we are frequently asked to justify why a certain class is a required part of the curriculum. It is interesting that this question arises frequently when discussing mathematics, but rarely when discussing music, philosophy, or literature—areas that are equally abstract. Many instructors try to respond to that inquiry with applications of the mathematics. Unfortunately, most applications that are accessible to beginning algebra students are so contrived that they weaken the argument.

The applications in this text are quite different from those contrived exercises in beginning algebra. Each of them gives the student a better understanding of some aspect of computer technology. An element of either the past, the present, or the future of technology is revealed in each application, providing the student with the background they will need in their computer science courses.

That relevance is key for a student who is trying to find the motivation to do the hard work of learning. It also provides something tangible to which an otherwise abstract idea can be related. This combination has allowed many students who had previously been only moderately successful in mathematics to excel in a class that contains many advanced topics.

The key to becoming newly successful in mathematics is to discard old prejudices and habits related to math courses and start afresh. Although you may have seen (and perhaps ignored) the following list of suggestions before, this is an opportunity to make them work for you.

1. Get an e-mail address and phone number from at least one other student in your class. This will be very useful throughout the term.
2. Read the section(s) to be covered before attending the lecture.
3. Write out a list of the topics you expect to hear about in the class discussion.

4. Use that list to add structure to your class notes. Use them as topic headings followed by notes or examples.
5. Ask questions when you are confused. There are undoubtedly other students with the same confusion.
6. Reread the section and start your homework while the ideas are still fresh in your mind.
7. Prepare for a test at least a week ahead of time.
8. Be curious.

Enjoy your learning!

Don Hutchison
Mark Yannotta

Computation

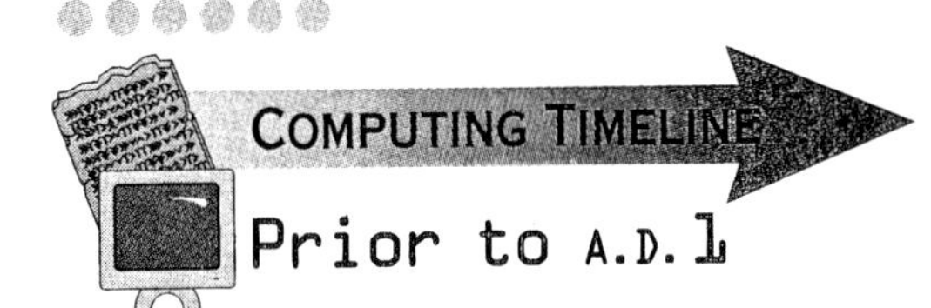

The concept of place value can be credited to the Babylonians, sometime before 2000 B.C. The Mayas also had an early place-value system, but there is no accurate way to estimate its date of origin. The Hindu-Arabic numeral system, from which we derive our base ten numbers, was invented by the Hindus and brought to Europe by the Arabs. The earliest preserved examples of the numerals are found in columns erected in India around 250 B.C.

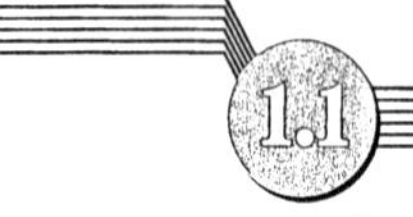

1.1 An Introduction

BITS of HISTORY

1900–1800 B.C. The first known positional number system is credited to the Babylonians. Although it would be several centuries before they had a symbol for zero, their sexigesimal, or base-sixty number system, is illustrated in modern society when measuring time.

Chapter 1 covers many topics that will be needed later in this text. We recommend that you closely examine and study every section. Even the fairly simple sections, such as this one, are designed to get you to see the relationship between numbers and operations more clearly than you might have in earlier mathematics classes. That different perspective will be very helpful both when we work with numbers in bases other than ten and when we operate on symbols other than numbers.

First, let us look closely at base ten. A number such as 537 can be expressed in base ten as

$537 = 5 \times 100 + 3 \times 10 + 7 \times 1$ (that is, 5 hundreds, 3 tens, and 7 ones).

A number such as 66 can be expressed as

$66 = 6 \times 10 + 6 \times 1.$

But what happens when we add these two numbers? This leads us to our first important idea from mathematics.

DEFINITION: Addition is an operation that takes two numbers and turns them into a single number.

The same can be said for subtraction, multiplication, or division. In this example we deal with adding 537 + 66.

$$\begin{array}{r} 537 \\ +\ 66 \\ \hline \end{array}$$

$$\begin{array}{r} {\scriptstyle 1} \\ 537 \\ +\ 66 \\ \hline 3 \end{array}$$

7 + 6 = 13, which is 3 ones and 1 ten. We carry the 1 ten over to the tens' place.

$$\begin{array}{r} {\scriptstyle 1\,1} \\ 537 \\ +\ 66 \\ \hline 03 \end{array}$$

3 tens + 6 tens + 1 ten = 10 tens, which is the same as no tens and 1 hundred.

Finally,

$$\begin{array}{r} {\scriptstyle 1\,1} \\ 537 \\ +\ 66 \\ \hline 603 \end{array}$$

500 + 100 = 600.

EXAMPLE 1 **How many tens are in each of the following sums?**

(a)
$$\begin{array}{r} 37 \\ +22 \\ \hline \end{array}$$
The sum has 5 tens.

(b)
$$\begin{array}{r} 58 \\ +27 \\ \hline \end{array}$$
The sum has 8 tens.

In this text, we offer you a chance to practice on problems similar to those you have just seen in the example. You can find all practice problem answers at the end of the section.

PRACTICE PROBLEMS 1 **How many tens are in each of the following sums?**

(a)
$$\begin{array}{r} 26 \\ +31 \\ \hline \end{array}$$

(b)
$$\begin{array}{r} 47 \\ +47 \\ \hline \end{array}$$

Many important measured quantities, such as temperature, altitude, and one's bank balance, can have positive or negative values. The next example gives you practice working with negative numbers. If your skills with negative numbers are rusty, we recommend that you work on those skills immediately; they will be important later in the text.

EXAMPLE 2 **Simplify each expression.**

(a) $-23 + (-15) - 12 = -38 - 12 = -50$

(b) $-6(35 - 7\times6) = -6(35 - 42) = -6(-7) = 42$

(c) $-3 - 3(5 - 7) = -3 - 3(-2) = -3 + 6 = 3$

PRACTICE PROBLEMS 2 **Simplify each expression.**

(a) $-15 + -33 - (-3)$

(b) $3(-3 + 4 \times 2) - (3 - 12)$

ANSWERS TO PRACTICE PROBLEMS

1. **(a)** 5 tens **(b)** 9 tens (carrying the 1)

2. **(a)** $-15 + (-33) - (-3) = -15 + (-33) + 3 = -48 + 3 = -45$

(b) $3(-3 + 4 \times 2) - (3 - 12) = 3(-3 + 8) - (-9) =$

$3(5) + 9 = 15 + 9 = 24$

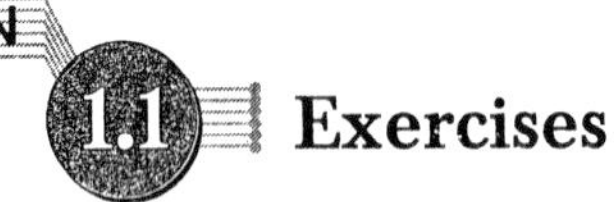

Exercises

How many *tens* and how many *ones* are in the following sums?

1. $\begin{array}{r} 86 \\ +\,12 \\ \hline \end{array}$

2. $\begin{array}{r} 49 \\ +\,30 \\ \hline \end{array}$

3. $\begin{array}{r} 27 \\ +\,94 \\ \hline \end{array}$

4. $\begin{array}{r} 53 \\ +\;\;8 \\ \hline \end{array}$

5. $\begin{array}{r} 85 \\ +\,76 \\ \hline \end{array}$

6. $\begin{array}{r} 98 \\ +\,96 \\ \hline \end{array}$

Simplify each expression.

7. $-8 + (-2)$

8. $-3 + (-14)$

9. $10 + (-6)$

10. $11 + (-14)$

11. $-2(5 + 12)$

12. $-3(-2 + 7)$

13. $(-5)(-4) + 7$

14. $(-3)(-6) - 5$

15. $5(2 + (-2)(7))$

Simplify each expression.

16. $2(-4 + 3 \times 2)$

17. $-8 + (-6) + 7 \times 4$

18. $-9 + (-5) + 2(-3)$

19. $5(2 + 3) - 3(4 \times 2)$

20. $-3(7 + (-1)) - 5(-2\times2)$

21. $6(-2) + 7(-8 + 5)$

22. $-11(4(-2) - 3\times7) - 6$

1.2 Exponents and Their Properties

BITS of HISTORY

600–500 B.C. The Greek mathematician Pythagoras and his followers are credited with formally studying numbers, their properties, and their relationship to society. Although references appear in several ancient cultures, even prior to Pythagoras, the famous Pythagorean theorem is attributed to him. The theorem states the square of the hypotenuse of a right triangle is equal to the sum of the squares of the other two legs. Algebraically, this equation is $c^2 = a^2 + b^2$.

This section examines the properties of exponents, both positive and negative, and zero. Much of the material in this section relates to the properties of exponents as they apply to powers of two. Base two will be very important in your continued studies in computer technology.

DEFINITION: Exponents are a shorthand version of repeated multiplication.

Instead of writing

$$a \cdot a \cdot a \cdot a \cdot a \cdot a \cdot a,$$

we write

$$a^7,$$

which we read as "a to the seventh power."

An expression of this type is said to be in **exponential form**. We call a the **base** of the expression and 7 the **exponent**, or **power**.

EXAMPLE 1 **Write each of the following using exponential notation.**

(a) $a \cdot a \cdot a \cdot a \cdot a = a^5$

(b) $2 \cdot 2 \cdot 2 \cdot 2 \cdot 2 \cdot 2 \cdot 2 \cdot 2 \cdot 2 = 2^9$ (which also equals 512)

PRACTICE PROBLEMS 1 **Write each of the following using exponential notation.**

(a) $b \cdot b \cdot b \cdot b \cdot b \cdot b \cdot b$

(b) $2 \cdot 2 \cdot 2 \cdot 2 \cdot 2 \cdot 2 \cdot 2 \cdot 2 \cdot 2 \cdot 2 \cdot 2 \cdot 2 \cdot 2 \cdot 2 \cdot 2$

PRACTICE PROBLEM Answers on page 9

In our next example, we use the product rule for exponents. When we multiply two expressions that have the same base, the resulting expression has the same base to the power that is the sum of the two exponents.

$$a^m \cdot a^n = a^{m+n}$$

EXAMPLE 2 **Simplify each product, first as a base to a single power, then, if possible, as a decimal number.**

(a) $a^5 \cdot a = a^6$ (b) $2^3 \times 2^5 = 2^8 = 256$

PRACTICE PROBLEMS 2 **Simplify each product, first as a base to a single power, then, if possible, as a decimal number.**

(a) $b^4 \cdot b^3$ (b) $2^8 \times 2^4$

In our next example, we use the **quotient rule for exponents.** When we divide one expression by another where both have the same nonzero base, the resulting expression has the same base to the power that is the difference of the two exponents.

$$\frac{a^m}{a^n} = a^{m-n}$$

EXAMPLE 3 **Simplify each quotient, first as a base to a single power, then, if possible, as a decimal number.**

(a) $\frac{a^9}{a^3} = a^6$ (b) $\frac{2^{12}}{2^3} = 2^9 = 512$

PRACTICE PROBLEMS 3 **Simplify each quotient, first as a base to a single power, then, if possible, as a decimal number.**

(a) $\frac{b^{12}}{b^4}$ (b) $\frac{2^{20}}{2^{10}}$

What does the quotient rule yield when m is equal to n? Consider the following:

$$\frac{a^m}{a^m} = a^{m-m} = a^0.$$

But we know that

$$\frac{a^m}{a^m} = 1 \text{ (because the numerator equals the denominator).}$$

Putting both equations together implies that $a^0 = 1$. In fact, any base (other than zero) raised to the zero power equals 1.

PRACTICE PROBLEM Answers on page 9

DEFINITION: Zero exponent: for any real number $a \neq 0$, $a^0 = 1$.

EXAMPLE 4 **Simplify each expression (assume that $a \neq 0$).**

(a) $12^0 = 1$

(b) $(-24)^0 = 1$

(c) $(5a)^0 = 1$

(d) $5a^0 = 5 \times 1 = 5$

Note that in (c), the expression $5a$ is raised to the zero power, whereas in (d), only the variable a is raised to the zero power.

PRACTICE PROBLEMS 4 **Simplify each expression (assume that $b \neq 0$).**

(a) 9^0

(b) $(-11)^0$

(c) $(11b)^0$

(d) $11b^0$

What if we allow one of the exponents to be negative and apply the power rule? Suppose that $m = 3$ and $n = -3$. Then

$$a^m \cdot a^n = a^3 \cdot a^{-3} = a^{3+(-3)} = a^0 = 1,$$

so

$$a^3 \cdot a^{-3} = 1.$$

Dividing both sides by a^3, we get $a^{-3} = \dfrac{1}{a^3}$.

This leads to the following definition.

DEFINITION: Negative integer exponents: for any nonzero real number a, and any number n,

$$a^{-n} = \frac{1}{a^n}.$$

Example 5 will illustrate this definition.

PRACTICE PROBLEM Answers on page 9

EXAMPLE 5 **Write the following expressions: y^{-3}, 2^{-5}, and $2^{32} \times 2^{-16}$, so that no answer includes negative exponents.**

(a) $y^{-3} = \frac{1}{y^3}$

(b) $2^{-5} = \frac{1}{2^5} = \frac{1}{32}$ (or, as a decimal number, 0.03125)

(c) $2^{32} \times 2^{-16} = 2^{16} = 65{,}536$

PRACTICE PROBLEMS 5 **Write the following expressions so that no answer includes negative exponents.**

(a) z^{-5}

(b) 2^{-7}

(c) $2^{-8} \times 2^{12}$

When working with exponents, you must also keep two other properties in mind. You have seen all of them in your beginning algebra course.

DEFINITION: Power rule: For any nonzero real number *a*, and integers *m* and *n*,

$$(a^m)^n = a^{mn}.$$

EXAMPLE 6 **Simplify each expression.**

(a) $(a^3)^5 = a^{15}$

(b) $(2^4)^5 = 2^{20} = 1{,}048{,}576$

PRACTICE PROBLEMS 6 **Simplify each expression.**

(a) $(b^4)^2$

(b) $(2^8)^2$

PRACTICE PROBLEM Answers
on page 9

DEFINITION: The product to a power rule: for any nonzero real numbers a and b, and integer n,

$$(ab)^n = a^n b^n.$$

EXAMPLE 7 **Simplify each expression.**

(a) $(ab)^5 = a^5b^5$ **(b)** $(2a)^5 = 2^5a^5 = 32a^5$

PRACTICE PROBLEMS 7 **Simplify each expression.**

(a) $(xy)^6$ **(b)** $(3x)^4$

ANSWERS TO PRACTICE PROBLEMS

1. (a) $b \cdot b \cdot b \cdot b \cdot b \cdot b \cdot b = b^7$
(b) $2 \cdot 2 \cdot 2 \cdot 2 \cdot 2 \cdot 2 \cdot 2 \cdot 2 \cdot 2 \cdot 2 \cdot 2 \cdot 2 \cdot 2 \cdot 2 \cdot 2 = 2^{15} = 32{,}768$
2. (a) $b^4 \cdot b^3 = b^7$ **(b)** $2^8 \times 2^4 = 2^{12} = 4096$ **3. (a)** $\dfrac{b^{12}}{b^4} = b^8$
(b) $\dfrac{2^{20}}{2^{10}} = 2^{10} = 1024$ **4. (a)** $9^0 = 1$ **(b)** $(-11)^0 = 1$
(c) $(11b)^0 = 1$ **(d)** $11b^0 = 11$ **5. (a)** $z^{-5} = \dfrac{1}{z^5}$
(b) $2^{-7} = \dfrac{1}{2^7} = .0078125$ **(c)** $2^{-8} \times 2^{12} = 2^4 = 16$ **6. (a)** $(b^4)^2 = b^8$
(b) $(2^8)^2 = 2^{16} = 65{,}536$ **7. (a)** $(xy)^6 = x^6y^6$ **(b)** $(3x)^4 = 81x^4$

Exercises

Write each of the following in exponential form.

1. $a \cdot a \cdot a \cdot a \cdot a \cdot a \cdot a \cdot a \cdot a$
2. $b \cdot b \cdot b \cdot b \cdot b$
3. $2 \cdot 2 \cdot 2 \cdot 2$
4. $2 \cdot 2 \cdot 2 \cdot 2 \cdot 2 \cdot 2 \cdot 2 \cdot 2$
5. $2 \cdot 2 \cdot 2 \cdot b \cdot b \cdot b \cdot b \cdot b \cdot b$
6. $2 \cdot 2 \cdot 2 \cdot 2 \cdot 2 \cdot a \cdot a$

Simplify each expression, first as a base to a single power, then, if possible, as a decimal. Assume the variables are nonzero.

7. $x^4 \cdot x^3$
8. $y^5 \cdot y^7$
9. $2^5 \times 2^7$
10. $2^4 \times 2^3$
11. $2^6 \times 2^{-1}$
12. $2^{-6} \times 2^{10}$
13. $\dfrac{x^8}{x^2}$
14. $\dfrac{x^{11}}{x^8}$
15. $\dfrac{2^{10}}{2^3}$
16. $\dfrac{2^{20}}{2^1}$
17. $\dfrac{x^{16}}{x^{-5}}$
18. $\dfrac{2^5}{2^{-5}}$
19. x^0
20. 6^0
21. $6x^0$
22. $(6x)^0$
23. $(-6x)^0$
24. $-6x^0$

Write the following expressions so that no answer includes negative exponents. Assume the variables are nonzero.

25. x^{-4}

26. x^{-1}

27. 2^{-2}

28. 2^{-6}

29. $2^6 \times 2^{-10}$

30. $2^3 \times 2^{-6}$

31. $\frac{x^4}{x^5}$

32. $\frac{2^5}{2^8}$

33. $\frac{x^{-6}}{x^5}$

34. $(x^6)^2$

35. $(x^3)^5$

36. $(2^3)^4$

37. $(2^2)^2$

38. $(2x)^2$

39. $(4x)^3$

40. $(2xy)^5$

41. $(2x^2y)^0 \times (2x)^4$

42. $(2x)^3 \times (2x^2)^2$

43. What is one-half of 2^{10}? What is one-half of 2^{100}?

44. What is one-third of 3^{10}? What is one-third of 3^{1000}?

1.3 Calculator Functions

BITS of HISTORY

500–400 B.C. Archaeological evidence of the first Greek mathematical counting machine dates to this era. The word *abacus* originates from the Greek *abakos* or *abax*, which translates to "board" or "tablet." These tablets were made out of wood or marble.

A somewhat wise person once said, "Calculators are like brains, everybody has one but few people know how to use one!" It is important that you become familiar with your calculator.

In this section, we offer many challenges but give little instruction. Because calculators are different, it will be incumbent on you to use past experience, trial and error, or the instruction manual to figure out how to use your calculator. Remember, a calculator is just a miniature, simple computer.

The first few exercises involve basic arithmetic skills.

PRACTICE PROBLEMS 1 **Simplify each expression.**

(a) $23 - 3 \times 4$ (b) $13.56 - 3.568$

(c) $56.7 \div 5.25$ (d) $4.24 \div 5$

To simplify an expression that includes negative numbers, you need to know how to enter a negative number on your calculator. Some calculators use a [+/-] key that is entered after the number; others use a [(–)] key that is entered before the number. After you have found how your calculator deals with negatives, solve the next set of exercises.

PRACTICE PROBLEMS 2 **Simplify each expression.**

(a) $7 - 6 \times 4$ (b) $12 + (-3.568)$

(c) $(-2.547) + (-5.41)$ (d) $(-2.547) - (-5.41)$

(e) $(-2.547) \times (-5.41)$

The most common ways to enter 2^5 into a calculator are

[2] [^] [5] and [2] [y^x] [5].

PRACTICE PROBLEMS 3 **Compute the decimal value of each expression.**

(a) 2^7 (b) 3×2^3 (c) $(3\times2)^3$

PRACTICE PROBLEM Answers on page 13

Most calculators have a square root key $\boxed{\sqrt{\ }}$. Find out whether you must enter the symbol before or after the number on your calculator.

PRACTICE PROBLEMS 4 **Approximate each square root. Round to the nearest hundredth.**

(a) $\sqrt{322}$ (b) $\sqrt{37.5}$ (c) $\sqrt{25{,}688}$

Here are a few more things that you should learn to do on your calculator. Other calculator functions that are introduced later in this text include the following:

- Store a number in the calculator memory.
- Recall a number from the calculator memory.
- Find a multiplicative inverse with a single keystroke.
- Retrieve a stored value for π.
- Find the value of e (usually as e^1).

ANSWERS TO PRACTICE PROBLEMS

1. **(a)** $23 - 3 \times 4 = 11$ **(b)** $13.56 - 3.568 = 9.992$
(c) $56.7 \div 5.25 = 10.8$ **(d)** $4.24 \div 5 = 0.848$
2. **(a)** $7 - 6 \times 4 = -17$ **(b)** $12 + (-3.568) = 8.432$
(c) $(-2.547) + (-5.41) = -7.957$ **(d)** $(-2.547) - (-5.41) = 2.863$
(e) $(-2.547) \times (-5.41) = 13.77927$ **3.** **(a)** $2^7 = 128$
(b) $3 \times 2^3 = 24$ **(c)** $(3 \times 2)^3 = 6^3 = 216$ **4.** **(a)** $\sqrt{322} \approx 17.94$
(b) $\sqrt{37.5} \approx 6.12$ **(c)** $\sqrt{25{,}688} \approx 160.27$

SECTION 1.3

Exercises

Use your calculator to evaluate the following.

1. 2×2^{-1}
2. $\frac{1}{5} + \frac{4}{13} + \frac{2}{7}$
3. -5^2
4. $(-5)^2$

Answer or approximate the following. Round your answer to four decimal places where appropriate.

5. $\sqrt{16}$
6. $\sqrt{-16}$
7. $\sqrt[3]{16^2}$
8. $16^2 + 16^{-2}$
9. $\frac{6.022 \times 1.6}{1.41}$
10. $12.7 - (3.5 \times 8^2)$
11. 2^{2^2}
12. $\sqrt{641} + 5$

Approximate the following. Round your answers after six decimal places.

13. π
14. e (Euler's number)
15. $\frac{1 + \sqrt{5}}{2}$

Evaluate first without using a calculator, and then use your calculator to check your answer.

16. $5000 \div [100 - 2 \times (24 + 1)]^2 \times 3 - 5$
17. $200 - ((100 \div 4 \times (5 - 1))^3 \times 2 - 9$

1.4 Scientific Notation

BITS of HISTORY

300–200 B.C. Considered by many to be one of the greatest mathematicians of all time, Archimedes of Syracuse extended the Greek number system enabling it to express extraordinarily large numbers, such as the number of grains of sand that could be fitted into the universe. Archimedes also estimated *pi* with great accuracy, stating that

$3.14084 < \pi < 3.14285$.

Using your calculator, multiply 2.3 by 1000. The display now reads 2300 (23 followed by two zeros).

Multiply by 1000 a second time. Now you see 2,300,000 (23 followed by five zeros).

Multiplying by 1000 a third time adds three zeros to the display (23 followed by eight zeros).

Now, multiply this result by 1000 again. The display reads

2.3 ^{12} **(some calculators will read 2.3 E12).**

What do you think the 12 means?

Yet another multiplication results in the display

2.3 ^{15} **(or 2.3 E15).**

Can you see what is happening? This is the way calculators display very large numbers. The number on the left is always at least 1 but less than 10, and the number on the right indicates the number of places the decimal is to be moved to the right to put the answer in standard (decimal) form. So, 2.3 ^{15} (or 2.3 E15) is really 2.3×10^{15}.

This notation is used frequently in science. In scientific applications of algebra, it is not uncommon to work with very large or very small numbers. Even in the time of Archimedes (287–212 B.C.), the study of such numbers was not unusual. Archimedes estimated that the universe was

23,000,000,000,000,000 meters in diameter.

In scientific notation, his estimate for the diameter of the universe is written

2.3×10^{16} meters

In **scientific notation**, the number to the left (called the **mantissa**) is always at least 1 but less than 10, and the exponent is always an integer.

EXAMPLE 1 **Write each number in scientific notation.**

(a) 57,000,000 is written 5.7×10^{7}

(b) 2,345,500,000,000 is written 2.3455×10^{12}

(c) 999 is written 9.99×10^{2}

PRACTICE PROBLEMS 1 **Write each number in scientific notation.**

(a) 2,770,000

(b) 43

(c) 91,211,000,000,000,000,000

To convert from scientific to decimal notation, move the decimal point in the opposite direction.

EXAMPLE 2 **Rewrite each number in decimal notation.**

(a) 2.666×10^7 is equivalent to 26,660,000 (the decimal point moves seven places to the right)

(b) 5.76×10^2 is equivalent to 576 (the decimal point moves two places to the right)

(c) 1.01×10^{22} is equivalent to 10,100,000,000,000,000,000,000

PRACTICE PROBLEMS 2 **Rewrite each number in decimal notation.**

(a) 3.696×10^4 (b) 8.0×10^{20} (c) 1.0101×10^8

Scientific notation is also used to store very small numbers. For example, the time it takes light to travel 1 meter is approximately 0.000000003 seconds. In scientific notation, we write this as

3.0×10^{-9} seconds.

The negative exponent indicates that the decimal moves to the left, rather than to the right. The following table will help you see why.

POWER	10^{-5}	10^{-4}	10^{-3}	10^{-2}	10^{-1}	10^0	10^1	10^2	10^3	10^4
DECIMAL	0.00001	0.0001	0.001	0.01	0.1	1	10	100	1000	10,000
FRACTION	$\frac{1}{100,000}$	$\frac{1}{10,000}$	$\frac{1}{1000}$	$\frac{1}{100}$	$\frac{1}{10}$	1	10	100	1000	10,000

PRACTICE PROBLEMS Answers on page 17

Example 3 **Rewrite each number in decimal notation.**

(a) 2.666×10^{-7} is equivalent to 0.0000002666

(b) 5.76×10^{-2} is equivalent to 0.0576

(c) 1.01×10^{-12} is equivalent to 0.00000000000101

Practice Problems 3 **Rewrite each number in decimal notation.**

(a) 3.696×10^{-4} (b) 8.0×10^{-20} (c) 1.0101×10^{-8}

In the previous section, we noted that you are expected to master several skills on your calculator. One of these is using scientific notation. You have seen that, when answers have too many digits, the calculator displays the result in scientific notation. You must now figure out how to enter numbers in scientific notation into your calculator.

You almost always enter the mantissa first, then press the appropriate key (usually [EE] or [EXP]), and then enter the power of ten.

Example 4 **Use your calculator to simplify each of the following. Write your answer in scientific notation.**

(a) $\dfrac{1.7152 \times 10^{27}}{6.7 \times 10^{58}} = 2.56 \times 10^{-32}$ (b) $\dfrac{2.6 \times 10^{27} - 9.8 \times 10^{26}}{5.4 \times 10^{26}} = 3$

Practice Problems 4 **Use your calculator to simplify each of the following. Write your answer in scientific notation.**

(a) $\dfrac{2.349 \times 10^{73}}{8.1 \times 10^{70}}$ (b) $\dfrac{9.4 \times 10^{22} - 9.8 \times 10^{21}}{2.105 \times 10^{22}}$

Answers to Practice Problems

1. (a) $2{,}770{,}000 = 2.77 \times 10^{6}$ (b) $43 = 4.3 \times 10$ (or 4.3×10^{1})
(c) $91{,}211{,}000{,}000{,}000{,}000{,}000 = 9.1211 \times 10^{19}$

2. (a) $3.696 \times 10^{4} = 36{,}960$
(b) $8.0 \times 10^{20} = 800{,}000{,}000{,}000{,}000{,}000{,}000$
(c) $1.0101 \times 10^{8} = 101{,}010{,}000$

3. (a) $3.696 \times 10^{-4} = 0.0003696$
(b) $8.0 \times 10^{-20} = 0.00000000000000000008$
(c) $1.0101 \times 10^{-8} = 0.000000010101$

4. (a) 2.9×10^{2} (b) 4

Exercises

Write each decimal number in scientific notation.

1. 120,000,000

2. 54,680,000,000,000

3. 0.0000000000024

4. 0.000010000100

5. 0.0000037 m
(The length of an infrared lightwave.)

6. 7,692,000 km^2
(The area of Australia.)

Convert each number into decimal notation.

7. 5.27×10^9

8. 1.99997×10^7

9. 2.217×10^{-6}

10. 9.011×10^{-11}

11. $\$5.87 \times 10^{10}$
(*Forbes* estimate of Bill Gates's net worth.)

12. 4.91×10^{-2} sec
(The transfer time of a 1K file via DSL.)

Use your calculator to simplify the following. Write your answers in scientific notation.

13. $\dfrac{3.14 \times 10^4}{2.72 \times 10^2}$

14. $\dfrac{8.7602 \times 10^{-5}}{5.81 \times 10^9}$

15. $\dfrac{4.36 \times 10^3 - 1.2 \times 10^{11}}{3.01 \times 10^7}$

16. $\dfrac{9.0 \times 10^{5} + 2.2 \times 10^{2}}{5.022 \times 10^{-2}}$

17. $\dfrac{6.022 \times 10^{23}}{4.174 \times 10^{-17}}$

18. $\dfrac{1.618 \times 10^{-9}}{3.236 \times 10^{-3}}$

For Exercises 19–24, translate each figure into scientific notation.

19. According to the *1999 CIA World Factbook,* the population of China was approximately one billion, two hundred forty-six million, eight hundred seventy-two thousand.

20. According to the *1999 CIA World Factbook,* the population of the United States was approximately two hundred seventy-two million, six hundred thirty-nine thousand, six hundred.

21. Nanotechnology is the manufacturing of materials and structures that can measure up to one hundred-billionths of a meter.

22. The Chicago Brand digital micrometer can measure an object as small as fifty millionths of a meter.

23. A $3\frac{1}{2}$-inch floppy disk has the capacity to store one million, four hundred fifty-seven thousand, six hundred sixty-four bytes of information.

24. An Internet transfer rate for sending 1 megabyte is clocked at forty-five thousand, five hundred forty millionths of a second.

For Exercises 25–27, convert each number into decimal notation.

25. According to the Internet Movie Database, the film *The Fellowship of the Ring* grossed $\$6.6114 \times 10^{7}$ during its opening weekend in the United States.

26. A JPEG image of da Vinci's painting *La Gioconda* (better known as the Mona Lisa) is 1.3859×10^{-3} megabytes.

27. According to the *1999 CIA World Factbook,* the United States produces 3.9629×10^{13} kilowatt-hours of electricity per year.

1.5 Error Analysis

BITS of HISTORY

300–200 B.C. The Greek mathematician and astronomer Erastothenes was probably the first person to accurately explain the process of the Nile River's flood patterns as well as catalog more than 600 stars. His most famous accomplishment is his estimate of the circumference of Earth using noontime shadows in two different cities. Because Erastothenes used a unit of measure called a *stadium*, which no longer exists, scholars question just how accurate it was, but most agree that it was one of the most accurate for its time.

Is it OK to round off an answer? When we write a number like

234,573,469,781,693,466

in scientific notation, we sometimes write

2.34×10^{17}.

This is not precisely the same as the original number, because it represents

234,000,000,000,000,000.

Try the following with your calculator. Start multiplying

$9 \times 9 \times 9 \times \ldots$

until your calculator goes into scientific notation. Note that the calculator can display only 10 digits, and no power of nine can ever end in zero. (Do you understand why?) The answer you see is not precise. Now, with the calculator display at

3.138105961^{10}, start dividing by 9.

Eventually, you return to the starting point, 9. (You can take this exercise to any extreme if you wish. Most calculators can continue as long as the power of ten is less than 100.)

The calculator stores far more data than it displays. There is much information in the calculator that we can lose if we just write down what we see on the display, which is a rounded off version of what the calculator stores.

How does rounding affect the final answer? Try the following two problems:

$237 + 396$ and 237×396.

Your answers should be 633 and 93,852. Now, rounding off the answers to the nearest 100, we get 600 and 93,900.

Now, rounding each number in the original problem to the nearest 100, approximate the sum and product. We have

$200 + 400 = 600$ and $200 \times 400 = 80{,}000$.

Our estimate in the addition problem was quite useful, but in the multiplication problem we were off by 13,900! What happened?

The mathematical lesson here is *never round your numbers before you have a final answer.* Always do as much of your computation as you can in the calculator. Your calculator is more accurate than the display may suggest because it takes and stores the result of the computation to several decimal places, even if it cannot display all of the digits.

The digits that define a numerical value are called **significant digits**. For a number with no decimal part, the significant digits can be counted starting at the left and counting up to the last nonzero digit. That gives the minimum number of significant digits.

EXAMPLE 1 **Determine the number of significant digits.**

(a) 47,500 **This number has a minimum of three significant digits. We say it is the minimum because we are unsure whether the zeros are a result of computation, measurement, or rounding.**

(b) 100,200 **This number has a minimum of four significant digits. We count starting with the 1 and ending with the 2, the last nonzero digit.**

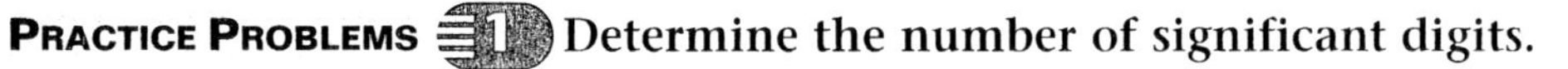

PRACTICE PROBLEMS 1 **Determine the number of significant digits.**

(a) 50,050,000 **(b)** 23,000,000,000

If a number has a decimal part, the significant digits are determined by counting the total number of displayed digits, starting with the first nonzero digit on the left.

EXAMPLE 2 **Determine the number of significant digits.**

(a) 12.034 has five significant digits.

(b) 0.00000034 has two significant digits.

(c) 0.0000003400 has four significant digits. We do not write zeros at the end of a decimal unless they mean something.

PRACTICE PROBLEMS 2 **Determine the number of significant digits.**

(a) 57.020 **(b)** 0.000369 **(c)** 0.0200200

PRACTICE PROBLEM Answers on page 25

The number of significant digits usually determines the way a number is represented in scientific notation. In the next example, we look at the scientific notation representation for the numbers used in Example 2.

EXAMPLE 3 **Write each of the following numbers in scientific notation.**

(a) 12.034 is written 1.2034×10^1

(b) 0.00000034 is written 3.4×10^{-7}

(c) 0.0000003400 is written 3.400×10^{-7}

PRACTICE PROBLEMS 3 **Write each of the following numbers in scientific notation.**

(a) 57.020 (b) 0.000369 (c) 0.0200200

The number of significant digits is important in determining how we report an answer in a real-world situation. Any time you multiply two numbers, report your answer only to the minimum number of significant digits of the two numbers multiplied. One reason for this is that *every measurement is an approximation.*

EXAMPLE 4 **Find the area of a room that has been measured (electronically) at 12.5 ft. by 18.3456 ft.**

Using a calculator, we find that $12.5 \times 18.3456 = 229.32$, but because the first measurement has only three significant digits, we report the answer as 229 square feet.

PRACTICE PROBLEM 4 **Find the volume of a cube that measures 5.25 cm on each side. (The volume of a cube is the cube of the length of one side.)**

When we speak of the **accuracy** of a measurement, we are interested in the error of the measurement. There are two ways to measure the error (assuming we know the exact value).

DEFINITION: The **absolute error** is the (absolute value of the) difference between a measurement and the exact value.

PRACTICE PROBLEM Answers on page 25

EXAMPLE 5 **Find the absolute error, a, if a pipe that is actually 15.32 inches long is measured at 15.375 inches.**

$$a = |15.32 - 15.375| = 0.055 \text{ inches}$$

PRACTICE PROBLEM 5 **Find the absolute error if a piece of molding that is actually 4.36 meters long is measured at 4.25 meters.**

DEFINITION: The **relative error** is the ratio that results when the absolute error is divided by the exact measurement. We usually express this as a percentage.

EXAMPLE 6 **Find the relative error if a pipe that is actually 15.32 inches long is measured at 15.375 inches.**

$$r = |15.32 - 15.375| \div 15.32 = 0.055 \div 15.32 \approx 0.0036$$

or 0.36%

PRACTICE PROBLEM 6 **Find the relative error if a piece of molding that is actually 4.36 meters long is measured at 4.25 meters.**

Why do we use two different measurements for error? Consider the next example.

EXAMPLE 7 **Suppose you are formatting your computer's hard drive. You estimate formatting time to be about 15 minutes, and it actually takes 20 minutes. Find the absolute error.**

Computing the absolute error, we find $a = |15 - 20| = 5$ minutes for the formatting task.

Now you are estimating the time it takes to load a Web page. You estimate the time to be 30 seconds, and it actually takes 5 minutes and 30 seconds (330 seconds). Find the absolute error.

Computing the absolute error, we find $a = |30 - 330| = 300$ seconds $= 5$ minutes for the Web-page load time.

PRACTICE PROBLEM Answers on page 25

In both cases, the absolute error is 5 minutes, but the context is very different.

Now we compute the relative error for each:

$r = |15 - 20| \div 20 = 0.25 = 25\%$ error for the formatting task, and

$r = |30 - 330| \div 330 \approx 0.9091 = 90.91\%$ error for the Web-page load time.

In both cases, the error is undesirable, but the relative error for the Web-page load time is almost 91%, which is much greater than for the formatting task.

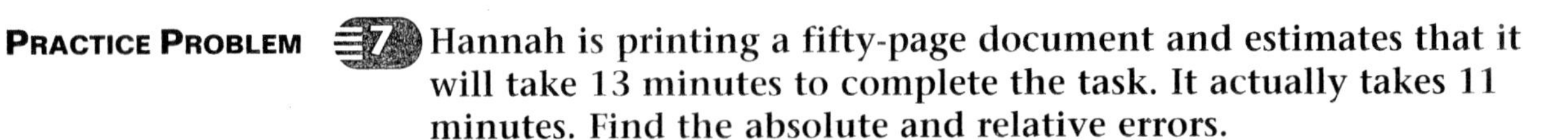

PRACTICE PROBLEM 7 **Hannah is printing a fifty-page document and estimates that it will take 13 minutes to complete the task. It actually takes 11 minutes. Find the absolute and relative errors.**

Many times, an error is measured between an **expected** outcome and the actual measurement. The expected outcome is usually derived by finding the **mean** (sometimes simply called the average) from some previously collected data.

DEFINITION: To find the **mean** for a set of numeric data values, add all the values, and then divide that sum by the number of data values.

EXAMPLE 8 **Find the mean for the following set of six computer installation times:**

23 min 37 min 35 min 72 min 28 min 44 min

Adding the values, we get $23 + 37 + 35 + 72 + 28 + 44 = 239$.

Dividing 239 by 6, we get $39\frac{5}{6} \approx 39.83$ minutes. Because we have only two significant digits in the original data, we report a mean of 40 minutes.

PRACTICE PROBLEM 8 **Find the mean for the following set of software installation times:**

57 sec 93 sec 88 sec 102 sec 63 sec 77 sec 68 sec

In the next example, we combine several of the ideas from this section.

PRACTICE PROBLEM Answers on page 25

Example 9 Assume that you have been given the data of Example 8 to help you determine how long it should take to install a computer. If it actually takes 46 minutes, find the relative and absolute errors.

The absolute error is $|40 - 46| = 6$.

The relative error is $\frac{6}{46} \approx 0.1304 = 13.04\%$

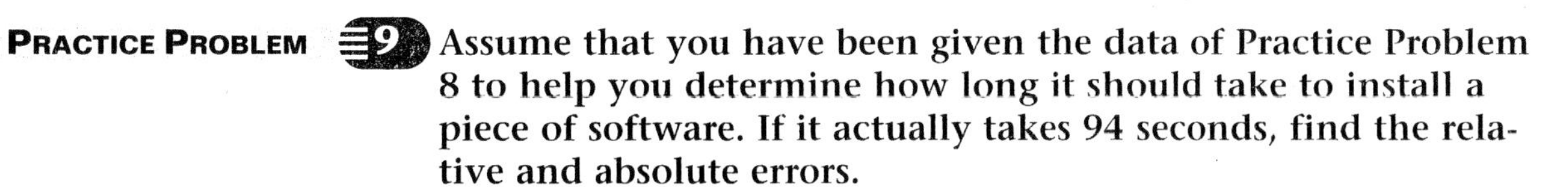

Practice Problem 9 Assume that you have been given the data of Practice Problem 8 to help you determine how long it should take to install a piece of software. If it actually takes 94 seconds, find the relative and absolute errors.

Answers to Practice Problems

1. **(a)** a minimum of 4 **(b)** a minimum of 2 **2.** **(a)** 5 **(b)** 3 **(c)** 6
3. **(a)** 5.7020×10^1 **(b)** 3.69×10^{-4} **(c)** 2.00200×10^{-2}
4. 145 cubic cm **5.** 0.11 m

6. 0.0252 or 2.52% **7.** absolute error = 2 min, relative error = 18%

8. 78 sec

9. absolute error = 16 sec, relative error = $\frac{16}{94} \approx 0.1702 = 17.02\%$

Exercises

Determine the number of significant digits.

1. 120,000,000
2. 54,680,000,000,000
3. 0.0000000000024
4. 0.000010000100
5. 0.0000037
6. 7,692,000

Compute the following to the correct number of significant digits.

7. Find the area of a rectangular rug that has been measured (electronically) at 5.145 ft by 7.92 ft.

8. Find the volume of a Rubik's cube that measures 3.125 inches on each side.

In Exercises 9–12, compute the absolute error and relative error.

9. The exact value $\frac{3}{7}$ is approximated with the value 0.43.

10. The exact value $\frac{5}{6}$ is approximated with the value 0.8.

11. The exact value $\frac{1}{11}$ is approximated with the value 0.1.

12. In the U.S., a can of Coke contains exactly 355 mL of liquid. It is approximated in Australia with a can that contains 375 mL of liquid.

13. Find the mean for the following set of data.
32 min 45 min 57 min 49 min 72 min 66 min 63 min

14. Find the mean for the following set of data.
125 sec 116 sec 131 sec 261 sec 149 sec 155 sec

15. Assume the data for Exercise 13 is used to predict an average time for a Web-page design. If the actual time is 61 minutes, find the absolute and relative errors.

16. Assume the data for Exercise 14 is used to predict an average time for an online purchase. If the actual time is 120 seconds, find the absolute and relative errors.

17. Find the mean for each of the following data sets.
(a) 5.0 10.0 15.0 20.0 25.0 30.0 35.0 40.0 45.0

(b) 22.0 23.0 24.0 25.0 26.0 27.0 28.0 29.0

18. Given the data in Exercise 17, is there any difference in the means? Is there any difference in the data sets? Describe the difference.

19. Given the data in Exercise 17, for which set would you expect an actual measurement to be further from the estimate? Explain your answer.

1.6 Dimensional Analysis

Some of the more common measurements in the ancient world were cubit, ped, millarium, and stadium. Our modern-day mile is derived from millarium, which was equal to 6002 peds.

Almost any application of mathematics involves not just numbers, but units. Units are the labels after a number. Units behave very much like mathematical quantities. Here is an example.

EXAMPLE 1 **Haifez is driving at 66 $\frac{\text{miles}}{\text{hour}}$. How far will he travel in 5.5 hours?**

$$66\,\frac{\text{miles}}{\text{hour}} \times 5.5\text{ hours} = 66\,\frac{\text{miles}}{\cancel{\text{hour}}} \times 5.5\,\cancel{\text{hours}} = 363\text{ miles}$$

Because we had only two significant digits in the original data, we would report an answer of 360 miles.

Note that the units are treated exactly like numeric fractions. "Canceling" the hour in the denominator and the hour in the numerator leaves the unit miles.

PRACTICE PROBLEMS 1 **Manuela gets 53 $\frac{\text{miles}}{\text{gallon}}$ in her new Toyota Prius. How far can she travel if the gas tank holds 10.5 gallons?**

In the applications of mathematics in general, and in this text in particular, the relationship between conversions and ratios is very important. The next example illustrates such a conversion.

EXAMPLE 2 **Earlier in this chapter, we mentioned that Archimedes estimated the diameter of the universe to be 23,000,000,000,000,000 meters. In scientific notation, we wrote this as 2.3×10^{16} m.**

How long would it take light to travel across this universe?

Light travels at 300,000 $\frac{\text{kilometers}}{\text{second}}$.

We write this as a ratio:

$$\frac{300{,}000\text{ kilometers}}{1\text{ second}}.$$

PRACTICE PROBLEM Answers on page 30

If a ratio is valid (and this one certainly is), then its reciprocal is equally valid. We write the reciprocal:

$$\frac{1 \text{ second}}{300{,}000 \text{ kilometers}}.$$

The unit of length used in the initial statement is meters, whereas the unit used in expressing the speed of light is kilometers. We need a conversion factor, which we can express as either of two ratios:

$$\frac{1000 \text{ meters}}{1 \text{ kilometer}} \text{ or } \frac{1 \text{ kilometer}}{1000 \text{ meters}} \text{ (both are equivalent to 1).}$$

Multiplying these three together, we get

$$2.3\times10^{16} \text{ meters} \times \frac{1 \text{ kilometer}}{1000 \text{ meters}} \times \frac{1 \text{ second}}{300{,}000 \text{ kilometers}}.$$

Before solving such a problem, look first at the algebra of the units. Treat the units as fractions and multiply.

$$2.3\times10^{16} \cancel{\text{meters}} \times \frac{1 \cancel{\text{kilometer}}}{1000 \cancel{\text{meters}}} \times \frac{1 \text{ second}}{300{,}000 \cancel{\text{kilometers}}}.$$

When the meters and the kilometers are canceled, the remaining units are seconds. Multiplying the three quantities results in the number of seconds it would take light to traverse Archimedes's universe. The answer is approximately

$$7.7 \times 10^7 \text{ seconds.}$$

PRACTICE PROBLEM 2 **Example 2 concludes that it would take 7.7×10^7 seconds for light to travel across Archimedes's universe. How many years would that be? Leave your answer in decimal form. Round to the nearest tenth of a year. Assume 1 year = 365 days.**

Knowing the units of the answer can often help us set up the correct equation. Look at the following example.

EXAMPLE 3 **Roger can type 120 $\frac{\text{words}}{\text{minute}}$. How many characters can he type per second? Assume an average of 5 characters per word.**

We know that the answer has the units $\frac{\text{characters}}{\text{second}}$.

PRACTICE PROBLEM Answers on page 30

The units of the solution come from an equation in the following form:

$$\frac{\text{words}}{\text{minute}} \times \frac{\text{characters}}{\text{word}} \times ? = \frac{\text{characters}}{\text{second}}.$$

The missing element must be a ratio, so we write

$$\frac{\text{words}}{\text{minute}} \times \frac{\text{characters}}{\text{word}} \times \frac{?}{?} = \frac{\text{characters}}{\text{second}}.$$

To get an answer with the desired units, you must use minutes in the numerator and seconds in the denominator of the missing ratio. The ratio of minutes to seconds is

$$\frac{1 \text{ minute}}{60 \text{ seconds}}.$$

We are now ready to set up the equation.

$$120 \frac{\text{words}}{\text{minute}} \times 5 \frac{\text{characters}}{\text{word}} \times \frac{1 \text{ minute}}{60 \text{ seconds}} = ? \frac{\text{characters}}{\text{second}}$$

$$120 \frac{\cancel{\text{words}}}{\cancel{\text{minute}}} \times 5 \frac{\text{characters}}{\cancel{\text{word}}} \times \frac{1 \cancel{\text{minute}}}{60 \text{ seconds}} = 10 \frac{\text{characters}}{\text{second}}$$

Roger types at a rate of $10 \frac{\text{characters}}{\text{second}}$.

PRACTICE PROBLEM  **A 100-MHz processor averages 5 cycles per machine instruction $\left(\frac{\text{5 cycles}}{\text{1 machine instruction}}\right)$. How many machine instructions are performed each second?** $\left(\textit{Hint: } 100 \text{ MHz} = 100{,}000{,}000 \frac{\text{cycles}}{\text{second}}.\right)$

ANSWERS TO PRACTICE PROBLEMS

1. 556.5 miles 2. 2.4 years 3. $20{,}000{,}000 \frac{\text{machine instructions}}{\text{second}}$

SECTION

1.6 Exercises

Use dimensional analysis to answer the following.

1. A Toyota 4Runner gets 19 $\frac{\text{miles}}{\text{gallon}}$ when driven on the highway. How many miles can the 4Runner travel if its fuel capacity is 18.5 gallons?

2. A Ford Ranger gets 24 $\frac{\text{miles}}{\text{gallon}}$ when driven on the highway. How many miles can the Ranger travel if its fuel capacity is 16.5 gallons?

3. How many hours are there in 1 year? How many seconds are there in 1 year? (Assume there are 365 days in 1 year.)

4. How many yards are there in 1 mile? How many inches are there in 1 mile? (There are 5280 feet in 1 mile and 3 feet in 1 yard.)

5. Given that 1 U.S. dollar = 1.660 Australian dollars,

1 Australian dollar = 1.119 Canadian dollars, and

1 Canadian dollar = 71.108 Japanese yen. How many Japanese yen are in 1 U.S. dollar?

6. Given that 1 U.S. dollar = .6536 British pounds,

1 British pound = 3.1517 New Zealand dollars, and

1 New Zealand dollar = .8262 Australian dollar. Determine how many U.S. dollars are in 10 Australian dollars.

7. Given that 1 kilometer $\approx$ 0.62 mile, 36 inches = 1 yard, and

   1 mile = 1760 yards, approximately how many inches are in 1 km?

8. Given that the average pulse is 70 heartbeats per minute, determine the number of heartbeats per lifetime of a person whose life expectancy is 80 years.

9. The average number of words per page for a Microsoft Word document formatted with default margins and a 12-point Times New Roman font is 260. Estimate the number of pages needed for a 10,000-word document.

10. Changing the font to 14-point Times New Roman reduces the average number of words per page by approximately 8%. Estimate the number of pages needed for a 10,000-word document using this larger font (refer to Exercise 9).

11. Changing the side margins from 1.25" to 1" increases the average number of words per page by approximately 15% when 12-point Times New Roman is used. Estimate the number of pages needed for a 10,000-word document using these narrower margins (refer to Exercise 9).

12. The average number of words per page for a Microsoft Word document formatted with the default margins and 12-point Arial is 245. Estimate the number of pages needed for a 10,000-word document.

13. An instructor assigns you a 12-page paper and specifies a 12-point font and default margins. Approximately how many fewer words would you need to type if you used Arial instead of Times New Roman (refer to Exercises 9 and 12)?

14. When we computed the absolute error in Section 1.5, the units were significant. When we computed relative error, there were no units. Explain why this occurs.

CHAPTER 1 SUMMARY

1.1 An Introduction

- A number such as 765 can be expressed in base ten as

$$765 = 7 \times 100 + 6 \times 10 + 5 \times 1$$

- The Order of Operations (Parentheses, Exponents, Multiplication and Division, Addition and Subtraction) applies to any arithmetic expression.

1.2 Exponents and Their Properties

- Exponents are a shorthand version of repeated multiplication.

$$a \cdot a \cdot a \cdot a \cdot a \cdot a \cdot a = a^7$$

- The **product rule for exponents** is written

$$a^m \cdot a^n = a^{m+n}$$

- The **quotient rule for exponents** is written

$$\frac{a^m}{a^n} = a^{m-n}$$

- Any base (other than zero) raised to the zero power equals one.

$$a^0 = 1$$

- For any nonzero real number a, and any number n,

$$a^{-n} = \frac{1}{a^n}$$

- For any nonzero real numbers a and b, and integer n,

$$(ab)^n = a^n b^n$$

1.3 Calculator Functions

- There are two common procedures for entering a negative number in a calculator.

  (a) If a calculator has a [+/-] key, that key is pressed after the number.

  (b) If a calculator has a [(–)] key, that key is pressed before the number.

- There are two common procedures for entering exponents in a calculator.

  (a) If a calculator has a [^] key, 2^5 is entered with the sequence
  [2] [^] [5]

  (b) If a calculator has a [y^x] key, 2^5 is entered with the sequence
  [2] [y^x] [5]

1.4 Scientific Notation

- In **scientific notation**, the number to the left (called the **mantissa**) is always at least 1 but less than 10 and the exponent is always an integer.

  $2{,}300{,}000{,}000 = 2.3 \times 10^9$

- A negative exponent indicates that the decimal moves to the left.

  $0.000000024 = 2.4 \times 10^{-8}$

1.5 Error Analysis

- For a number with no decimal part, the **significant digits** can be counted by starting at the left and counting up to the last nonzero digit.

- If a number has a decimal part, the significant digits are determined by counting the total number of displayed digits, starting with the first nonzero digit on the left.

- The number of significant digits usually determines the way a number will be represented in scientific notation.

- Every measurement is an approximation.

- The **absolute error** is the (absolute value of the) difference between a measurement and the exact value.

- The **relative error** is the ratio that results when the absolute error is divided by the exact measurement. We usually express this as a percentage.

- To find the **mean** for a set of numeric data values, add all the values, and then divide that sum by the number of data values.

1.6 Dimensional Analysis

- Almost any application of mathematics involves not just numbers, but units. Units are the labels after a number. Units behave very much like mathematical quantities.

- Before solving a problem that includes units, look first at the algebra of the units. Treat the units as fractions and multiply.

Glossary Quiz

Match each of the following terms from Chapter 1 with one of the definitions that follows. Write the definition's letter next to its term.

accuracy ______ absolute error ______ relative error ______

the mean ______ units ______ exponential form ______

base ______ the *exponent*, or *power* ______ scientific notation ______

significant digits ______ mantissa ______

(a) a shorthand version of repeated multiplication
(b) the a in the expression a^n
(c) the n in the expression a^n
(d) the form $a \times 10^n$ where $1 \leq a < 10$
(e) the a in $a \times 10^n$ where $1 \leq a < 10$
(f) the digits that define a numerical value without regard to the units
(g) the degree of correctness of a measurement
(h) the (absolute value of the) difference between a measurement and the exact value
(i) the ratio that results when the absolute error is divided by the exact measurement (usually expressed as a percentage)
(j) the sum of a set of values divided by the number of data values
(k) the labels after a number

CHAPTER 1 REVIEW EXERCISES

For help with these exercises, refer to the section number given in brackets.

[1.1] How many *hundreds*, *tens*, and *ones* are in the following sums?

1. $\begin{array}{r} 26 \\ +17 \\ \hline \end{array}$

2. $\begin{array}{r} 82 \\ +\ 9 \\ \hline \end{array}$

3. $\begin{array}{r} 113 \\ +\ 78 \\ \hline \end{array}$

4. $\begin{array}{r} 276 \\ +512 \\ \hline \end{array}$

5. $\begin{array}{r} 891 \\ +429 \\ \hline \end{array}$

6. $\begin{array}{r} 1409 \\ +2598 \\ \hline \end{array}$

[1.1] Simplify each expression.

7. $-10 + (-32)$

8. $5 + (-13) - 8$

9. $-6(8 + 4)$

10. $(-8) - (4 \times 2 - 5)$

11. $2(4 + 3) - 4(5 \times 2)$

[1.2] Write each in exponential form.

12. $a \cdot a \cdot a \cdot a \cdot a \cdot a \cdot a \cdot a$

13. $16 \times 16 \times 16 \times 16 \times 16$

14. $2 \cdot 2 \cdot 2 \cdot a \cdot a \cdot a \cdot a \cdot a$

[1.2] **Simplify each expression, first as a base to a single power, then if possible as a decimal. Assume the variables are nonzero.**

15. $a^5 \cdot a^{12}$

16. $b^7 \cdot b^{-2}$

17. $2^4 \times 2^9$

18. $16^3 \times 16^{-1}$

19. $\dfrac{x^{13}}{x^5}$

20. $\dfrac{2^{10}}{2^4}$

21. 2^0

22. $2x^0$

23. $(2x)^0$

[1.2] **Write the following expressions so that no answer includes negative exponents. Assume the variables are nonzero.**

24. x^{-12}

25. 5^{-2}

26. -2^5

27. $\dfrac{x^4}{x^9}$

28. $\dfrac{x^{-8}}{x^{-6}}$

29. $(x^4)^2$

30. $\dfrac{2^5}{2^8}$

31. $(2^{-2})^4$

32. $16(x^5)^0$

33. $(16x^5)^0$

34. $(2x^2)(2x)^3$

[1.3] Use your calculator to evaluate the following expressions.

35. $16^4 \times 16^{-2}$

36. $\frac{1}{3} + \frac{3}{7} + \frac{1}{2}$

37. -2^2

38. $(-2)^2$

[1.3] Approximate the following. Round your answers to four decimal places if the value is not exact.

39. $\sqrt{2}$

40. $\sqrt[3]{2}$

41. $\frac{4.78 \times 3.715}{8.21}$

42. $8.67 - (5 \times 3^0) + 9$

43. $\frac{\sqrt{5} - 1}{2}$

44. $\frac{\pi}{2}$

[1.4] Write each decimal number in scientific notation.

45. 0.00000552

46. 0.00000000000012305

47. 20,327,000 km^2
(area of the Southern Ocean)

48. 96,002 miles^2
(area of Oregon)

[1.4] Convert each number into decimal notation.

49. 8.78×10^{12}

50. 5.62×10^{-6}

51. 1.125×10^{-8}

52. 4.4471×10^6 people
(2000 U.S. census estimate of Alabama's population)

[1.4] **Use your calculator to simplify the following. Write your answers in scientific notation.**

53. $\dfrac{(9.537 \times 10^{-7})(1.024 \times 10^{3})}{8.0 \times 10^{-4}}$

54. $\dfrac{(1.04 \times 10^{14})(6.022 \times 10^{23})}{2.718 \times 10^{17}}$

[1.5] **55. Find the mean for the following set of data.**

4 min 7 min 4 min 3 min 5 min

Assume the data above is used to predict the average time to rip a 42-minute audio CD. Use the mean from the data set to compute the absolute error and relative error if the actual time is 5.25 minutes.

[1.5] **Determine the number of significant digits.**

56. 20,327,000

57. 0.00000000125

58. 0.00000552000

[1.5] **Compute the absolute error and relative error for each of the following.**

59. In Mexico, a bottle of hot sauce contains exactly 148 mL of liquid. That is equivalent to a 150 mL bottle in the United States.

60. In Europe, a 16-inch neck size is equivalent to 41 cm. After a measurement is taken, the exact length is determined to be 40.64 cm.

[1.6] Use dimensional analysis to answer the following.

61. According to Honda, a 2003 5-speed Civic Hybrid gets 51 $\frac{\text{miles}}{\text{gallon}}$ when driven on the highway. How many miles can the Hybrid travel if its fuel capacity is 13.2 gallons?

62. How many minutes are there in 1 year? (Assume there are 365 days in 1 year).

63. Given that 1 U.S. dollar = 10.0474 Mexican pesos, 1 Mexican peso = 3.14756 Russian rubles, and 1 Russian ruble = 0.236124 Norwegian kroner, how many kroner are in 1 U.S. dollar?

64. Given that the average American sleeps 7.5 hours a day, determine the total amount of sleep in minutes per lifetime for a person whose life expectancy is 75 years.

65. In the United States, a standard piece of paper is $8\frac{1}{2}$ inches $\times$ 11 inches. In Europe, the standard is called A4, which measures 210 mm $\times$ 297 mm. Given that 1 cm = 10 mm, and 1 cm = 0.3937 inches, find the dimensions of A4 paper in inches.

66. Given a standard piece of paper that is $8\frac{1}{2}$ inches $\times$ 11 inches, the average number of words per page for a Microsoft Word document formatted with default margins and a 12-point Times New Roman font is 260. Estimate the number of pages needed for a 12,000-word document.

67. Changing the document from $8\frac{1}{2}$ inches $\times$ 11 inches to A4 increases the average number of words per page by approximately 10%. Estimate the number of pages needed for a 12,000-word document using this larger paper size (refer to Exercise 66).

CHAPTER 1 SELF-TEST

1. **How many *hundreds*, *tens* and *ones* are in the following sum?**

   (a) $\begin{array}{r} 456 \\ +\,617 \\ \hline \end{array}$

   (b) Write in exponential form.

   $a \cdot a \cdot a \cdot a \cdot a \cdot a \cdot b \cdot b \cdot b \cdot b \cdot b \cdot b \cdot b \cdot b$

2. **Simplify each expression, first as a base to a single power, then if possible as a decimal. Assume the variables are nonzero.**

   (a) $2^2 \cdot y^5 \cdot y^{16}$ (b) $\dfrac{x^{14}}{x^9}$ (c) $\dfrac{2^{12}}{2}$

3. **Write the following expressions so that no answer includes negative exponents. Assume the variables are nonzero.**

   (a) x^{-17} (b) 2^{-5} (c) $\dfrac{y^4}{y^{15}}$

   (d) $(x^5)^{-2}$ (e) $5x^0$ (f) $(10x^3)(10x)^2$

4. Use your calculator to evaluate the following expressions. If the value is not exact, round your answer to four decimal places.

(a) $\frac{1}{4} + \frac{2}{5} + \frac{1}{6}$ (b) $\sqrt{3}$ (c) $\frac{\pi}{3}$

5. Write each decimal number in scientific notation.

(a) 2,015,000,000,000 (b) 0.00000136

6. Convert each number into decimal notation.

(a) 5.12×10^{-8} (b) 2.715×10^{9}

7. Simplify the following and write your answers in scientific notation.

(a) $\dfrac{(3.05 \times 10^{12})(8.356 \times 10^{-7})}{5.62 \times 10^{-2}}$ (b) $\dfrac{(5.7 \times 10^{-22})(1.024 \times 10^{3})}{7.02 \times 10^{8}}$

8. Find the mean for the following set of data.

42 sec 57 sec 61 sec 51 sec 57 sec

Assume the data above is used to predict the average time to download a 12-MB file with a corporate digital subscription line (DSL). Use the mean from the data set to compute the absolute error and relative error if the actual download time is 48 seconds.

9. Determine the number of significant digits.

(a) 0.007201 (b) 5,001,725,000,000,000

10. Use dimensional analysis to answer the following.

(a) According to Toyota, a 2003 Prius gets a combined average of $48 \frac{\text{miles}}{\text{gallon}}$ when driven on the highway and in the city. How many miles can the Prius travel if its fuel capacity is 11.9 gallons?

(b) The city of Tempe, Arizona, claims it gets nearly 300 days of sunshine per year. Approximately how many sunsets could a person living in Tempe see during her lifetime if her life expectancy is 85 years?

Binary Numbers

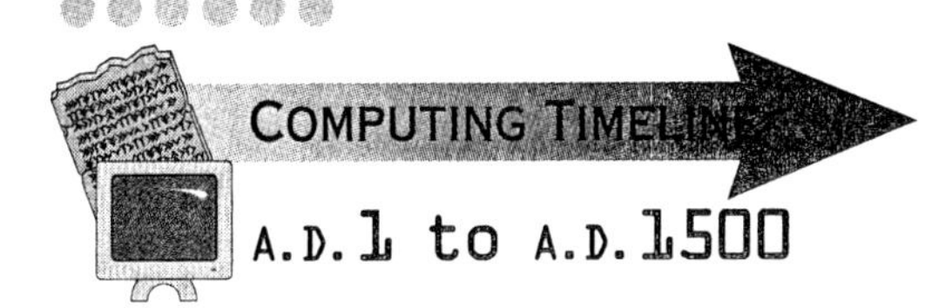

This era was characterized by a battle of the abacists (those who did arithmetic mechanically) and the algorists (those who did algebra). The algorists were able to persevere because of the development of the printing press. Both the symbols and the algorithms were standardized as a result of publishing books.

2.1 The Binary System

BITS of HISTORY

In addition to the decimal (base ten) system, other number systems have been developed throughout history. The Babylonians employed a base sixty system (sexigesimal), while the ancient Maya of Central America used a modified base twenty system (vigesimal). The Maya are also one of the first civilizations to have a symbol for zero, a remarkable accomplishment.

Information Storage

When we use a computer, we expect it to store and retrieve information in a variety of forms. Here are a few examples of the kinds of data one expects to find in a computer:

- A word-processing program that can be used to create, correct, and save text files.
- Text files that have been created on the computer.
- A spreadsheet program that can store and manipulate numeric-based data.
- Numeric files that have been created using a spreadsheet.
- Paintbrush, computer-aided design (CAD) programs, or other programs designed to create graphics.
- Graphics files created by these programs.
- Directions that allow the computer to locate stored files.
- Instructions for creating and digitizing sound.
- Games that involve graphics, numbers, words, and sounds.
- Sets of commands (programs) that will add to the utility of the computer.

To work with such data, the computer needs to use a **code** to represent the information being processed. We use codes every day of our lives. The set of all English words is a code used to communicate ideas. The set of twenty-six letters of the English alphabet is a code used to represent the pronunciation of those words. The set of musical notes is a code used to represent musical pitches. The set of ten digits (0 through 9) is a code used to represent quantities.

It is this last code, the set of digits, which comes closest to the computer code. The difference is that a computer knows only two basic code values (0 and 1), which we simplify as **on/off** or **positive/negative**. The on/off model is the original computer model. Early computers used electronic circuitry that connected vacuum tubes. The more modern silicon chips are a miniature version of this idea. Disk storage is an example of the positive/negative model. Floppy disks are covered with a thin layer of ferromagnetic material in which each cell can be polarized as positive or negative. A sequence of cells creates a code that is interpreted by the computer. Because the prefix *bi-* means two, we call the computer code a ***bi*nary code**.

Binary Code

In Chapter 1, you were reminded of the importance of place value in numeric calculations. Place value was built on a base of ten, because humans have ten fingers. Binary place values are built on a base of two.

In base ten, each place has a value that is ten times greater than the place to its right:

$$\begin{aligned} 11{,}111 &= 1 \times 10^4 + 1 \times 10^3 + 1 \times 10^2 + 1 \times 10^1 + 1 \times 10^0 \\ &= 10{,}000 + 1000 + 100 + 10 + 1. \end{aligned}$$

In base two, each place has a value that is twice as great as the place to its right. As in base ten, we begin at a place value of 1. The next place to the left has a value twice that, or 2. Again moving left we get a place value twice 2, which is 4. Moving left, subsequent place values are 8, 16, 32, 64, 128, 256, 512,... . Note that every place is a power of two.

The following model may help you start to think in base two. In Canada there are coins that represent 25¢ 50¢, $1.00, and $2.00. In the table below, those coins are labeled Q, H, D, and T.

T	D	H	Q	TOTAL QUARTERS
0	0	0	0	0
0	0	0	1	1
0	0	1	0	2
0	0	1	1	3
0	1	0	0	4
0	1	0	1	5
0	1	1	0	6
0	1	1	1	7
1	0	0	0	8
1	0	0	1	9
1	0	1	0	10
1	0	1	1	11
1	1	0	0	12
1	1	0	1	13
1	1	1	0	14
1	1	1	1	15

Note that with one of each of the four coins, we can create the equivalent of any number of quarters from 1 to 15. This is essentially the same concept used in the base two system.

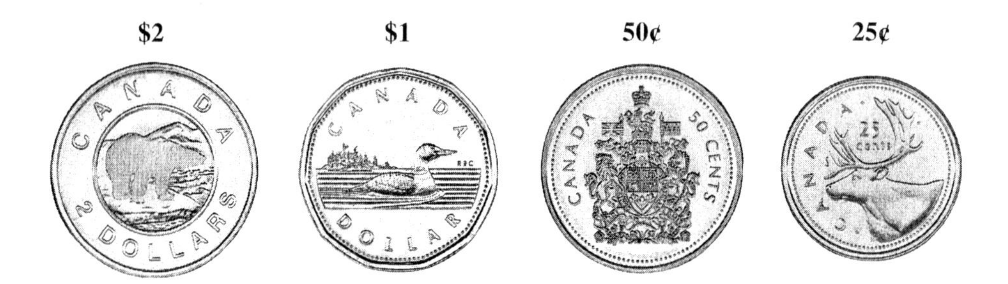

When we write a number in base two, we write a subscript 2 at the end of the number. If a number is written without a subscript, we assume that it is a decimal number, which is base ten.

$$\begin{aligned} 11111_2 &= 1 \times 2^4 + 1 \times 2^3 + 1 \times 2^2 + 1 \times 2^1 + 1 \times 2^0 \\ &= 16 + 8 + 4 + 2 + 1 \\ &= 31 \end{aligned}$$

Notice the pattern when you count in binary.

$0000_2 = 0$	$1000_2 = 8$
$0001_2 = 1$	$1001_2 = 9$
$0010_2 = 2$	$1010_2 = 10$
$0011_2 = 3$	$1011_2 = 11$
$0100_2 = 4$	$1100_2 = 12$
$0101_2 = 5$	$1101_2 = 13$
$0110_2 = 6$	$1110_2 = 14$
$0111_2 = 7$	$1111_2 = 15$

It is the same pattern we created with our coin table. This pattern will be useful later. You should practice writing these sixteen numbers in order until the pattern is obvious to you.

Now, we will examine place values for an eight-digit binary number. Refer to the following table for the next few examples.

BIT LOCATION	Eighth	Seventh	Sixth	Fifth	Fourth	Third	Second	First
POWER	2^7	2^6	2^5	2^4	2^3	2^2	2^1	2^0
VALUE	128	64	32	16	8	4	2	1

EXAMPLE 1 **Find the place value for the 1 in each binary representation.**

(a) 00000100_2 **Because the 1 is in the third place from the right, the place value is 4. (1, 2, 4,...)**

(b) 00010000_2 **The place value is 16.**

(c) 01000000_2 **The place value is 64.**

PRACTICE PROBLEMS 1 **Find the place value for the 1 in each binary representation.**

(a) 00000010_2 (b) 00100000_2 (c) 10000000_2

By summing the place values of a binary number, we can find its decimal equivalent. Let's first try it with some four-digit binary numbers.

EXAMPLE 2 **Find the decimal equivalent for each binary representation.**

BINARY	DECIMAL
0111_2	$0 + 4 + 2 + 1 = 7$
1010_2	$8 + 0 + 2 + 0 = 10$
1101_2	$8 + 4 + 0 + 1 = 13$

PRACTICE PROBLEMS 2 **Find the decimal equivalent for each binary representation.**

(a) 0011_2 (b) 1100_2 (c) 1111_2

In the next example, we look at the decimal equivalent for some eight-digit binary numbers.

PRACTICE PROBLEM Answers on page 51

EXAMPLE 3 **Find the decimal equivalent for each binary representation.**

BINARY	DECIMAL
01010111_2	0 + 64 + 0 + 16 + 0 + 4 + 2 + 1 = 87
10111000_2	128 + 0 + 32 + 16 + 8 + 0 + 0 + 0 = 184
11001101_2	128 + 64 + 0 + 0 + 8 + 4 + 0 + 1 = 205

PRACTICE PROBLEMS 3 **Find the decimal equivalent for each binary representation.**

(a) 01000011_2 (b) 01111000_2 (c) 11111111_2

How do we go about rewriting a decimal number as its binary equivalent? As we noted earlier, each place in a binary representation is a power of two. We find the binary equivalent for a decimal number by subtracting the largest power of two that is less than the decimal number, noting a 1 in that binary place, then continuing the process with the result until we are left with zero. Let's look at an example.

EXAMPLE 4 **Rewrite the decimal number 231 as a binary number.**

The largest power of two that is less than 231 is 128. 128 is represented in base two by a 1 in the eighth place to the left (128, 64, 32, 16, 8, 4, 2, 1).

231 − 128 = 103

128	64	32	16	8	4	2	1
1							

The largest power of two less than 103 is 64, which is represented in base two by a 1 in the seventh place to the left.

103 − 64 = 39

128	64	32	16	8	4	2	1
1	1						

The largest power of two less than 39 is 32, which is represented in base two by a 1 in the sixth place to the left.

39 − 32 = 7

128	64	32	16	8	4	2	1
1	1	1					

Note that 7 = 4 + 2 + 1. We also need 1s in the third, second, and first places.

128	64	32	16	8	4	2	1
1	1	1	0	0	1	1	1

PRACTICE PROBLEM Answers on page 51

Thus,

DECIMAL	BINARY
231	11100111_2

Note there are zeros in the fifth and fourth places. When we were subtracting, the decimal value was reduced from 39 to 7, so we skipped both 16 and 8 because they were both greater than 7.

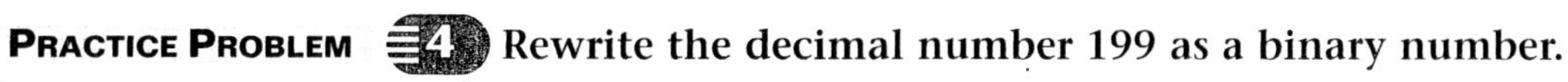

PRACTICE PROBLEM 4 **Rewrite the decimal number 199 as a binary number.**

When we use the term *digit*, we usually think of decimal digits. When we wish to refer to a binary digit, the common contraction **bit** is used. In a computer, bits are usually arranged in groups of eight. A collection of eight consecutive bits is so common that it is often referred to as a **byte**. It is because of the common grouping that most of the binary representations we have seen in this section were in byte form. For larger numbers two or more bytes are combined. Later on, we will look at some of these arrangements.

ANSWERS TO PRACTICE PROBLEMS

1. **(a)** 2 **(b)** 32 **(c)** 128 **2.** **(a)** 3 **(b)** 12 **(c)** 15

3. **(a)** 67 **(b)** 120 **(c)** 255 **4.** 11000111_2

SECTION 2.1

Exercises

Find the decimal equivalent for each of the following binary numbers.

1. 0101_2
2. 0011_2
3. 1101_2
4. 1010_2
5. 01101101_2
6. 11000110_2
7. 10101010_2
8. 11111110_2
9. 111100001111_2
10. 110011001100_2
11. 100000000001_2
12. 110000000011_2

Find the binary equivalent for each of the following decimal numbers.

13. 7
14. 2
15. 12
16. 10

Represents additionally challenging problems.

17. 15

18. 49

19. 57

20. 64

21. 107

22. 112

23. 1024

24. 2003

Translate the following binary phone numbers into their decimal equivalents. Each digit is separated by a hyphen.

25. $1000_2 - 0110_2 - 0111_2 - 0101_2 - 0011_2 - 0000_2 - 1001_2$

26. $0101_2 - 0101_2 - 0101_2 - 0010_2 - 0111_2 - 0100_2 - 0011_2$

27. $0001_2 - 1000_2 - 0000_2 - 0000_2 - 1000_2 - 0111_2 - 0110_2 - 0101_2 - 0011_2 - 0101_2 - 0011_2$

28. $0001_2 - 1000_2 - 0000_2 - 0000_2 - 0010_2 - 0110_2 - 0101_2 - 0101_2 - 0011_2 - 0010_2 - 1000_2$

Translate the following binary combination into its decimal equivalent.

29. $1001_2 - 1010_2 - 0100_2$

30. Explain the difference between your answer in Exercise 29 and the decimal equivalent of the binary number: 100110100100_2.

Binary numbers can also be represented as groups of blocks. Instead of taking values of 1 and 0, the blocks can be filled or unfilled. For example, the binary number 1101_2 (equivalent to the decimal number 13) could be expressed as the following group of blocks:

2^3	2^2	2^1	2^0
8	4	2	1
1	1	0	1

Write the binary number represented by the following sets of blocks, then translate each binary number into its decimal equivalent.

31. 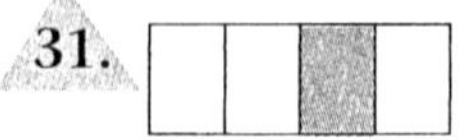

32. 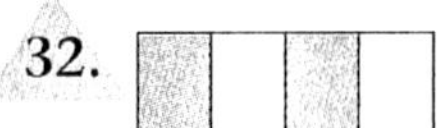

33. 

34. 

35. 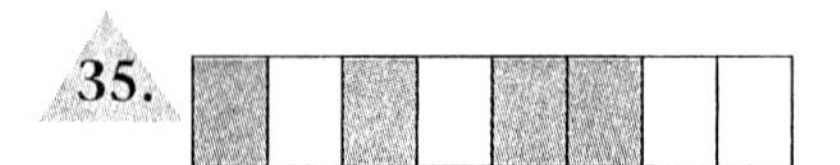

36. 

2.2 Base Two Arithmetic: Addition and Multiplication

Bits of History

A.D. 300–400 The birth of the modern decimal system can be traced back to India. While many cultures used some sort of base ten system, the Brâhmî system of nine digits was modified to include a small circle or dot, which represented zero. The beginning of our modern numerical notation and modern arithmetic can be attributed to both Indian and Arabic scholars, but it would still be almost a thousand years before Western Europe would employ the Hindu-Arabic numerals.

Adding in Base Two

The principles of adding binary numbers are exactly the same as the principles of adding decimal numbers. We work right to left and carry when we do not have a digit large enough to express the sum. First, as a reminder, look at this decimal sum.

$$\begin{array}{r} 357 \\ +\,172 \\ \hline \end{array}$$

$$\begin{array}{r} 357 \\ +\,172 \\ \hline 9 \end{array}$$

The sum of the digits on the right is 9.

$$\begin{array}{r} {\scriptstyle 1} \\ 357 \\ +\,172 \\ \hline 29 \end{array}$$

The sum of the next pair of digits is 12. We write the 2 as part of the sum and carry the 1 (which is actually 10 tens, or 1 hundred).

$$\begin{array}{r} {\scriptstyle 1} \\ 357 \\ +\,172 \\ \hline 529 \end{array}$$

In our first example, we use the same process to add two binary numbers. Note that, when we add two binary numbers, we get the following sums.

$0_2 + 0_2 = 0_2$	**In decimal notation, this is $0 + 0 = 0$.**
$0_2 + 1_2 = 1_2$	**This is the same as in decimal notation.**
$1_2 + 0_2 = 1_2$	**Again, it's the same as the decimal form.**
$1_2 + 1_2 = 10_2$	**In decimal notation, we say $1 + 1 = 2$. The 10_2 indicates 1 in the twos' place and 0 in the ones' place.**

EXAMPLE 1 Add the two binary numbers.

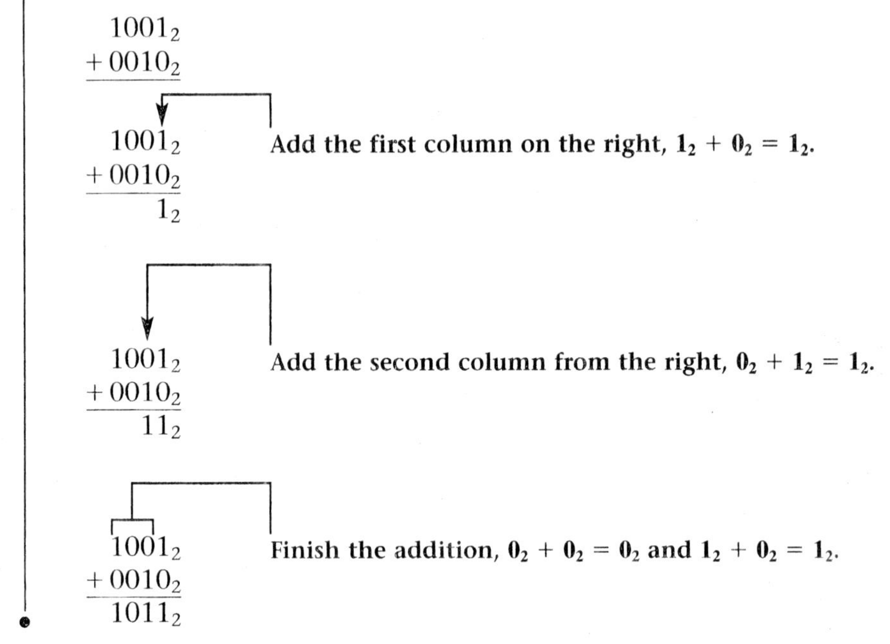

We can verify our result by converting all three numbers to their decimal equivalent:

$$1001_2 = 9, \qquad 0010_2 = 2, \qquad 1011_2 = 11.$$

Because 9 + 2 = 11, we have verified the result.

PRACTICE PROBLEM 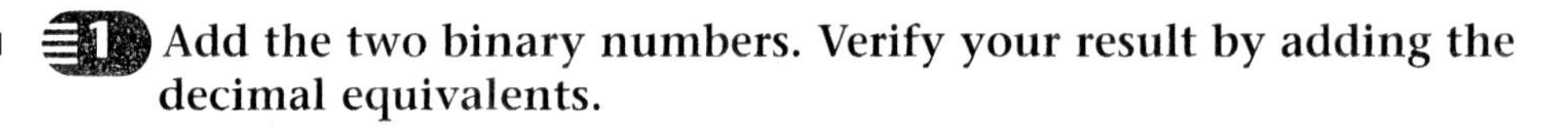
1 Add the two binary numbers. Verify your result by adding the decimal equivalents.

$$\begin{array}{r} 1010_2 \\ +\,0101_2 \\ \hline \end{array}$$

In decimal addition, what happens when we add 2 two-digit numbers such as 67 and 84? Because we carry in the tens' place, we end up with a three-digit answer, 151. Addition of binary numbers is similar.

PRACTICE PROBLEM Answers
on page 61

Example 2 **Add the binary numbers 1101_2 and 1011_2.**

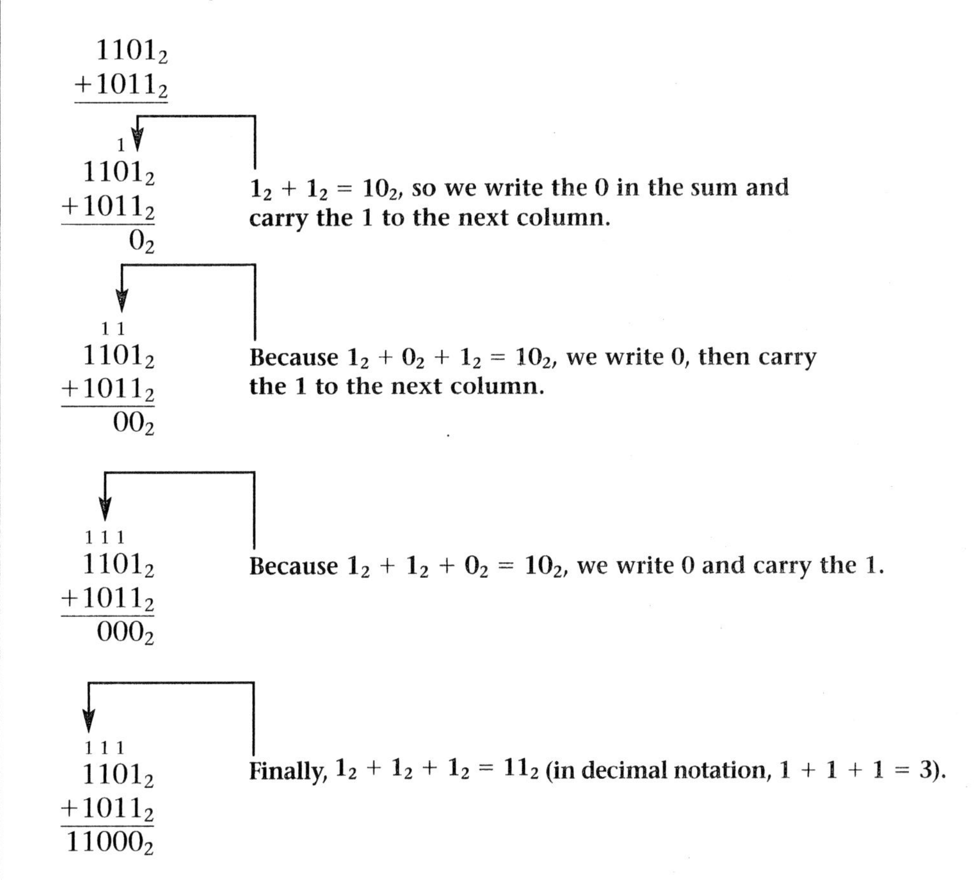

Now, let's convert all three numbers to the decimal equivalent to check our result:

$$1101_2 = 13, \qquad 1011_2 = 11, \qquad 11000_2 = 24.$$

Indeed, $13 + 11 = 24$. We have verified our result.

If you recall the Canadian coins discussed in Section 2.1, you may be able to better relate to the idea of carrying in binary addition. Let $\mathbf{0101}_C$ represent (from right to left) one quarter, no halves, one dollar and no two-dollar coins.

By the same model, **0011**$_{C}$ would represent one quarter and one half.

What happens when we add these two together (**0011**$_{C}$ + **0101**$_{C}$)?

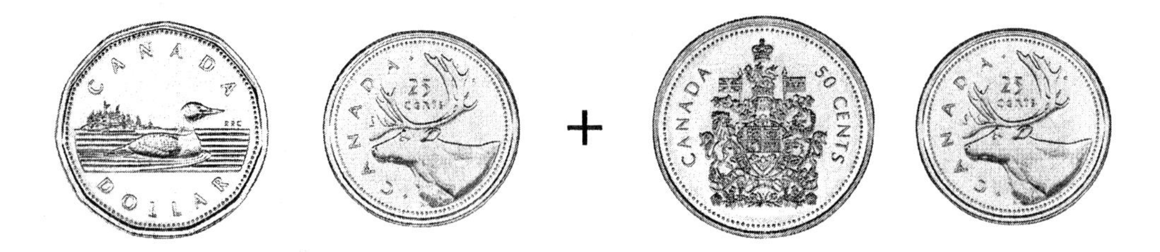

The 2 quarters become a half.

The 2 halves become a dollar.

The 2 one-dollar coins become a two-dollar coin.

We have demonstrated that $\mathbf{0101_C} + \mathbf{0011_C} = \mathbf{1000_C}$.

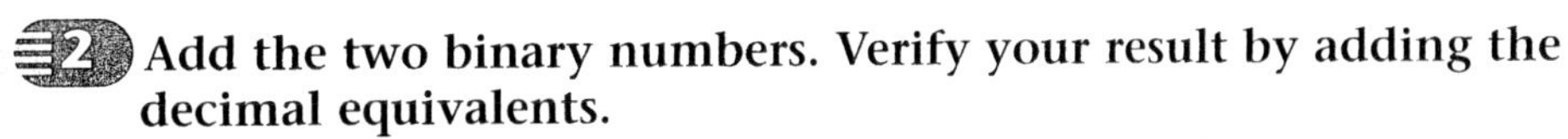

PRACTICE PROBLEM 2 **Add the two binary numbers. Verify your result by adding the decimal equivalents.**

$$\begin{array}{r} 1110_2 \\ +\ 1101_2 \\ \hline \end{array}$$

EXAMPLE 3 **Add the two bytes. Verify your result by adding the decimal equivalents.**

$$\begin{array}{r} {\scriptstyle 1111} \\ 10110110_2 \\ +\ 00111101_2 \\ \hline 11110011_2 \end{array}$$

Now let's convert each byte to its decimal equivalent to check our results.

$$10110110_2 = 128 + 0 + 32 + 16 + 0 + 4 + 2 + 0 = 182$$
$$00111101_2 = 0 + 0 + 32 + 16 + 8 + 4 + 0 + 1 = 61$$
$$11110011_2 = 128 + 64 + 32 + 16 + 0 + 0 + 2 + 1 = 243$$

Because $182 + 61 = 243$, we have verified the results.

Once you are confident in your ability to add two binary numbers, there will be no need to convert to base ten to verify your result. In fact, you may want to convert decimals to binary numbers to verify the result of adding two decimals!

PRACTICE PROBLEM 3 **Add the 2 bytes. Verify your result by adding the decimal equivalents.**

$$\begin{array}{r} 10101001_2 \\ +\ 01001101_2 \\ \hline \end{array}$$

PRACTICE PROBLEM Answers on page 61

Multiplying in Base Two

When you work in base ten, the easiest multiplications involve multiplying by a power of ten. Look at the following products.

$53 \times 10 = 530$
$53 \times 100 = 5300$
$53 \times 1000 = 53000$
$53 \times 10000 = 530000$

The same principle works in the binary system, except that you are multiplying by powers of two.

$11_2 \times 10_2 = 110_2$
$11_2 \times 100_2 = 1100_2$
$11_2 \times 1000_2 = 11000_2$

This principle is used in the next example.

EXAMPLE 4 **Find the product.**

$1011_2 \times 100_2 = 101100_2$

PRACTICE PROBLEM 4 **Find the product.**

$1011_2 \times 1000_2$

In Example 5, we will learn to multiply any two binary numbers. First, let's look at a decimal multiplication model.

Because $(123 \times 12) = (123 \times 10) + (123 \times 2) = 1230 + 246$, we can multiply in the following way:

$$\begin{array}{r} 123 \\ \times \quad 12 \\ \hline 246 \end{array}$$

First, we multiply by the 2.

$$\begin{array}{r} 123 \\ \times \quad 12 \\ \hline 246 \\ 0 \end{array}$$

We write a 0 to indicate that we are multiplying by tens.

$$\begin{array}{r} 123 \\ \times \quad 12 \\ \hline 246 \\ 1230 \\ \hline 1476 \end{array}$$

We multiply by 1 and then add.

PRACTICE PROBLEM Answers on page 61

EXAMPLE 5 **Find the product.**

$1011_2 \times 111_2$

Recall that $111_2 = 100_2 + 10_2 + 1_2$.

When you multiply binary numbers, you use a process that is sometimes called "shift and add."

$$\begin{array}{r} 1011_2 \\ \times\ \ 111_2 \\ \hline 1011_2 \end{array}$$

Multiply by the 1.

$$\begin{array}{r} 1011_2 \\ \times\ \ 111_2 \\ \hline 1011 \\ 10110 \end{array}$$

Multiply by the 10_2.
Write the result (10110) under the 1011.

$$\begin{array}{r} 1011_2 \\ \times\ \ 111_2 \\ \hline 1011 \\ 10110 \\ 101100 \\ \hline 1001101_2 \end{array}$$

Multiply by the 100_2 and then add.

PRACTICE PROBLEM 5 **Find the product.**

$$\begin{array}{r} 1101_2 \\ \times\ \ 101_2 \\ \hline \end{array}$$

ANSWERS TO PRACTICE PROBLEMS:

1. 1111_2; $10 + 5 = 15$ **2.** 11011_2; $14 + 13 = 27$

3. 11110110_2; $169 + 77 = 246$ **4.** 1101000_2 **5.** 1000001_2

SECTION

2.2 Exercises

Add the following binary numbers. Rewrite each problem in decimal notation to check your work.

1. $0100_2 + 0010_2$

2. $0101_2 + 0010_2$

3. $0101_2 + 0110_2$

4. $1010_2 + 0011_2$

5. $1101_2 + 0111_2$

6. $1011_2 + 0111_2$

7. 1010_2, 1101_2, $+ 1011_2$

8. 1011_2, 1101_2, $+ 1111_2$

9. $10010010_2 + 10101010_2$

10. $10001110_2 + 10101010_2$

11. $10110111_2 + 01100100_2$

12. $00011010_2 + 11010111_2$

Represents additionally challenging problems.

Add the following binary numbers in block form. Shade each answer in the empty set of blocks below.

13.

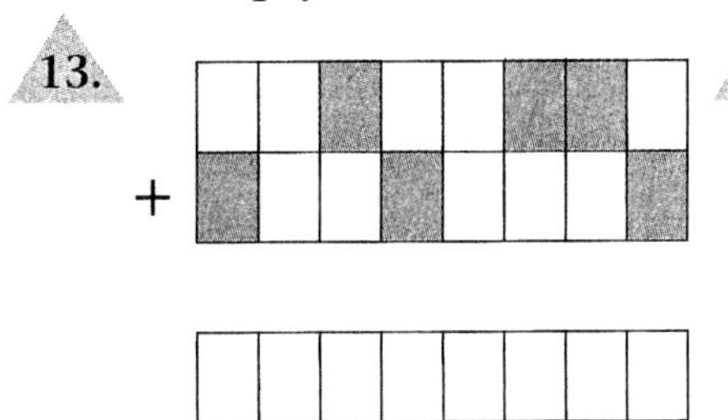

14.

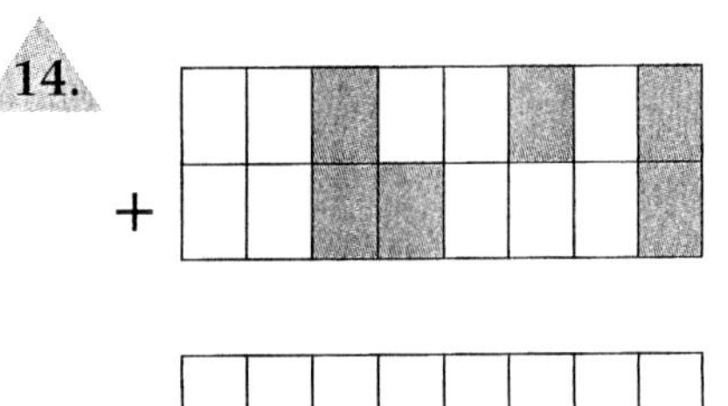

15.

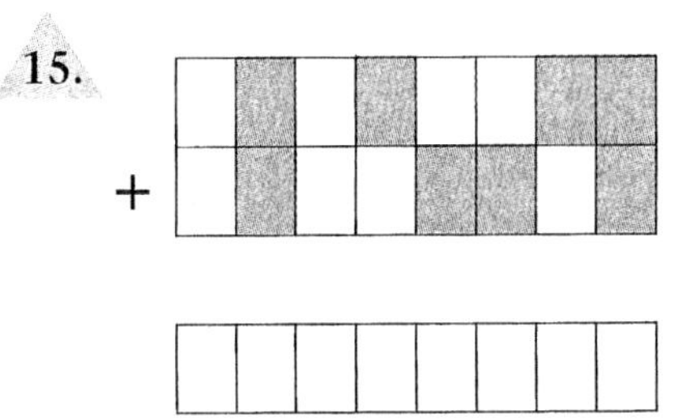

Multiply each pair of binary numbers. Rewrite each problem in decimal notation to check your work.

16. $100_2 \times 10_2$

17. $1101_2 \times 10_2$

18. $101_2 \times 110_2$

19. $1110_2 \times 11_2$

20. $1101_2 \times 111_2$

21. $1011_2 \times 111_2$

22. Is the sum of 2 bits always a bit? Justify your answer.

23. Is the product of 2 bits always a bit? Justify your answer.

2.3 Base Two Arithmetic: Subtraction and Division

Subtracting in Base Two

As was the case with addition, the principles of subtracting binary numbers are exactly the same as the principles of subtracting decimal numbers. First, as a reminder, look at this decimal difference.

$$\begin{array}{r} 247 \\ -\ 162 \\ \hline \end{array}$$

$$\begin{array}{r} 247 \\ -\ 162 \\ \hline 5 \end{array}$$ **The difference of the digits on the right is 5.**

$$\begin{array}{r} {}^{1}\not{2}{}^{1}47 \\ -\ 162 \\ \hline 85 \end{array}$$ **We cannot subtract 6 from 4. We borrow from the hundreds' column, and the 4 tens is now 14 tens.**

$$\begin{array}{r} {}^{1}\not{2}47 \\ -\ 162 \\ \hline 85 \end{array}$$ **Finally, in the hundreds' column, 1 − 1 = 0.**

In our first two examples, we use the same process to subtract two binary numbers. Note that when we subtract two binary numbers, we have the following possibilities:

$0_2 - 0_2 = 0_2$	**In decimal notation, this is 0 − 0 = 0.**
$1_2 - 1_2 = 0_2$	**It is the same as in decimal notation.**
$1_2 - 0_2 = 1_2$	**Again, it is the same as the decimal form.**
$0_2 - 1_2 = ?$	**We must borrow from the column to the left to perform this operation.**

EXAMPLE 1 **Subtract the binary numbers.**

$$\begin{array}{r} 1011_2 \\ -\,0010_2 \\ \hline \end{array}$$

$$\begin{array}{r} 1011_2 \\ -\,0010_2 \\ \hline 1_2 \end{array}$$ **Subtract the first column on the right, $1_2 - 0_2 = 1_2$.**

$$\begin{array}{r} 1011_2 \\ -\,0010_2 \\ \hline 01_2 \end{array}$$ **Subtract the second column from the right, $1_2 - 1_2 = 0_2$.**

$$\begin{array}{r} 1011_2 \\ -\,0010_2 \\ \hline 1001_2 \end{array}$$ **Finish the subtraction, $0_2 - 0_2 = 0_2$ and $1_2 - 0_2 = 1_2$.**

We can verify our result by converting all three numbers to their decimal equivalent:

$1011_2 = 11, \quad 0010_2 = 2, \quad 1001_2 = 9.$

Because $11 - 2 = 9$, so we have verified the result.

PRACTICE PROBLEM 1 **Subtract the binary numbers. Verify your result by subtracting the decimal equivalents.**

$$\begin{array}{r} 1111_2 \\ -\,0101_2 \\ \hline \end{array}$$

In our second example, we look at a situation in which we must borrow from a column on the left to complete the subtraction.

EXAMPLE 2 **Subtract the binary number 1101_2 from 10011_2.**

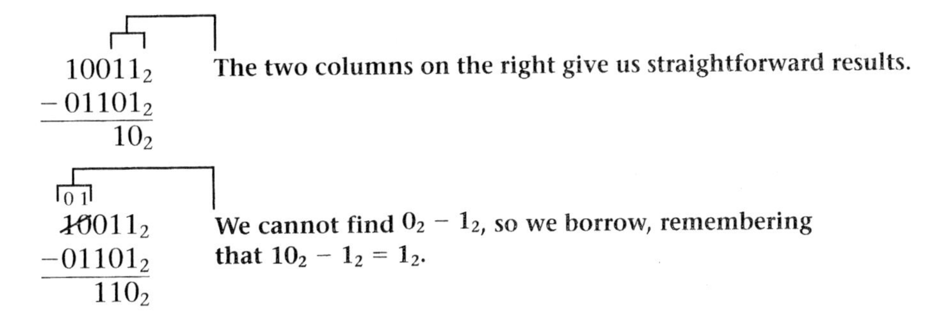

$$\begin{array}{r} 10011_2 \\ -\,01101_2 \\ \hline 10_2 \end{array}$$ **The two columns on the right give us straightforward results.**

$$\begin{array}{r} {}^{0\,1} \\ \not{1}\not{0}011_2 \\ -\,01101_2 \\ \hline 110_2 \end{array}$$ **We cannot find $0_2 - 1_2$, so we borrow, remembering that $10_2 - 1_2 = 1_2$.**

PRACTICE PROBLEM Answers on page 68

As in base ten arithmetic, we can check our results through addition. Note that $110_2 + 01101_2 = 10011_2$ (in base ten, $6 + 13 = 19$).

PRACTICE PROBLEM

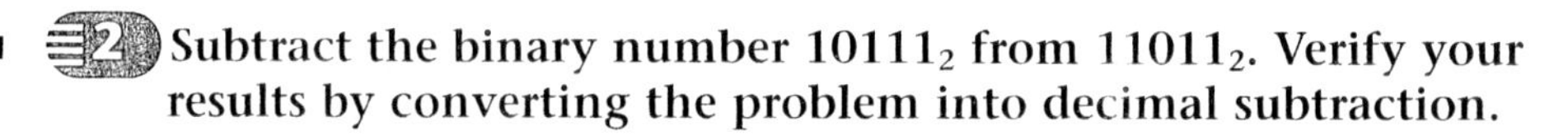

2 Subtract the binary number 10111_2 from 11011_2. Verify your results by converting the problem into decimal subtraction.

Once again we refer to the coins discussed in Section 2.1. This time we demonstrate the idea of borrowing. Let $\mathbf{1001_C}$ represent (from right to left) 1 quarter, no halves, no dollars, and 1 two-dollar coin.

From this collection, we wish to subtract 1 dollar, which is represented as $\mathbf{100_C}$.

To accomplish this subtraction, we break the two-dollar coin into 2 one-dollar coins and use one of them to make the payment. Essentially, we borrow from the fourth place to the left. We have a dollar and the quarter remaining.

We have demonstrated that $\mathbf{1001_C} - \mathbf{0100_C} = \mathbf{101_C}$.

Dividing in Base Two

In Chapter 1, we looked at multiplication as repeated addition and exponents as repeated multiplication. In much the same way, we now look at division as repeated subtraction.

PRACTICE PROBLEM Answers
on page 68

When dividing 20 by 5, you find that there are four groups of 5 in the number 20.

$$
\begin{aligned}
20 - 5 &= 15\\
15 - 5 &= 10\\
10 - 5 &= 5\\
5 - 5 &= 0
\end{aligned}
$$

This is similar to one process by which a computer accomplishes division.

When you actually started doing division, you learned an algorithm we refer to as *long division*. We will examine that algorithm as it applies to binary division, but before we move to that example, you should review a base ten division problem.

We suggest that you recreate the steps used to divide 5,016 by 11 using long division.

$$
\begin{array}{r}
456\\
11\overline{)5016}
\end{array}
$$

EXAMPLE 3 **Divide 10010_2 by 11_2.**

$$
11_2\overline{)10010_2}
$$

11_2 does not go into 10_2, but 11_2 *does* go into 100_2 one time.

$$
\begin{array}{r}
1\\
11_2\overline{)10010_2}\\
-11\\
\hline
1
\end{array}
$$

Multiply 1_2 times 11_2 to get 11_2, and then subtract 11_2 from 100_2 to get 1_2

$$
\begin{array}{r}
1\\
11_2\overline{)10010_2}\\
-11\\
\hline
11
\end{array}
$$

Bring down the 1.

$$
\begin{array}{r}
11\\
11_2\overline{)10010_2}\\
-11\\
\hline
11\\
-11\\
\hline
00
\end{array}
$$

11_2 goes into 11_2 one time. Bring down the 0.

$$
\begin{array}{r}
110\\
11_2\overline{)10010_2}\\
-11\\
\hline
11\\
-11\\
\hline
00\\
-00\\
\hline
0
\end{array}
$$

11_2 goes into 0_2 zero times. There is no remainder.

We can check our answer using binary multiplication.

$$\begin{array}{r} 110_2 \\ \times\ \ 11_2 \\ \hline 110_2 \\ 1100_2 \\ \hline 10010_2 \end{array}$$

PRACTICE PROBLEM 3 **Divide 10010_2 by 10_2, and then verify your answer using binary multiplication.**

ANSWERS TO PRACTICE PROBLEMS

1. 1010_2; $15 - 5 = 10$ **2.** 100_2; $27 - 23 = 4$

3. 1001_2; $\begin{array}{r} 1001_2 \\ \times\ \ 10_2 \\ \hline 0000_2 \\ 10010_2 \\ \hline 10010_2 \end{array}$

SECTION

2.3 Exercises

Perform the following binary subtractions. Rewrite each problem in decimal notation to check your work.

1. $0110_2 - 0010_2$

2. $0111_2 - 0010_2$

3. $1111_2 - 0110_2$

4. $0100_2 - 0010_2$

5. $0101_2 - 0010_2$

6. $1001_2 - 0110_2$

7. $10011010_2 - 00011111_2$

8. $10101101_2 - 01000111_2$

9. Use repeated subtraction to find the number of times 10_2 goes into 1010_2.

10. Use repeated subtraction to find the number of times 11_2 goes into 1001_2.

11. Use long division to find the number of times 10_2 goes into 1010_2.

12. Use long division to find the number of times 11_2 goes into 1001_2.

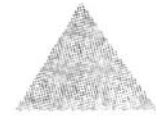 **Represents additionally challenging problems.**

Simplify each of the following by using long division. Convert the values to decimal notation to verify your answer.

13. $11000110_2 \div 110_2$

14. $11011101_2 \div 1101_2$

Simplify each of the following by using long division. Once you have found the quotient, use it to rewrite the problem as a multiplication problem. (*Hint:* There will be a remainder. Practice with base ten first so that you remember what to do with the remainder.)

15. $11000110_2 \div 111_2$

16. $11000101_2 \div 1101_2$

2.4 Two's Complement Notation

BITS of HISTORY

After the fall of the Roman Empire, much of Western Europe entered a period known as the Dark Ages, where scholarly activity was limited primarily to monks who worked in isolation. In the East, however, mathematics continued to flourish. The great Muslim mathematician and astronomer, Al Khuwārzmi, improved upon previous numerical and algebraic contributions, which eventually would be disseminated to the Western World. The Latin version of his name, Algorismi, would eventually translate into our modern word *algorithm*.

To this point, we have seen only non-negative integers represented in binary notation, yet we know that the computer must be able to represent many other kinds of numbers, including decimals, fractions, and negative numbers. We will now examine the way in which bits can be used to represent negative integers.

In a computer, signed numbers are represented using something called **two's complement notation**. This form is used to ensure that addition and multiplication work efficiently. The method used in Examples 1 and 2 to add two binary numbers must work for any two integers, positive or negative.

To write an integer in two's complement notation, we must first identify whether the given number is positive, negative, or zero. If the number is a positive integer or zero, its binary representation begins with a 0. No matter how many bits we have available for non-negative integer representation, the bit farthest to the left is zero. This bit is called the **sign bit**, because it tells us the sign of the number. If we have 4 bits available, the non-negative numbers will have the following binary representation.

BINARY	DECIMAL
0111_2	7
0110_2	6
0101_2	5
0100_2	4
0011_2	3
0010_2	2
0001_2	1
0000_2	0

Earlier in Chapter 2, we used 4 bits to represent the decimal numbers 0–15. In two's complement notation, 4 bits allows us to represent only the non-negative decimal numbers 0–7, because the sign bit (now set to zero) is in the fourth position. You may have guessed that the binary representation for a negative number will begin with a 1, and you are correct. Given a negative integer, its representation can be found by using the following algorithm.

Finding the Two's Complement Representation for a Negative Integer

To find the two's complement representation for negative integer x,

Step 1. Find the binary representation for $|x|$.

Step 2. Find the complement of that binary number (change the 1s to 0s and the 0s to 1s).

Step 3. Add 1 to the result of step 2.

The result is the two's complement representation for x.

EXAMPLE 1 **(a) Find the two's complement notation for −1, using 4 bits.**

1. Find the binary representation for $|-1|$.

$$|-1| = 1$$
$$1 = 0001_2$$

2. Find the complement of that binary number (change the 1s to 0s and the 0s to 1s).

   The complement of 0001_2 is 1110_2.

3. Add 1 to the result of step 2.

   Adding 1, we get 1111_2, so

   $-1 = 1111_{2*}$.

We will use the subscript 2* to indicate a binary number that uses two's complement representation. Note that $\mathbf{1111_2 = 15}$, whereas $\mathbf{1111_{2*} = -1}$ (two very different numbers).

(b) Find the two's complement notation for −4, using 4 bits.

1. Find the binary representation for $|-4|$.

$$4 = 0100_2$$

2. Find the complement of that binary number (change the 1s to 0s and the 0s to 1s).

   The complement of 0100_2 is 1011_2.

3. Add 1 to the result of step 2.

   Adding 1, we get

$$\begin{array}{r} {\scriptstyle 1\,1} \\ 1011_2 \\ +\quad 1_2 \\ \hline 1100_2 \end{array}$$

   so $-4 = 1100_{2*}$.

Again, notice that $1100_2 = 12$, whereas $1100_{2*} = -4$.

Practice Problem  **Find the two's complement notation for −6 and −5, using 4 bits.**

It is sometimes useful to find the decimal equivalent for a given two's complement representation. The following algorithm will help you do this.

Finding a Negative Integer for a Two's Complement Representation

To find a negative integer when you are given its two's complement representation,

Step 1. Find the complement of the given binary number (change the 1s to 0s and the 0s to 1s).

Step 2. Add 1 to the result of step 1.

Step 3. Convert the result of step 2 into an integer.

Step 4. The negative of the integer from step 3 is the decimal equivalent of the two's complement representation.

Example 2 **Convert the two's complement number 1110_{2*} into its decimal equivalent.**

Step 1. Find the complement of 1110_2, which is 0001_2.

Step 2. Add 1, so $0001_2 + 1_2 = 0010_2$.

Step 3. $0010_2 = 2$

Step 4. $1110_{2*} = -2$

Practice Problem  **Convert the two's complement number 1001_{2*} into its decimal equivalent.**

Two's Complement Addition

You may be asking yourself why we should learn the two's complement notation when it seems so complicated. The answer lies in the elegance and efficiency with which the computer can now add two signed numbers. To add in two's complement notation, we simply use the same rule that we used when we added two binary numbers in Section 2.2. The next example illustrates this process.

Practice Problem Answers
on page 75

EXAMPLE 3 **Using two's complement notation, add the following numbers.**

(a) 0010_{2^*} **2 in decimal notation.**
$+0100_{2^*}$ **4 in decimal notation.**
0110_{2^*} **6 in decimal notation.**

(b) 1100_{2^*} **−4 in decimal notation.**
$+0010_{2^*}$ **2 in decimal notation.**
1110_{2^*} **As we saw in the first example, this is −2 in decimal notation.**

PRACTICE PROBLEM 3 **Using two's complement notation, add the following numbers.**

$$
\begin{array}{r}
0001_{2^*} \\
+\,1100_{2^*} \\
\hline
\end{array}
$$

Any binary addition that involves the sum of a negative and a positive can be done using two's complement representation.

EXAMPLE 4 **Using two's complement notation, add the following numbers.**

1100_{2^*} **−4 in decimal notation.**
$+0110_{2^*}$ **6 in decimal notation.**
$(1)0010_{2^*}$ **If you ignore the 1 in the fifth bit, 0010_{2^*} is positive 2 in decimal notation.**

Any time we are adding a positive number and a negative number using two's complement notation, we can ignore a 1 in the bit to the left of the sign bit.

PRACTICE PROBLEM 4 **Using two's complement notation, add the following numbers.**

$$
\begin{array}{r}
1110_{2^*} \\
+\,0111_{2^*} \\
\hline
\end{array}
$$

When you are adding two positive or two negative numbers using four bits, a 1 in the fifth bit may create a problem. The next example illustrates such a case.

PRACTICE PROBLEM Answers on page 75

EXAMPLE 5 **Using two's complement notation, add the following numbers.**

1010_{2*}	**−6 in decimal notation.**
$+\ 1101_{2*}$	**−3 in decimal notation.**
$(1)0111_{2*}$	**If you ignore the 1 in the fifth bit, the result in decimal notation is 7, clearly not the sum of −3 and −6.**

Here, we have what is called an **overflow error**. Such an error occurs whenever the information is too large for the storage that is allotted. Note that the answer (−9) is correct if we use 5 bits, but we are restricted to 4 bits in this example. The size of the numbers we can add (and of the answers we can reasonably interpret) is determined by the number of bits we have to work with. In the 4-bit notation used in the previous example, we are limited to $2^4 = 16$ different numbers. That is to say, we can represent only sixteen different numbers, −8, −7, −6, −5, −4, −3, −2, −1, 0, 1, 2, 3, 4, 5, 6, 7.

If we had 8 bits, we could represent $2^8 = 256$ different numbers (−128 to 127). With 32 bits, we can represent $2^{32} = 4{,}294{,}967{,}296$ different numbers. No matter how many bits we use, there are numbers that are too large or too small to be represented in our system. When we try to represent such numbers, or we end up with answers that are too large, we encounter an overflow error. Some overflow errors are caught by the computer, some disable the computer, and (perhaps the most dangerous) some just return an incorrect answer. In subsequent courses, you will learn more about overflow errors.

ANSWERS TO PRACTICE PROBLEMS

1. $1010_{2*} = -6$ and $1011_{2*} = -5$ **2.** −7

3. $1101_{2*} = -3$ **4.** $(1)0101_{2*} = 5$

SECTION 2.4

Exercises

Translate each of the following decimal numbers into two's complement notation. Use a 4-bit representation.

1. 2
2. 6
3. −6
4. −1

Each of the following numbers is in 4-bit, two's complement notation. Translate each into its decimal equivalent.

5. 0101_2
6. 1101_2
7. 1011_2
8. 0100_2

Translate each of the following decimal numbers into two's complement notation. Use an 8-bit representation.

9. 6
10. −13
11. −62
12. 95
13. −108
14. −121

Each of the following numbers is in 8-bit, two's complement notation. Translate each into its decimal equivalent.

15. 00011101_2
16. 11111011_2
17. 10111010_2
18. 01100010_2
19. 11111111_2
20. 10000010_2

▲ Represents additionally challenging problems.

Add the following binary numbers in 4-bit, two's complement notation. Identify any overflow errors.

21. $0101_{2^*} + 0010_{2^*}$

22. $1010_{2^*} + 1111_{2^*}$

23. $1010_{2^*} + 0110_{2^*}$

24. $0110_{2^*} + 0100_{2^*}$

25. $1001_{2^*} + 1011_{2^*}$

26. $1011_{2^*} + 0111_{2^*}$

Add the following binary numbers in 8-bit, two's complement notation. Rewrite each problem in decimal notation to check your work.

27. $10001110_{2^*} + 10101010_{2^*}$

28. $10110111_{2^*} + 01100100_{2^*}$

29. $00011010_{2^*} + 11010111_{2^*}$

30. How many different numbers can be represented in 4-bit, two's complement notation? What is the minimum value? What is the maximum value?

31. How many different numbers can be represented in 8-bit, two's complement notation? What is the minimum value? What is the maximum value?

32. Many computers use 32-bit, two's complement notation. How many different numbers can be represented in 32-bit, two's complement notation? What is the minimum value? What is the maximum value?

2.5 Binary Fractions

BITS of HISTORY

Although fractions were used during ancient times, their notation was cumbersome. It was not until the late fifteenth century that advances in notation for decimal fractions would become apparent. The Swiss mathematician and watchmaker Jost Bürgi made a significant improvement on previous notation by placing a ring sign above the digit in the unit's position. For example, our modern $73\frac{25}{100}$ would have been written $7\mathring{3}\ 25$. The Italian Magini is credited with replacing the ring with a dot we now call a decimal point between the units and the tenths position, thus giving us our modern convention of 73.25.

Before we introduce binary fractions, let's review some concepts you have encountered while learning about decimal fractions.

The following diagram shows the value of each place of the decimal number 452,976.

DIGIT	4	5	2	9	7	6
PLACE VALUE	10^5	10^4	10^3	10^2	10^1	10^0

What happens to place value as we move to the right of a decimal point?

First, let's look at the names associated with the place values for the decimal fraction 0.3574, which can also be written as $\frac{3574}{10,000}$.

NUMBER	.3	5	7	4
DIGIT	3	5	7	4
PLACE VALUE NAME	tenths	hundredths	thousandths	ten thousandths
PLACE VALUE	$\frac{1}{10}$	$\frac{1}{100}$	$\frac{1}{1000}$	$\frac{1}{10,000}$

Now, look at the same decimal fraction if we label the place values as powers of ten.

NUMBER	.3	5	7	4
DIGIT	3	5	7	4
PLACE VALUE NAME	tenths	hundredths	thousandths	ten thousandths
PLACE VALUE	10^{-1}	10^{-2}	10^{-3}	10^{-4}

Note that the pattern established to the left of the decimal continues to the right. The same pattern holds true when we write binary fractions. There are a couple of obvious differences when the fraction is binary. Although the pattern of the powers is the same, the base is two instead of ten. Also, we cannot call the period a decimal point, so we call it a **binary point**.

Look at the place values for the binary fraction 111.1111_2.

DIGIT	1	1	1	.	1	1	1	1
PLACE VALUE NAME	fours	twos	ones	BINARY POINT	halfs	fourths	eighths	sixteenths
PLACE VALUE	2^2	2^1	2^0		2^{-1}	2^{-2}	2^{-3}	2^{-4}
DECIMAL EQUIVALENT	4	2	1		$\frac{1}{2}$	$\frac{1}{4}$	$\frac{1}{8}$	$\frac{1}{16}$

Do you see the pattern? Each place value is one-half the place value to the left. We will use this decimal equivalence to rewrite binary fractions as the decimal equivalent.

EXAMPLE 1 **Find the place value for the 1 in the binary fraction 0.001_2.**

The place value is $2^{-3} = \frac{1}{8}$.

PRACTICE PROBLEM 1 **Find the place value for the 1 in the binary fraction 0.0001_2.**

EXAMPLE 2 **Rewrite the binary fraction 110.0101_2 as its decimal equivalent.**

Summing the place values, we find

$$110.0101_2 = 4 + 2 + 0 + 0 + \frac{1}{4} + \frac{1}{16} = 6\frac{5}{16} = 6.3125.$$

(We work from the binary point to the left, and then from the binary point to the right.)

PRACTICE PROBLEM 2 **Rewrite the binary fraction 11.101_2 as its decimal equivalent.**

PRACTICE PROBLEM Answers on page 84

Using the same technique that allows us to write 0.3574 as $\frac{3574}{10{,}000}$ rather than as $\frac{3}{10} + \frac{5}{100} + \frac{7}{1000} + \frac{4}{10{,}000}$, we can more quickly compute a binary fraction's decimal equivalent. The important thing to note from the decimal fraction is this:

To write a decimal as a decimal fraction, we use the number after the decimal point as the numerator, and we write the denominator as a 1 followed by a zero for each place after the decimal point.

EXAMPLE 3 **Write each binary as a binary fraction.**

(a) $0.011_2 = \frac{11_2}{1000_2}$

Note that we use the same process that we use to write 0.3574 as $\frac{3574}{10{,}000}$.

(b) $0.00000001_2 = \frac{1_2}{100000000_2}$

(c) $0.01001011_2 = \frac{1001011_2}{100000000_2}$

PRACTICE PROBLEMS 3 **Write each binary as a binary fraction.**

(a) 0.0101_2 (b) 0.00000011_2 (c) 0.11111111_2

In the next example, we translate binary fractions into decimal fractions.

EXAMPLE 4 **Using the fractions from Example 3, write each binary fraction as a decimal fraction.**

(a) $\frac{11_2}{1000_2} = \frac{3}{8}$ **or we could say that** $\mathbf{0.011_2 = 0 + 0 + \frac{1}{4} + \frac{1}{8} = \frac{3}{8}}$

(b) $\frac{1_2}{100000000_2} = \frac{1}{256}$ **or** $\mathbf{0.00000001_2 = \frac{1}{256}}$

(c) $\frac{1001011_2}{100000000_2} = \frac{75}{256}$ **or** $\mathbf{0.01001011_2 = \frac{75}{256}}$

PRACTICE PROBLEM Answers on page 84

Note that any time we have 8 zeros after the binary point, the decimal equivalent fraction has a denominator of 256. Because there are 8 bits in a byte, this is a commonly occurring denominator.

PRACTICE PROBLEMS 4 **Write each binary fraction as a decimal fraction.**

(a) 0.0101_2 **(b)** 0.00000011_2 **(c)** 0.11111111_2

To rewrite a decimal as a binary fraction, we first check to see if we can express the decimal with a denominator that is a power of two. As stated earlier, we usually use 8 bits to represent the fraction, which results in a denominator of 256.

EXAMPLE 5 **Find the binary equivalent for each decimal fraction.**

(a) $\frac{79}{256} = \frac{1001111_2}{100000000_2}$ or 0.01001111_2

(b) $\frac{5}{256} = \frac{101_2}{100000000_2}$ or 0.00000101_2

PRACTICE PROBLEMS 5 **Find the binary equivalent for each decimal fraction.**

(a) $\frac{93}{256}$ **(b)** $\frac{11}{256}$

Sometimes we need to rewrite a decimal fraction so that the denominator is 256. To accomplish that, we multiply the decimal by $\frac{256}{256}$, which equals 1. Example 6 illustrates this technique.

EXAMPLE 6 **Rewrite each decimal as a fraction with denominator 256, and then convert that fraction to its binary equivalent.**

(a) 0.01171875

$$0.01171875 \times \frac{256}{256} = \frac{3}{256} = \frac{11_2}{100000000_2} \text{ or } 0.00000011_2$$

(b) 0.1875

$$0.1875 \times \frac{256}{256} = \frac{48}{256} = \frac{110000_2}{100000000_2} \text{ or } 0.00110000_2$$

PRACTICE PROBLEM Answers on page 84

PRACTICE PROBLEMS 6 **Rewrite each decimal as a fraction with denominator 256, and then convert that fraction to its binary equivalent.**

(a) 0.01953125 (b) 0.3125

Can every decimal fraction be rewritten as a binary fraction? Any rational number can be written as a binary fraction. But if the denominator of the fraction is not a power of two, the fraction can be approximated only as a terminating binary.

EXAMPLE 7 **Find the 8-bit binary fraction that most closely approximates each of the following.**

(a) 0.1

$$0.1 \times \frac{256}{256} = \frac{25.6}{256} \approx \frac{26}{256} = 0.00011010_2$$

(Note that $\frac{26}{256}$ actually equals 0.1015625. As we said, we are approximating 0.1.)

(b) $\frac{2}{3}$

$$\frac{2}{3} \approx 0.66666667$$

$$0.66666667 \times \frac{256}{256} \approx \frac{170.6}{256} \approx \frac{171}{256}$$

$$\frac{171}{256} = 0.10101011_2$$

(Again, note that $\frac{171}{256} = 0.10101011_2 = 0.66796875$, which is not exactly $\frac{2}{3}$.)

PRACTICE PROBLEMS 7 **Find the 8-bit binary fraction that most closely approximates each of the following.**

(a) 0.2 (b) $\frac{1}{3}$

PRACTICE PROBLEM Answers on page 84

In Chapter 1, we discussed the concepts of absolute and relative errors. These ideas are used in the next example.

EXAMPLE 8 **Find the absolute and relative errors for each of the following approximations.**

(a) 0.0010_2 for the decimal value 0.1

(Recall that the **absolute error** is the absolute value of the difference.)

Because $0.0010_2 = 0 + 0 + \frac{1}{8} + 0 = 0.125$, the absolute error $= |0.125 - 0.1| = 0.025$.

(Recall that the **relative error** is the ratio of the absolute error to the actual value, usually expressed as a percentage.)

$$\text{The relative error} = \frac{|0.125 - 0.1|}{0.1} = \frac{0.025}{0.1} = 0.25 = 25\%.$$

(b) 0.1011_2 for $\frac{2}{3}$

Because $0.1011_2 = 0 + \frac{1}{2} + 0 + \frac{1}{8} + \frac{1}{16} = \frac{11}{16} = 0.6875$, the absolute

$$\text{error} = \left|.6875 - \frac{2}{3}\right| = \left|\frac{11}{16} - \frac{2}{3}\right| = \left|\frac{33}{48} - \frac{32}{48}\right| = \frac{1}{48}$$

and the

$$\text{relative error} = \frac{\frac{1}{48}}{\frac{2}{3}} = \frac{1}{48} \times \frac{3}{2} = \frac{3}{96} = \frac{1}{32} = 0.03125 = 3.125\%.$$

PRACTICE PROBLEMS 8 **Find the absolute and relative errors for each of the approximations you found in Practice Problems 7a and 7b.**

The concepts of this chapter have many applications in both mathematics and computer science. The binary representation that you have seen is the foundation for (although not identical to) floating-point notation. You will encounter further discussion of floating-point notation in your computer science course.

ANSWERS TO PRACTICE PROBLEMS

1. 0.0625 or $\frac{1}{16}$ 2. $3.625 = 3\frac{5}{8}$

3. (a) $\frac{101_2}{10000_2}$ (b) $\frac{11_2}{100000000_2}$ (c) $\frac{11111111_2}{100000000_2}$

4. (a) $\frac{5}{16}$ (b) $\frac{3}{256}$ (c) $\frac{255}{256}$ 5. (a) $\frac{1011101_2}{100000000_2}$ (b) $\frac{1011_2}{100000000_2}$

6. (a) $\frac{5}{256} = \frac{101_2}{100000000_2}$ (b) $\frac{80}{256} = \frac{1010000_2}{100000000_2}$

7. (a) $0.00110011_2 = \frac{51}{256}$ (b) $0.01010101_2 = \frac{85}{256}$

8. (a) absolute error = 0.00078125, relative error = 0.39%
   (b) absolute error = 0.0013021, relative error = 0.39%

SECTION

2.5 Exercises

Find the decimal equivalent for each of the following binary numbers.

1. 0.1000_2
2. 0.0010_2
3. 0.0110_2
4. 0.1011_2
5. 0.01101100_2
6. 0.11000110_2
7. 1011.1010_2
8. 1101.1110_2
9. 1111.01100100_2
10. $\frac{110_2}{1000_2}$
11. $\frac{1010_2}{10000_2}$
12. $\frac{1001010_2}{10000000_2}$
13. $\frac{11111110_2}{100000000_2}$
14. $\frac{111010_2}{10000000_2}$
15. $\frac{10101010_2}{100000000_2}$

Represents additionally challenging problems.

Find the binary equivalent for each of the following decimal numbers.

16. 0.25

17. 0.03125

18. 0.6875

19. 0.9375

20. 0.1640625

21. 0.33203125

22. 8.75

23. 12.625

24. 14.25390625

Use long division to convert each decimal fraction into a binary expansion.

25. $\frac{1}{8}$

26. $\frac{3}{4}$

27. $\frac{2}{3}$

28. $\frac{3}{5}$

29. Compare your answers in Exercises 25–28. Which fractions terminate in base ten? Which fractions terminate in base two? How can you tell?

30. Suppose we want to approximate the decimal fraction $\frac{1}{3}$ using a binary representation. Find the absolute and relative errors of each approximation. Which of the following is closest to the actual value?

(a) 0.0100_2

(b) 0.0101_2

(c) 0.01010100_2

31. Suppose we want to approximate the decimal fraction $\frac{1}{5}$ using a binary representation. Find the absolute and relative errors of each approximation. Which of the following is closest to the actual value?

(a) 0.0011_2 **(b)** 0.00110011_2 **(c)** 0.00110010_2

32. What is the best approximation of the decimal fraction $\frac{2}{3}$, using a denominator of 8, of 16, and of 256?

33. What is the best approximation of the decimal fraction $\frac{1}{6}$ using a denominator of 8, of 16, and of 256?

Each block of three represents a single binary number. Translate each binary number into its decimal equivalent.

34.

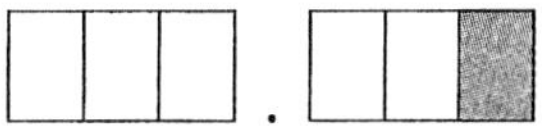

35.

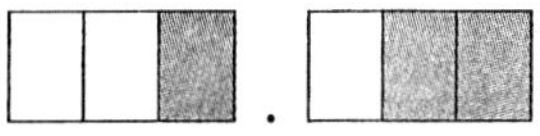

36.

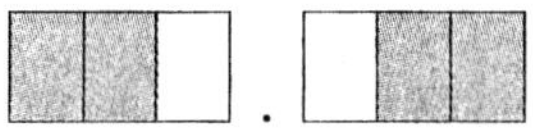

37.

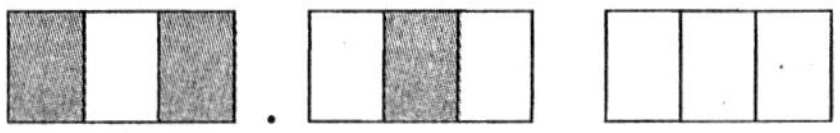

38.

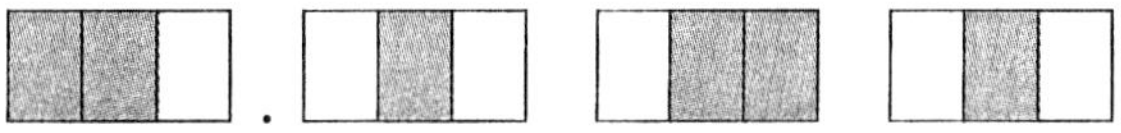

2.6 Computer Memory and Quantitative Prefixes

BITS of HISTORY

In 1202, Leonardo Fibonacci formally introduced the Western World to the Hindu-Arabic number system as well as posing many problems in both arithmetic and algebra.

Earlier in this chapter, we defined a byte as a basic computer memory unit, consisting of 8 bits. We are now prepared to delve further into the topic of computer memory. Much of our work in this section depends on the material covered in Section 1.6, Dimensional Analysis.

What is meant by the phrase "4K of memory"? The K is generally accepted to mean approximately 1000 bytes (8 bits unless otherwise specified), although, in fact, 1 K is actually 2^{10} (1024) bytes.

EXAMPLE 1 **How many bytes and how many bits are in 64 K of memory?**

$$64\text{K} = 64 \times 2^{10} \text{ bytes} = 2^6 \times 2^{10} \text{ bytes} = 2^{16} \text{ bytes} = 65{,}536 \text{ bytes}$$
$$2^{16} \text{ bytes} = 2^{16} \times 8 \text{ bits} = 2^{16} \times 2^3 \text{ bits} = 2^{19} \text{ bits} = 524{,}288 \text{ bits}$$

Note that, using dimensional analysis, we could also write

$$64 \text{ K} = 2^{16} \text{ bytes} \times \frac{2^3 \text{ bits}}{1 \text{ byte}} = 2^{19} \text{ bits.}$$

Note further that our use of unit analysis has helped us keep track of the appropriate units in the answer.

PRACTICE PROBLEM  **How many bytes and how many bits are in 128 K of memory?**

When we speak of 64 K of memory, K is an abbreviation for kilobyte. *Kilo* is the metric prefix that means 1000. A kilometer is 1000 meters, and a kilogram is 1000 grams. As we said earlier, when referring to memory, the 1000 is only an approximation. The actual value is 2^{10}, which equals 1024.

All of the prefixes used to describe computer memory come from the list of standard metric prefixes. The following table summarizes the units commonly used for computer memory.

PREFIX	METRIC MEANING	MEMORY USAGE	ACTUAL NUMBER OF BYTES	INTEGER EQUIVALENT
KILO-	1,000	Kilobyte (K)	2^{10}	1,024
MEGA-	1,000,000	Megabyte (MB)	2^{20}	1,048,576
GIGA-	1,000,000,000	Gigabyte (GB)	2^{30}	1,073,741,824
TERA-	1,000,000,000,000	Terabyte (TB)	2^{40}	1,099,511,627,776

PRACTICE PROBLEM Answers on page 93

EXAMPLE 2 **How many bytes are there in 128 GB?**

This is an example of a conversion problem. As we did in Section 1.6, we will use unit ratios. Here the unit ratio is $\frac{2^{30} \text{ bytes}}{1 \text{ GB}}$ from the preceding table.

$$128 \text{ GB} = 2^7 \text{ GB} = 2^7 \text{ GB} \times \frac{2^{30} \text{ bytes}}{1 \text{ GB}} = 2^{37} \text{ bytes}$$

PRACTICE PROBLEM 2 **How many bytes are there in 512 TB?**

A typical computer monitor display might have 1024 columns and 768 rows of **pixels** (this word derives from *picture elements*). In a monochrome (black and white) monitor, each pixel can be represented by a single bit. If the pixel is on (white), the bit has a value of 1. If the pixel is off (black), it has a value of 0. Different configurations of white pixels next to black ones can produce various shades of gray on a monochrome monitor.

EXAMPLE 3 **How many bytes are required to store a monochrome display that uses 1024 columns and 768 rows of pixels? (Note that we could say we have 1024 columns with 768 pixels per column.)**

The important conversion factors are $\frac{1 \text{ bit}}{1 \text{ pixel}}$ and $\frac{1 \text{ byte}}{8 \text{ bits}}$.

We have the expression

$$1024 \text{ columns} \times \frac{768 \text{ pixels}}{1 \text{ column}} \times \frac{1 \text{ bit}}{1 \text{ pixel}} \times \frac{1 \text{ byte}}{8 \text{ bits}}.$$

Simplifying the units, we have

$$\frac{1024 \times 768}{8} \text{ bytes} = 98{,}304 \text{ bytes}.$$

PRACTICE PROBLEM Answers
on page 93

PRACTICE PROBLEM 3 **How many bytes are required to store a monochrome display that uses 2048 columns and 1536 rows of pixels?**

In the previous example, we assumed that we were using a monochrome monitor. In most cases, the monitor has a palette of colors from which to choose. Such a selection requires more than 1 bit per pixel, as we see in the next example.

EXAMPLE 4 **Assume that 2 bytes are used to store the color for each pixel on a display screen. How many kilobytes are required to store a display that uses 1024 columns and 768 rows of pixels?**

The conversion factors are $\frac{2 \text{ bytes}}{1 \text{ pixel}}$ and $\frac{1 \text{ kilobyte}}{1024 \text{ bytes}}$.

We have the expression

$$1024 \text{ columns} \times \frac{768 \text{ pixels}}{1 \text{ column}} \times \frac{2 \text{ bytes}}{1 \text{ pixel}} \times \frac{1 \text{ kilobyte}}{1024 \text{ bytes}}.$$

Simplifying the units, we have

$$\frac{1024 \times 768 \times 2}{1024} \text{ kilobytes} = 1536 \text{ kilobytes}.$$

PRACTICE PROBLEM 4 **Assume that 2 bytes are used to store the color for each pixel on a display screen. How many megabytes are required to store a display that uses 2048 columns and 1536 rows of pixels?**

In Example 5, we examine the computer memory required for computer animation.

EXAMPLE 5 **Computer animation requires an ever-increasing amount of computer memory. One form of animation uses a series of pictures that is stored in memory. Assume that a picture is represented on a monitor screen by a rectangular array containing 1024 columns and 768 rows of pixels. How many bytes of memory are required to store an animation that consists of 32 pictures, if each pixel requires 1 byte of storage? After you find the number of bytes, convert the answer to megabytes.**

First, let's find the number of pixels required for each picture. There are 1024 columns and 768 rows of pixels, so

$$1024 \times 768 = 786{,}432 \frac{\text{pixels}}{\text{picture}}.$$

PRACTICE PROBLEM Answers on page 93

Alternatively, we could factor the 1024 and the 768 and get

$$1024 \times 768 = 2^{10} \times (3 \cdot 2^{8}) = 3 \times 2^{18} \frac{\text{pixels}}{\text{picture}}.$$

We want to find the number of pixels per animation.

$$32 \frac{\text{pictures}}{\text{animation}} = 2^{5} \frac{\text{pictures}}{\text{animation}}$$

$$2^{5} \frac{\text{pictures}}{\text{animation}} \times 3 \times 2^{18} \frac{\text{pixels}}{\text{picture}} = 3 \times 2^{23} \frac{\text{pixels}}{\text{animation}}$$

The problem indicates that each pixel takes 1 byte of memory, so the memory needed is 3×2^{23} bytes.

Converting that to MB, we have

$$3 \times 2^{23} \text{ bytes} \times \frac{1 \text{ MB}}{2^{20} \text{ bytes}} = 3 \times 2^{3} \text{ MB} = 24 \text{ MB}.$$

We use MB for abbreviating megabyte. Mb is commonly used for megabits.

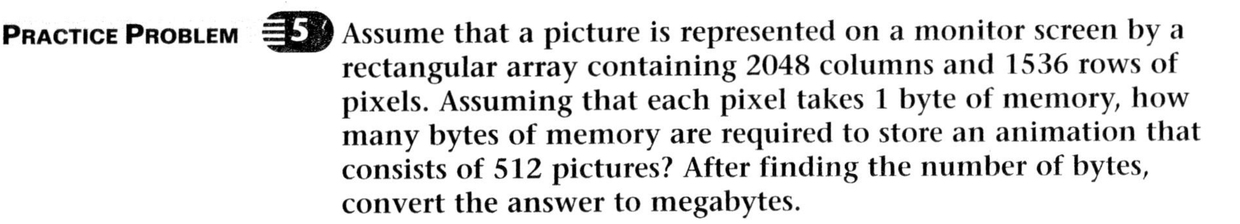

Practice Problem 5 **Assume that a picture is represented on a monitor screen by a rectangular array containing 2048 columns and 1536 rows of pixels. Assuming that each pixel takes 1 byte of memory, how many bytes of memory are required to store an animation that consists of 512 pictures? After finding the number of bytes, convert the answer to megabytes.**

To communicate very large and very small numbers, scientists throughout the world developed and agreed upon a set of prefixes that could be used with any unit of measure. Although these prefixes were not developed with the computer in mind, no other product or process comes close to the computer in terms of their usefulness. As computers become faster and memory larger, there is already a prefix designed to describe the change.

This set of prefixes is generally referred to as the SI, the International System of Units. The SI communicates the magnitude of the units by the use of appropriate prefixes. For example, the electrical unit of a watt is not a big unit, even in terms of ordinary household use, so it is generally used in terms of 1000 watts at a time. The prefix for 1000 is *kilo-*, so we use kilowatts (kW) as our unit of measurement. For makers of electricity, or bigger users such as industry, it is common to use megawatts (MW) or even gigawatts (GW).

Practice Problem Answers on page 93

The following table includes the full range of prefixes, their symbols or abbreviations, and their multiplying factors. These prefixes can be used to describe very large and very small numbers. We refer to this table later in this text.

The Prefixes of the SI (International System of Units)

PREFIX	SYMBOL	FACTOR	
Yotta-	Y	1 000 000 000 000 000 000 000 000	$= 10^{24}$
Zetta-	Z	1 000 000 000 000 000 000 000	$= 10^{21}$
Exa-	E	1 000 000 000 000 000 000	$= 10^{18}$
Peta-	P	1 000 000 000 000 000	$= 10^{15}$
Tera-	T	1 000 000 000 000	$= 10^{12}$
Giga-	G	1 000 000 000 (a thousand millions = a billion)	$= 10^{9}$
Mega-	M	1 000 000 (a million)	$= 10^{6}$
Kilo-	k	1 000 (a thousand)	$= 10^{3}$
Hecto-	h	100	$= 10^{2}$
Deca-	da	10	$= 10^{1}$
		1	$= 10^{0}$
Deci-	d	0.1	$= 10^{-1}$
Centi-	c	0.01	$= 10^{-2}$
Milli-	m	0.001 (a thousandth)	$= 10^{-3}$
Micro-	μ	0.000 001 (a millionth)	$= 10^{-6}$
Nano-	n	0.000 000 001 (a thousand millionth)	$= 10^{-9}$
Pico-	p	0.000 000 000 001	$= 10^{-12}$
Femto-	f	0.000 000 000 000 001	$= 10^{-15}$
Atto-	a	0.000 000 000 000 000 001	$= 10^{-18}$
Zepto-	z	0.000 000 000 000 000 000 001	$= 10^{-21}$
Yocto-	y	0.000 000 000 000 000 000 000 001	$= 10^{-24}$

EXAMPLE 6 **What is the multiplying factor associated with each of the following metric prefixes?**

(a) nano (b) yotta

(a) From the table above, we see that **nano** means "a thousand millionth" and has a corresponding multiplying factor of 10^{-9}.

(b) From the table, we see that **yotta** is "a trillion trillions" and has a multiplying factor of 10^{24}.

PRACTICE PROBLEMS 6 **What is the multiplying factor associated with each of the following metric prefixes?**

(a) deca (b) pico

It is important to note that because these metric prefixes are multiples or submultiples of 10, they are not used precisely in the context of computers, which relies upon base two. For example, 1 kilometer = 1000 meters, whereas 1 kilobyte = 1024 bytes.

EXAMPLE 7 **Compute the absolute and relative errors when 1000 bytes are used to approximate 1 KB.**

The actual value of 1 KB is 1024 bytes, and the approximation is 1000 bytes. Hence the absolute error is given by $a.e. = |1000 - 1024| = |-24| = 24$ bytes.

The relative error is given by $r.e. = \frac{|1000 - 1024|}{1024} = \frac{|-24|}{1024} = \frac{24}{1024} = \frac{3}{128} \approx 2.34\%$.

PRACTICE PROBLEM 7 **Compute the absolute and relative errors when 1,000,000 bytes are used to approximate 1 MB.**

ANSWERS TO PRACTICE PROBLEMS

1. 128 K = 2^{17} bytes = 2^{20} bits 2. 512 TB = 2^{49} bytes = 2^{52} bits

3. 393,216 bytes 4. $\approx$ 6.1 MB 5. 3×2^{29} bytes = 3×2^{9} MB

6. (a) 10^1 or 10 (b) 10^{-12} 7. a.e. = 48,576 bytes, r.e. $\approx$ 4.63%

SECTION

2.6 Exercises

Exactly how many *bytes* are in each of the following?

1. 1 K

2. 1 MB

3. 5 MB

4. 10 TB

5. 64 K

6. 60 MB

Exactly how many *bits* are in each of the following?

7. 1 K

8. 1 GB

9. 20 MB

10. 40 GB

11. 640 K

12. 5 TB

Give the multiplying factor associated with each of the following metric prefixes.

13. kilo-

14. deci-

15. centi-

16. giga-

17. micro-

18. tera-

19. An 8-page document consumes 72.5 K of disk space. How many documents of this size can fit onto a 1.44-MB floppy disk?

Represents additionally challenging problems.

20. A computer modem is connected at 2.4 $\frac{\text{Kb}}{\text{sec}}$. What is the modem's transfer rate in $\frac{\text{Mb}}{\text{hr}}$?

21. A computer modem is connected at 9 $\frac{\text{Mb}}{\text{hr}}$. What is the modem's transfer rate in $\frac{\text{Kb}}{\text{sec}}$?

22. How many bytes are required to store a monochrome display (assume 1 bit is needed for each pixel) that uses 640 columns and 480 rows of pixels?

23. The Game Boy Advance uses a 15-bit color scheme for its display. Assuming each pixel requires 15 bits, how many bits are needed if the display is 240 columns and 160 rows?

24. Many color monitors use 24-bit true color for their displays. Assuming each pixel needs 3 bytes for storage, find the number of bytes required for a color display of 640 columns and 480 rows.

25. Some color monitors use 32-bit true color for their displays. Assuming each pixel requires 4 bytes for storage, find the number of bytes required for a color display of 1024 columns and 768 rows.

26. Assume that a picture is represented on a monitor screen, which requires 1 byte per pixel (gray scale) by a rectangular array containing 1024 columns and 768 rows of pixels. How many bytes of memory are required to store an animation that consists of 64 pictures? After finding the number of bytes, convert the answer to megabytes.

27. Assume the animation described in Exercise 26 is done in 24-bit true color, which requires 3 bytes per pixel. How many megabytes are required to store the same animation in color?

28. A computer-animated GIF of a dog that is 127 by 77 pixels is designed in 16-bit color (two bytes per pixel). If the animation consists of 10 pictures, how many bytes are required to store the image? After finding the number of bytes, convert the answer to kilobytes.

29. A computer-animated short film that is 640 by 480 pixels is designed in 16-bit color (two bytes per pixel). The film is produced at a rate of 10 frames (pictures) per second and it is two minutes in length. How many megabytes are required to store this film?

30. Compute the absolute and relative error when 1,000,000,000 bytes are used to approximate 1 GB.

31. Compute the absolute and relative error when 1,000,000,000,000 bytes are used to approximate 1 TB.

32. Compute the absolute and relative error when 100 KB is used to approximate 1 Mb (megabit).

SUMMARY

2.1 The Binary System

- A code that contains only two basic code values, which we can simplify as **on/off** or **positive/negative**, is called a **binary code**.
- When a number has a subscript of 2 at the end, it is a base two number. If is written without a subscript, it is intended to be a decimal number, which is base ten.
- The following table indicates the place values for an eight-digit binary number.

BIT LOCATION	Eighth	Seventh	Sixth	Fifth	Fourth	Third	Second	First
POWER	2^7	2^6	2^5	2^4	2^3	2^2	2^1	2^0
VALUE	128	64	32	16	8	4	2	1

- When we use the term "digit," we usually think of decimal digits. When we wish to refer to a binary digit, the common contraction **bit** is used.
- In a computer, bits are usually arranged in groups of eight. A collection of eight consecutive bits is so common that it is often referred to as a **byte**.

2.2 Base Two Arithmetic: Addition and Multiplication

- When we add two binary numbers, we get the following sums:

$$0_2 + 0_2 = 0_2$$
$$0_2 + 1_2 = 1_2$$
$$1_2 + 0_2 = 1_2$$
$$1_2 + 1_2 = 10_2$$

- Addition of two binary numbers follows the same basic algorithm used to add two decimal numbers.

$$\begin{array}{r} {\scriptstyle 1\,1\,1\ \ \ } \\ 1101_2 \\ +\,1011_2 \\ \hline 11000_2 \end{array}$$

- To multiply by a binary number that is a one followed by a series of zeros, rewrite the number to be multiplied and attach the zeros after the final digit.

$$11_2 \times 100_2 = 1100_2$$

$$11_2 \times 1000_2 = 11000_2$$

- Multiplication of two binary numbers follows the same basic algorithm used to multiply two decimal numbers.

$$\begin{array}{r} 1011_2 \\ \times\ \ 111_2 \\ \hline 1011 \\ 10110 \\ \hline 1001101_2 \end{array}$$

2.3 Base Two Arithmetic: Subtraction and Division

- When we subtract two binary numbers, we have the following possibilities:

$$\begin{array}{l} 0_2 - 0_2 = 0_2 \\ 1_2 - 1_2 = 0_2 \\ 1_2 - 0_2 = 1_2 \\ 0_2 - 1_2 = ? \end{array}$$

We must borrow from the column to the left to perform this operation.

- The principles of subtracting binary numbers are exactly the same as the principles of subtracting decimal numbers.

$$\begin{array}{r} {\scriptstyle 0\,1\,1} \\ \not{1}\not{0}011_2 \\ -\,01101_2 \\ \hline 110_2 \end{array}$$

- As in base ten arithmetic, we can check our results through addition. Note that $110_2 + 01101_2 = 10011_2$ (in base ten, $6 + 13 = 19$).

  Division of two binary numbers follows the same basic algorithm used to divide two decimal numbers.

$$\begin{array}{r} 110 \\ 11_2\overline{)10010_2} \\ -11 \\ \hline 11 \\ -11 \\ \hline 00 \\ -00 \\ \hline 0 \end{array}$$

- We can check the result of binary division using binary multiplication:

$$\begin{array}{r} 110_2 \\ \times\ \ 11_2 \\ \hline 110_2 \\ 1100_2 \\ \hline 10010_2 \end{array}$$

2.4 Two's Complement Notation

In a computer, signed numbers are represented using **two's complement notation**.

In two's complement notation, the bit furthest to the left is called the **sign bit**. If that bit is zero, the number is positive. If that bit is one, the number is negative.

To find the two's complement representation for a negative integer, x:

1. Find the binary representation for $|x|$.
2. Find the complement of that binary number.
3. Add one to the result of step 2.

The result will be the two's complement representation for x.

The subscript 2* indicates a binary number that uses two's complement representation.

Any time we are adding a positive number and a negative number using two's complement notation, we can ignore a one in the bit to the left of the sign bit.

When adding two positive or two negative numbers using four bits, a one in the fifth bit indicates an **overflow error**. Such an error occurs whenever the information is too large for the storage that is allotted.

With n bits, we can represent 2^n different numbers.

2.5 Binary Fractions

- The table below indicates the place values for the binary fraction 111.1111_2.

DIGIT	1	1	1	• (BINARY POINT)	1	1	1	1
PLACE VALUE NAME	fours	twos	ones		halfs	fourths	eighths	sixteenths
PLACE VALUE	2^2	2^1	2^0		2^{-1}	2^{-2}	2^{-3}	2^{-4}
DECIMAL EQUIVALENT	4	2	1		$\frac{1}{2}$	$\frac{1}{4}$	$\frac{1}{8}$	$\frac{1}{16}$

- To write a binary in fraction form, use the number after the binary point as the numerator and write the denominator as a one followed by the number of zeros that is identical to the number of places after the binary point.

$$0.01001011_2 = \frac{1001011_2}{100000000_2}$$

- Any time we have eight zeros after the binary point, the decimal equivalent fraction has a denominator of 256.
- Can every decimal fraction be rewritten as a binary fraction? Any rational number can be written as a binary fraction. But if the denominator of the fraction is not a power of two, the fraction can be approximated only as a terminating binary.

$$0.1 \times \frac{256}{256} = \frac{25.6}{256} \approx \frac{26}{256} = 0.00011010_2$$

2.6 Computer Memory and Quantitative Prefixes

- When referring to computer memory, one K is generally accepted as meaning approximately 1000 bytes. In fact, one K is actually 2^{10} (1024) bytes.
- The following table summarizes the commonly used units for computer memory.

PREFIX	METRIC MEANING	MEMORY USAGE	NUMBER OF BYTES	INTEGER EQUIVALENT
Kilo-	1,000	Kilobyte (K)	2^{10}	1024
Mega-	1,000,000	Megabyte (MB)	2^{20}	1,048,576
Giga-	1,000,000,000	Gigabyte (GB)	2^{30}	1,073,741,824
Tera-	1,000,000,000,000	Terabyte (TB)	2^{40}	1,099,511,627,776

- A typical computer monitor display might have 1024 columns and 768 rows of **pixels** (this word derives from the term *picture elements*).
- We use MB for abbreviating megabyte. Mb is commonly used for megabits.

CHAPTER 2 **GLOSSARY QUIZ**

The following terms were introduced in Chapter 2 of the text. Match each with one of the definitions that follows:

pixel ______
decimal representation ______
digit ______
bit ______
overflow error ______
MB ______
micro- ______
monochrome ______
gigabyte ______
kilobyte ______
byte ______
nano- ______
binary representation ______
two's complement notation ______
binary code ______
Mb ______
megabyte ______
pico- ______
milli- ______

(a) a system using combinations of two symbols to represent letters or numbers
(b) a value written in base ten
(c) a value written in base two
(d) a finger, toe, or a symbol used to represent a whole number
(e) a binary digit
(f) a collection of eight
(g) a form used to allow binary representation of negative numbers
(h) an error that occurs whenever the information is too large for the storage that is allotted
(i) 2^{10} bytes
(j) 2^{20} bytes
(k) 2^{30} bytes
(l) picture elements
(m) having only one color
(n) an abbreviation for megabyte
(o) an abbreviation for megabit
(p) a prefix meaning a thousandth
(q) a prefix meaning a millionth
(r) a prefix meaning a billionth
(s) a prefix meaning a trillionth

CHAPTER 2 REVIEW EXERCISES

For help with exercises, refer to the section number given in the brackets.

[2.1] Find the decimal equivalent for each of the following binary numbers.

1. 0100_2
2. 1111_2
3. 10110001_2
4. 11100100_2
5. 101100111111_2

[2.1] Find the binary equivalent for each of the following decimal numbers.

6. 9
7. 14
8. 29
9. 172
10. 509

[2.1] Translate the following binary phone numbers into their decimal equivalents. Each digit is separated by a hyphen.

11. $0101_2 - 0000_2 - 0011_2 - 0110_2 - 0101_2 - 0111_2 - 0110_2 - 1001_2 - 0101_2 - 1000_2$

12. $0111_2 - 1000_2 - 0001_2 - 1001_2 - 0100_2 - 0100_2 - 0011_2 - 0111_2 - 0000_2 - 0000_2$

[2.1] Write the binary number represented by the following sets of blocks, then translate each binary number into its decimal equivalent.

13.

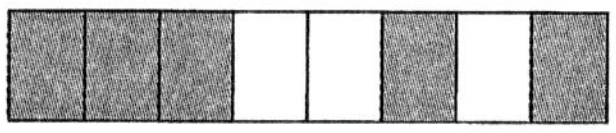

14.

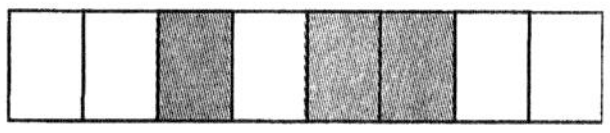

15.

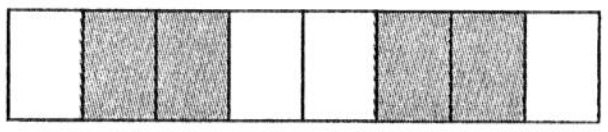

[2.2] Add the following binary numbers. Rewrite each problem in decimal notation to check your work.

16. $\begin{array}{r} 0110_2 \\ +\,0001_2 \\ \hline \end{array}$

17. $\begin{array}{r} 1011_2 \\ +\,0110_2 \\ \hline \end{array}$

18. $\begin{array}{r} 1101_2 \\ 0110_2 \\ +\,1001_2 \\ \hline \end{array}$

19. $\begin{array}{r} 1011_2 \\ 1001_2 \\ +\,1001_2 \\ \hline \end{array}$

20. $\begin{array}{r} 10111110_2 \\ +\,10101010_2 \\ \hline \end{array}$

[2.2] Multiply each pair of binary numbers. Rewrite each problem in decimal notation to check your work.

21. $\begin{array}{r} 101_2 \\ \times\ \ 10_2 \\ \hline \end{array}$

22. $\begin{array}{r} 1101_2 \\ \times\ \ \ 11_2 \\ \hline \end{array}$

23. $\begin{array}{r} 1010 \\ \times\ \ 101_2 \\ \hline \end{array}$

[2.3] **Perform the following binary subtractions. Rewrite each problem in decimal notation to check your work.**

24. $\begin{array}{r} 1011_2 \\ -\,0011_2 \\ \hline \end{array}$

25. $\begin{array}{r} 1101_2 \\ -\,0110_2 \\ \hline \end{array}$

26. $\begin{array}{r} 10010110_2 \\ -\,00001010_2 \\ \hline \end{array}$

27. $\begin{array}{r} 10000000_2 \\ -\,01111110_2 \\ \hline \end{array}$

[2.3] **28.** Use repeated subtraction to find the number of times 11_2 goes into 1111_2.

29. Use long division to find the number of times 11_2 goes into 1111_2.

[2.3] **Simplify each of the following by using long division. Convert the values to decimal to verify your answer. Some answers may have a remainder.**

30. $10110_2 \div 10_2$

31. $11001_2 \div 101_2$

32. $10100011_2 \div 100_2$

33. $1011010110_2 \div 101_2$

[2.4] **Translate each of the following decimal numbers into two's complement notation. Use an 8-bit representation.**

34. 3 **35.** –3 **36.** 29

37. –75 **38.** 102 **39.** –115

[2.4] **Each of the following numbers are in 8-bit, two's complement notation. Translate each into its decimal equivalent.**

40. 00000010_{2*} **41.** 11111101_{2*} **42.** 10010001_{2*}

43. 01011111_{2*} **44.** 10110101_{2*} **45.** 11011101_{2*}

[2.4] **Add the following binary numbers in 4-bit, two's complement notation. Rewrite each problem in decimal notation to check your work. Identify any overflow errors.**

46. $1101_{2*} + 1011_{2*}$ **47.** $0011_{2*} + 1111_{2*}$ **48.** $0110_{2*} + 0111_{2*}$

[2.5] **Find the decimal equivalent for each of the following binary numbers.**

49. 0.0101_2

50. 0.1101_2

51. 0.11010110_2

52. 1110.1011_2

53. $\frac{111_2}{1000_2}$

54. $\frac{100100_2}{10000000_2}$

55. $\frac{1010101_2}{10000000_2}$

56. $\frac{1110111_2}{10000000_2}$

[2.5] **Find the binary equivalent for each of the following decimal numbers.**

57. 0.125

58. 0.34375

59. 13.296875

60. 9.2734375

[2.5] Use long division to convert each decimal fraction into a binary expansion.

61. $\frac{1}{16}$ 62. $\frac{7}{8}$ 63. $\frac{4}{5}$

64. What is the best approximation of the decimal fraction $\frac{3}{5}$ using a denominator of 8, of 16, and of 256?

[2.6] Exactly how many *bytes* are in each of the following?

65. 1 GB 66. 12 MB 67. 640 K

[2.6] Exactly how many *bits* are in each of the following?

68. 1 MB 69. 120 GB 70. 1 TB

[2.6] 71. A 12-page document consumes 200 K of disk space. How many documents of this size can fit onto a 1.44 MB floppy disk?

72. According to Samsung, the Syncmaster 151S monitor uses 24-bit True Color for its display. Assuming each pixel needs 3 bytes for storage, find the number of bytes required for the Syncmaster's display of 1024 columns and 768 rows.

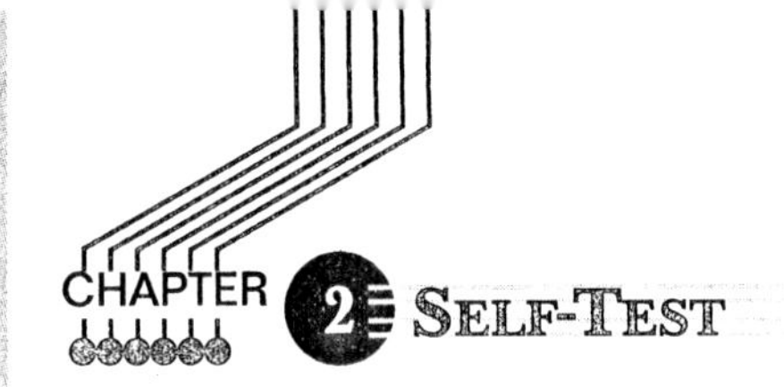

2 SELF-TEST

1. Find the decimal equivalent for each of the following binary numbers.

   (a) 1101_2 (b) 10111111_2 (c) 10100001_2

2. Find the binary equivalent for each of the following decimal numbers.

   (a) 2   (b) 27   (c) 113

3. Add the following binary numbers. Rewrite each problem in decimal notation to check your work.

   (a) $\begin{array}{r} 1101_2 \\ +0110_2 \\ \hline \end{array}$ (b) $\begin{array}{r} 10011101_2 \\ +01110110_2 \\ \hline \end{array}$

4. Multiply each pair of binary numbers. Rewrite each problem in decimal notation to check your work.

   (a) $\begin{array}{r} 111_2 \\ \times\ 10_2 \\ \hline \end{array}$ (b) $\begin{array}{r} 1011_2 \\ \times\ 11_2 \\ \hline \end{array}$ (c) $\begin{array}{r} 11010_2 \\ \times\ 101_2 \\ \hline \end{array}$

5. Perform the following binary subtractions. Rewrite each problem in decimal notation to check your work.

   (a) $\begin{array}{r} 1011_2 \\ -0011_2 \\ \hline \end{array}$ (b) $\begin{array}{r} 1101_2 \\ -0110_2 \\ \hline \end{array}$

6. **Simplify each of the following by using long division. Convert the values to decimal to verify your answer. Some answers may have a remainder.**

   (a) $11010_2 \div 10_2$ (b) $10101101_2 \div 11_2$ (c) $1101101_2 \div 10_2$

7. **Translate each of the following decimal numbers into two's complement notation. Use an 8-bit representation.**

   (a) 98 (b) –13 (c) –116

8. **Each of the following numbers are in 8-bit, two's complement notation. Translate each into its decimal equivalent.**

   (a) 00001110_{2*} (b) 11111101_{2*} (c) 100010001_{2*}

9. **Add the following binary numbers in 4-bit, two's complement notation. Rewrite each problem in decimal notation to check your work. Identify any overflow errors.**

   (a) $\begin{array}{r} 1011_{2*} \\ +\,0111_{2*} \\ \hline \end{array}$ (b) $\begin{array}{r} 1001_{2*} \\ +\,1010_{2*} \\ \hline \end{array}$ (c) $\begin{array}{r} 1111_{2*} \\ +\,1100_{2*} \\ \hline \end{array}$

10. Find the decimal equivalent for each of the following binary numbers.

(a) 0.11001_2

(b) 110.1001_2

(c) $\dfrac{10001_2}{10000000_2}$

(d) $\dfrac{1101101_2}{10000000_2}$

11. Find the binary equivalent for each of the following decimal numbers.

(a) 0.0625

(b) 5.34375

(c) 11.34375

12. Use long division to convert each decimal fraction into a binary expansion.

(a) $\dfrac{1}{32}$

(b) $\dfrac{3}{64}$

(c) $\dfrac{4}{9}$

13. Exactly how many *bytes* are in each of the following?

(a) 1 MB

(b) 5 K

14. Exactly how many *bits* are in each of the following?

(a) 40 GB

(b) 200 K

15. A 25-slide PowerPoint presentation consumes 8550 K of disk space. How many 1.44 MB floppy disks are required to store the presentation?

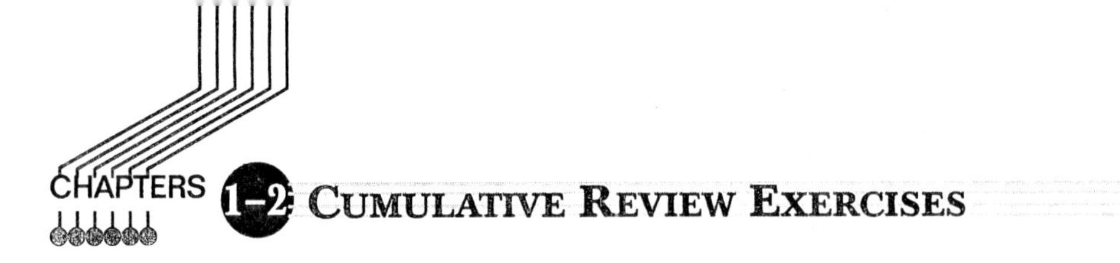

Chapters 1–2 Cumulative Review Exercises

1. Evaluate each of the following. Write each answer in either decimal form or scientific notation (rounded to 5 significant digits). If the operation is not defined, so state.

   (a) $5^2 + 5^{-2}$ (b) $5^2 \times 5^{-2}$ (c) $10^{800} \times 10^{-1400}$

   (d) $(3.14 \times 10^2)(2.71 \times 10^{-6})$ (e) $\frac{12}{0}$ (f) $\sqrt{-25}$

2. Put the following binary numbers in order from smallest to largest.

   (a) 1111111_2 (b) 101111011_2 (c) 11011111_2 (d) 011010101_2

3. The following numbers are in 4-bit, two's complement notation. Put them in order from smallest to largest.

   (a) 1011_{2*} (b) 0011_{2*} (c) 1101_{2*} (d) 1111_{2*}

4. Find the sum of the following binary numbers. Verify your result by showing the decimal equivalent problem.

   (a) $1111011_2 + 0111011_2$ (b) $1101_2 + 1100_2 + 1000_2 + 1001_2$

5. In the U.S., a football field (including the end zones) was measured to be 120 yards. A surveyor approximates this length with 110 meters. Given that 1 yard = 0.9144 meters, find the absolute and relative error associated with this approximation.

6. (a) Given that the average heartbeat rate is 70 beats/minute, use unit analysis to determine the number of heartbeats/lifetime of a person whose life expectancy is seventy-eight years.

   (b) Given that 1 U.S. dollar = 1.660 Australian dollar, 1 Australian dollar = 1.119 Canadian dollar, 1 Canadian dollar = 71.108 Japanese yen, use unit analysis to determine how many Japanese yen are in 1 U.S. dollar.

7. Use unit analysis to determine how many **bits** are in 6400 KB.

8. Multiply each pair of binary numbers. Rewrite each problem in decimal notation to check your work.

   (a) $\begin{array}{r} 100_2 \\ \times\ 11_2 \\ \hline \end{array}$

   (b) $\begin{array}{r} 1001_2 \\ \times\ 101_2 \\ \hline \end{array}$

9. What is one-half of 2^{1000}?

10. Translate each of the following binary numbers to its decimal equivalent.

(a) 0.101011_2

(b) 11101.011001_2

11. Translate the following decimal number to its binary equivalent.

(a) 0.09375

(b) 124.4375

(c) 1066

12. Given two's complement notation, what is the range of numbers (the smallest number and the largest) that can be represented with 8 bits?

13. Convert the following numbers into scientific notation.

(a) 24,813,600,000,000 miles

(Estimated distance to Proxima Centauri, Earth's closest star next to the Sun.)

(b) 0.00000002 meters

(Estimated size of an Ebola virus cell.)

14. Exactly how many bytes are in the following?

(a) 1.44 MB (b) 16 GB

15. Using 8-bit, two's complement notation, add the following decimal numbers. Give your answer in 8-bit, two's complement notation. If there is an overflow, so state.

(a) $-107 + 116$ (b) $107 + 116$

16. Use long division to convert each decimal fraction into a binary expansion.

(a) $\frac{3}{64}$ (b) $\frac{5}{6}$

Octal and Hexadecimal Numbers

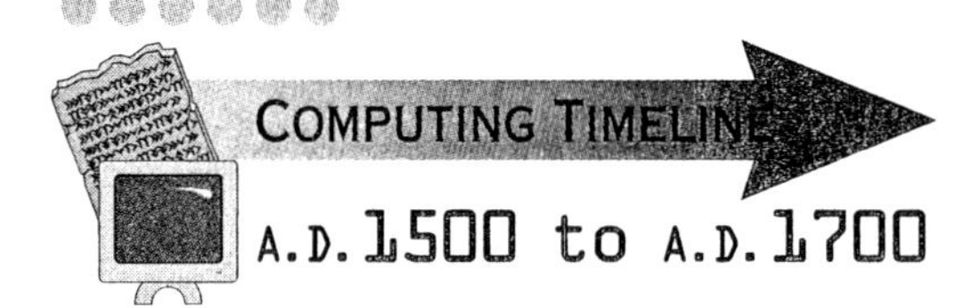

There were three significant mechanical computing machines invented in this era. The first, in 1525, was a mechanical pedometer that estimated distances by counting footsteps. The second was Blaise Pascal's computing clock. It had the theoretical ability to do multiplication and division but needed the user to do interim calculations to make it work. Finally, Gottfried Leibniz constructed a machine in 1694 that did multiplication and division by repeated addition and subtraction.

3.1 The Octal System

BITS of HISTORY

A.D. 1600–1700 One of the most important achievements in mathematics occurred in the early seventeenth century when John Napier about developed logarithms. Logarithms, allowed for multiplication and division by adding and subtracting exponents. In addition to the practical applications of logarithms in mathematics and astronomy at the time, logarithms and exponents were key ingredients in the emergence of modern computers.

In the first two chapters, we worked with two different number bases, base ten and base two. The base ten system is a natural choice for human beings because most people have ten fingers. The base two system is a nice model for computers because machine information is stored in a manner that accepts one of two states. Are these the only number systems that exist?

Reprinted with permission.

The characters of Matt Groening's *The Simpsons* all have eight digits (fingers), so how do they count to our decimal equivalent of ten? Lisa would start just as we would, 1, 2, 3, 4, 5, 6, 7, and 8, then she would need to use two digits on a hand she already used. In other words, she uses an entire group of hands and then needs two more digits to finish counting.

In this section, we introduce a third number base, base eight. This is called the **octal number system**. The place value pattern is the same as the two bases above.

OCTAL NUMBER	10000_8	1000_8	100_8	10_8	1_8
BASE TEN VALUE	4096	512	64	8	1
POWER OF EIGHT	8^4	8^3	8^2	8^1	8^0

In base ten, we use ten different symbols, 0, 1, 2, 3, 4, 5, 6, 7, 8, and 9. In base two, we use only two symbols, 0 and 1. In base eight, we use eight symbols, 0, 1, 2, 3, 4, 5, 6, and 7. Notice the counting pattern in base eight.

Note: As before, when there is no subscript, we are dealing with base ten.

00_8	01_8	02_8	03_8	04_8	05_8	06_8	07_8	10_8	11_8	12_8	13_8	14_8	15_8	16_8	17_8	20_8
0	1	2	3	4	5	6	7	8	9	10	11	12	13	14	15	16

In our first example, we determine the value of given octal digits.

EXAMPLE 1 **Find the decimal equivalent value for the given octal digit. Note that the given place is counted from right to left.**

(a) 1352_8 *The first digit*:
The first digit has a value of $2 \times 8^0 = 2 \times 1 = 2$.

(b) 1352_8 *The second digit*:
The second digit has a value of $5 \times 8^1 = 5 \times 8 = 40$.

(c) 1352_8 *The third digit*:
The third digit has a value of $3 \times 8^2 = 3 \times 64 = 192$.

PRACTICE PROBLEMS 1 **Find the decimal equivalent value for the given octal digit.**

(a) 0276_8 The first digit: ________

(b) 0276_8 The second digit: ________

(c) 0276_8 The third digit: ________

As is the case with binary numbers, we can find an octal number's decimal equivalent by converting and summing its place values. In our next example, we convert from octal to decimal numbers.

EXAMPLE 2 **Find the decimal equivalent for the octal representation 1352_8.**

OCTAL DIGIT LOCATION	fourth	third	second	first
POWER	8^3	8^2	8^1	8^0
VALUE	512	64	8	1

$$\begin{aligned} 1352_8 &= 1 \times 512 + 3 \times 64 + 5 \times 8 + 2 \times 1 \\ &= 512 + 192 + 40 + 2 \\ &= 746 \end{aligned}$$

PRACTICE PROBLEM Answers
on page 122

PRACTICE PROBLEMS 2 **Find the decimal equivalent for each octal representation.**

(a) 0276_8 (b) 2133_8

How do we go about rewriting a decimal number as its octal equivalent? One possibility is a process that is nearly identical to the one we use for converting decimal numbers to their binary equivalent.

EXAMPLE 3 **Rewrite the decimal number 231 as an octal number.**

The largest power of eight that is less than 231 is 64 (the next power of eight, 8^3, equals 512). Unlike base two, we must find the *number* of 64s in 231 (between 0 and 7), so we divide. Dividing 231 by 64, we find that 64 goes in three times with a remainder of 39.

So far, we have $231 = 3\_\_\;\_\__8$ ← **That is, 3 in the 64s' place and unknown amounts in the 8s' and 1s' places.**

Next, we look at the number of 8s in 39. There are 4 eights with a remainder of 7. From this information, we can find the octal equivalent of 231.

$231 = 347_8$ ← **There are 3 sixty-fours, 4 eights, and 7 ones.**

PRACTICE PROBLEM 3 **Rewrite the decimal number 199 as an octal number.**

We have two more conversions to master in this section, binary to octal and octal to binary. In our next example, we convert a binary number to its octal equivalent.

PRACTICE PROBLEM Answers on page 122

EXAMPLE 4 **Rewrite the binary number 1010011011_2 as its octal equivalent.**

We could convert the binary number to its decimal equivalent, and then convert that decimal number to its octal equivalent, but there is no need to do that much work! All we need to do is break the binary number into groups of 3 bits, starting on the right. We call these groups of 3 bits **tribbles**.

$1010011011_2 = 001\ 010\ 011\ 011_2$

We add the 2 zeros on the left so that we have groups of three.

Just convert each tribble to its octal equivalent, and you have made the conversion!

$$\begin{aligned} 1010011011_2 &= 001 \quad 010 \quad 011 \quad 011_2 \\ &= \ \ 1 \quad\ \ 2 \quad\ \ 3 \quad\ \ 3_8 \end{aligned}$$

$1010011011_2 = 1233_8$

PRACTICE PROBLEM 4 **Rewrite the binary number 11101101110_2 as an octal number.**

To convert octal numbers to their binary equivalent, we reverse the process.

EXAMPLE 5 **Convert the octal number 3715_8 to its binary equivalent.**

Convert each digit to its tribble. The series of tribbles becomes the binary equivalent.

$$\begin{array}{cccccc} 3 & 7 & 1 & 5_8 & & \\ 011 & 111 & 001 & 101_2 & = & 011111001101_2 \end{array}$$

PRACTICE PROBLEM 5 **Rewrite the octal number 5624_8 as a binary number.**

Our goal in this section was to help you understand the octal number system and to convert between the different bases we have discussed. When converting from base ten into other bases, there is a nice algorithm that uses division rather than repeated subtraction. We illustrate this in Example 6.

PRACTICE PROBLEM Answers on page 122

EXAMPLE 6 **Rewrite the decimal number 231 as an octal number.**

We start by taking 231 and dividing it by 8. We could set up the problem using our traditional long division, but the algorithm requires information about the remainders, so we use a modified division.

```
8|231
  28 R7   <----  First, 8 goes into 231 twenty-eight times, with a
                 remainder of 7.
                 Next, we divide the 28 by 8 and again keep track of
                 the remainder

8|231
 8|28 R7  <----  Then, 8 goes into 28 three times, with a remainder of 4.
    3 R4
    ^            Now, starting with the 3 and working upward, we
    |________    write the digits 347.
```

So, $231 = 347_8$.

PRACTICE PROBLEM 6 **Use the division algorithm above to rewrite the decimal number 199 as an octal number.**

ANSWERS TO PRACTICE PROBLEMS

1. (a) $6 \times 8^0 = 6$ (b) $7 \times 8^1 = 56$ (c) $2 \times 8^2 = 128$
2. (a) $2 \times 8^2 + 7 \times 8^1 + 6 \times 8^0 = 128 + 56 + 6 = 190$

(b) $2 \times 8^3 + 1 \times 8^2 + 3 \times 8^1 + 3 \times 8^0 = 1024 + 64 + 24 + 3 = 1115$

3. $199 = 307_8$

8^3	8^2	8^1	8^0
0	3	0	7

4. $11101101110_2 = 3556_8$

011	101	101	110
3	5	5	6

5. $5624_8 = 101110010100_2$

5	6	2	4
101	110	010	100

6.
```
8|199
 8|24 R7
    3 R0
```
Hence, $199 = 307_8$.

SECTION

3.1 Exercises

Find the decimal equivalent for each of the following octal numbers.

1. 7_8
2. 2_8
3. 11_8
4. 25_8
5. 40_8
6. 76_8
7. 61_8
8. 124_8
9. 256_8
10. 423_8
11. 1066_8
12. 1776_8

Find the octal equivalent for each of the following decimal numbers.

13. 5
14. 2
15. 10
16. 25
17. 32
18. 46
19. 65
20. 79
21. 128
22. 501
23. 1925
24. 2004

Find the octal equivalent for each of the following binary numbers.

25. 00101010_2
26. 00111011_2
27. 01001000_2

28. 11101111_2

29. 10101010_2

30. 11111110_2

31. 111100101010_2

32. 101010101111_2

33. 111010110101_2

Find the binary equivalent for each of the following octal numbers.

34. 26_8

35. 31_8

36. 75_8

37. 126_8

38. 255_8

39. 307_8

40. 624_8

41. 1057_8

42. 3120_8

43. What is true of the octal equivalent of any odd number?

44. What is true of the octal equivalent of any even number?

45. Suppose you want to work in base nine. What digits would you use?

46. Suppose you want to work in base twelve. What digits would you use? How would you write the decimal number 11 in base twelve?

3.2 Hexadecimal Representation

BITS of HISTORY

A.D. 1600–1700 Wilhelm Schickard, who was formally educated in mathematics, religion, and language, was a true Renaissance man. He possessed a remarkable balance of both theoretical knowledge and practical skill that enabled him to build the first mechanical calculator. Schickard's invention, known as the calculating clock, aided the astronomer Johannes Kepler in his work. Schickard's machine disappeared into history until the middle of the twentieth century, when some of his personal letters were discovered and his clock was successfully rebuilt to its original specifications.

In Chapter 2, we presented the binary numbering system, which is the system used within every computer. While computers work well in base two, you may have noticed that it is somewhat difficult for humans to sort through long strings of 1s and 0s, and that it is somewhat easier to look at the octal representation of that number. Most computer applications use one of two systems to make it easier for users to read and represent these strings. The octal system is one of those systems. However, the octal system can still be cumbersome when it is used to represent large base ten numbers. It is also the case that octal numbers, which represent three-bit segments, do not easily translate the byte, which has 8 bits. The other system that computer applications use, the hexadecimal system, can easily represent a byte and has become more standard in recent years.

Binary numbers have a base of two. Octal numbers have a base of eight. Decimal numbers have a base of ten. Hexadecimal numbers have a base of sixteen. We compare the four systems in the following table.

DECIMAL	BINARY	OCTAL	HEXADECIMAL
0	0000_2	0_8	0_{16}
1	0001_2	1_8	1_{16}
2	0010_2	2_8	2_{16}
3	0011_2	3_8	3_{16}
4	0100_2	4_8	4_{16}
5	0101_2	5_8	5_{16}
6	0110_2	6_8	6_{16}
7	0111_2	7_8	7_{16}
8	1000_2	10_8	8_{16}
9	1001_2	11_8	9_{16}
10	1010_2	12_8	A_{16}
11	1011_2	13_8	B_{16}
12	1100_2	14_8	C_{16}
13	1101_2	15_8	D_{16}
14	1110_2	16_8	E_{16}
15	1111_2	17_8	F_{16}
16	10000_2	20_8	10_{16}

The first thing you probably notice is the letters used in the base sixteen representations. When this system was designed, it was obvious that some new symbol needed to be used for the digits 10 to 15. In base two, we need only two different symbols (0 and 1) in each place. In base ten, we need ten different symbols for each place. In base sixteen, we need sixteen different symbols. The symbols A, B, C, D, E, and F were chosen for their familiarity.

Converting from Binary to Hexadecimal Notation

It may sound intimidating, but this process is similar to the binary to octal conversion discussed earlier. All that is needed is the preceding table. The next example illustrates the conversion process.

EXAMPLE 1 **Convert each binary number to its hexadecimal equivalent.**

(a) 0101_2 — From the table on page 125, we see that $0101_2 = 5_{16}$.

(b) 1101_2 — Again, from the table on page 125, we have $1101_2 = D_{16}$.

(c) 10100111_2 — To convert a string of bits to hexadecimal notation, we group the bits in sets of four (sometimes called a nibble) starting at the right. We then convert each nibble by using the table on page 125.

$10100111_2 = 1010\ 0111_2 = A7_{16}$

(d) $11001011_2 = 1100\ 1011_2 = CB_{16}$

(e) $11010110101101010_2 = 0001\ 1010\ 1101\ 0110\ 1010_2 = 1AD6A_{16}$

Note that in (e) we add extra zeros to the left to complete the nibble, just as we added them when we worked with the tribbles in the previous section.

PRACTICE PROBLEMS 1 **Convert each binary to its hexadecimal equivalent.**

(a) 0111_2

(b) 1110_2

(c) 10110000_2

(d) 1011011_2

(e) 110100000_2

PRACTICE PROBLEM Answers on page 129

Converting from Hexadecimal to Binary Notation

To convert from hexadecimal to binary notation, we again use the table on page 125.

EXAMPLE 2 **Convert the hexadecimal number $A4C_{16}$ to its binary equivalent.**

Using the table on page 125, we see that $A_{16} = 1010_2$, $4_{16} = 0100_2$, and $C_{16} = 1100_2$.

Putting these together, we see that $A4C_{16} = 101001001100_2$.

PRACTICE PROBLEM  **Convert the hexadecimal number $0C29_{16}$ to its binary equivalent.**

Hexadecimal Place Values

We have seen the pattern of place values for decimal and binary numbers. The pattern for hexadecimal numbers uses the same exponential pattern.

HEXADECIMAL NUMBER	1000_{16}	100_{16}	10_{16}	1_{16}
BASE TEN VALUE	16^3	16^2	16^1	16^0
DECIMAL EQUIVALENT	4096	256	16	1

Converting from Hexadecimal to Decimal Notation

The preceding table is used in the next example.

EXAMPLE 3 **Convert each hexadecimal number to its decimal equivalent.**

(a) $3A4_{16}$

We have $3 \times 16^2 + \text{A (which is 10)} \times 16^1 + 4 \times 16^0$.
$= 3 \times 256 + 10 \times 16 + 4 \times 1$
$= 768 + 160 + 4$
$= 932$ in decimal

(b) $B52C_{16} = 11 \times 16^3 + 5 \times 16^2 + 2 \times 16^1 + 12 = 46{,}380$

PRACTICE PROBLEMS **3** **Convert each hexadecimal number to its decimal equivalent**

(a) $B21_{16}$ (b) $9F2C_{16}$

PRACTICE PROBLEM Answers
on page 129

Converting from Decimal to Hexadecimal Notation

We discussed the conversion of decimals to binary numbers in Chapter 2. When we need to convert from decimal to hexadecimal notation, we use a similar process.

EXAMPLE 4 **Convert the decimal number 3728 to its hexadecimal equivalent.**

3728 is less than 4096, so the value of the fourth digit is 0. Moving to the next place (value 256), we find the number of 256s in 3728.

$3728 \div 256 = 14$, with a remainder of 144.

The 14 is E in hexadecimal, so now we have $0E\_\_\_\_\_\_{}_{16}$ as our hexadecimal equivalent.

The remainder of 144 is now divided by the 16 of the next place value to the right.

$144 \div 16 = 9$, with no remainder.

We have $3728 = 0E90_{16}$.

PRACTICE PROBLEM 4 **Convert the decimal number 5999 to its hexadecimal equivalent.**

Again, we can use the division algorithm discussed in Section 3.1 to convert a base ten number into its hexadecimal equivalent.

EXAMPLE 5 **Use the division algorithm to convert the decimal number 3728 to its hexadecimal equivalent.**

16|3728
233 R0

First, 16 goes into 3728 two hundred thirty-three times, with a remainder of 0.

Next, we divide the 233 by 16 and again keep track of the remainder.

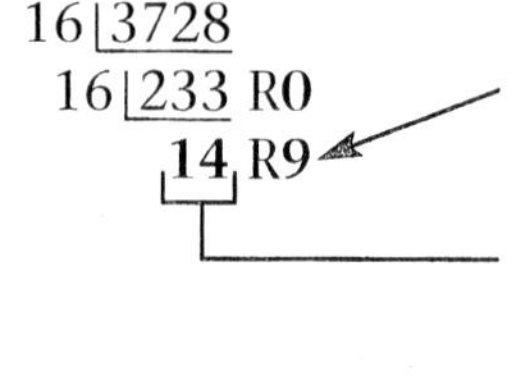

Then, 16 goes into 233 fourteen times, with a remainder of 9.

We must be careful when we write down the digits of our answer. If we write 1490, this is incorrect. Why? Because in base sixteen the 14 represents a single digit. We need to convert 14 to its hexadecimal equivalent, which is E.

So, $3728 = E90_{16}$.

PRACTICE PROBLEM Answers on page 129

PRACTICE PROBLEM 5 **Use the division algorithm to convert the decimal number 2607 to its hexadecimal equivalent.**

ANSWERS TO PRACTICE PROBLEMS

1. (a) 7_{16} (b) E_{16} (c) $B0_{16}$ (d) $5B_{16}$ (e) $1A0_{16}$

2.

0	C	2	9
0000	1100	0010	1001

$0C29_{16} = 0000110000101001_2$

3. (a) $B21_{16} = 2849$
   (b) $9F2C_{16} = 40{,}748$

4.

16^3	16^2	16^1	16^0
1	7	6	F

$5999 = 176F_{16}$

5. $16\underline{|2607}$
   $16\underline{|162}$ R15
   10 R2, hence $2607 = A2F_{16}$

SECTION 3.2

Exercises

Find the decimal equivalent for each of the following hexadecimal numbers.

1. 5_{16}

2. E_{16}

3. 13_{16}

4. 16_{16}

5. $A1_{16}$

6. $C4_{16}$

7. 70_{16}

8. ACE_{16}

9. $D12_{16}$

10. $AD8_{16}$

11. $ABBA_{16}$

12. $F150_{16}$

Find the hexadecimal equivalent for each of the following decimal numbers.

13. 1

14. 8

15. 12

16. 31

17. 47

18. 65

19. 72

20. 108

21. 121

22. 340

23. 1783

24. 5294

Find the hexadecimal equivalent for each of the following binary numbers.

25. 01001000_2 26. 00111011_2 27. 00101010_2

28. 10101010_2 29. 11101111_2 30. 11111110_2

31. 111010110101_2 32. 101010101111_2 33. 111100101010_2

Find the binary equivalent for each of the following hexadecimal numbers.

34. 25_{16} 35. 37_{16} 36. $4A_{16}$

37. FF_{16} 38. $B07_{16}$ 39. $C29_{16}$

40. $E41_{16}$ 41. $369C_{16}$ 42. $A2D2_{16}$

Convert each binary fraction into both its octal and hexadecimal equivalents by grouping into tribbles and nibbles working away from the binary point.

43. 0.1010_2 44. 0.1110_2 45. 0.01010100_2

46. 0.01101101_2 47. 0.110101101001_2 48. 0.100111101011_2

3.3 Base Sixteen Arithmetic

BITS of HISTORY

A.D. 1600–1700 Blaise Pascal, is often referred to as the inventor of the first digital calculator. His machine, called the Pascaline was a five-digit calculator about the size of a shoebox, that used a nine's complement (similar to the modern computer's two's complement) for subtraction. Although often noted as a mathematician, Pascal is immortalized in computer science by the computer language PASCAL.

We have seen that, in hexadecimal notation, the sixteen digits are

0, 1, 2, 3, 4, 5, 6, 7, 8, 9, A, B, C, D, E, F

Keeping this order in mind can help us add hexadecimal numbers. What is 4 + 5 in decimal notation? You probably answer "9" almost immediately. When you first learned to add, you probably started at 4, then counted 5 more on your fingers. It would be easiest to teach you to add this way in base sixteen, if only you had sixteen fingers! Instead, you will have to use the list of digits (figuratively, "fingers") above.

EXAMPLE 1 Add each pair of hexadecimal numbers.

(a) $4_{16} + 5_{16}$ — **If we start at 4 and move five places, we get 9. In base sixteen, as in base ten, $4_{16} + 5_{16} = 9_{16}$.**

(b) $8_{16} + 3_{16}$ — **Starting at 8, we move three digits. We end at B (this would be 11 in base ten). $8_{16} + 3_{16} = B_{16}$.**

(c) $C_{16} + 2_{16}$ — **By the same method, $C_{16} + 2_{16} = E_{16}$.**

PRACTICE PROBLEMS 1 Add each pair of hexadecimal numbers.

(a) $7_{16} + 3_{16}$

(b) $7_{16} + 7_{16}$

(c) $C_{16} + 3_{16}$

What do we do when our sum takes us "off the chart"? As we did in decimal notation, we add one in the place to the left each time we get to the end of the digit list. Recall from the last section that the place to the left of the 1s' place was the 16s' place. Every time we get to the end of the digits list, we have accumulated another 16, so we add 1 to that place. This is sometimes easier to see if we draw a base sixteen clock face.

PRACTICE PROBLEM Answers on page 140

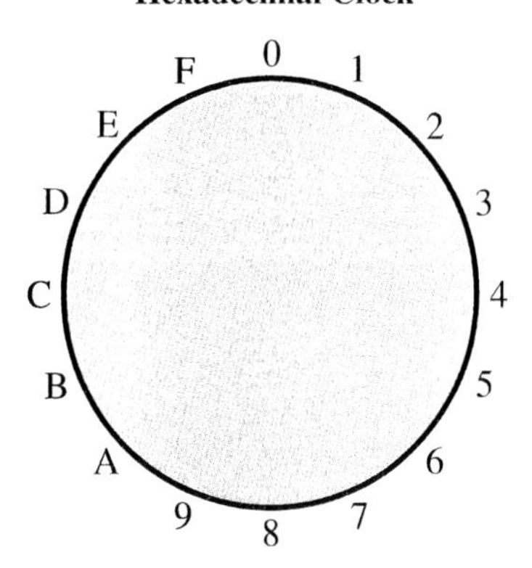

Every time we get back to 0, we add 1 in the 16s' column. We will refer to this clock in the following example.

EXAMPLE 2 **Find the sum for each pair of hexadecimal numbers.**

(a) $9_{16} + 9_{16}$

Using the following figure, we start at the 9, count off nine digits. Note that, after getting to the 0 on the clock face, you must continue two more digits to finish. The resulting sum is $9_{16} + 9_{16} = 12_{16}$. You could think of this as "$9_{16} + 9_{16}$ is one 16 and one 2."

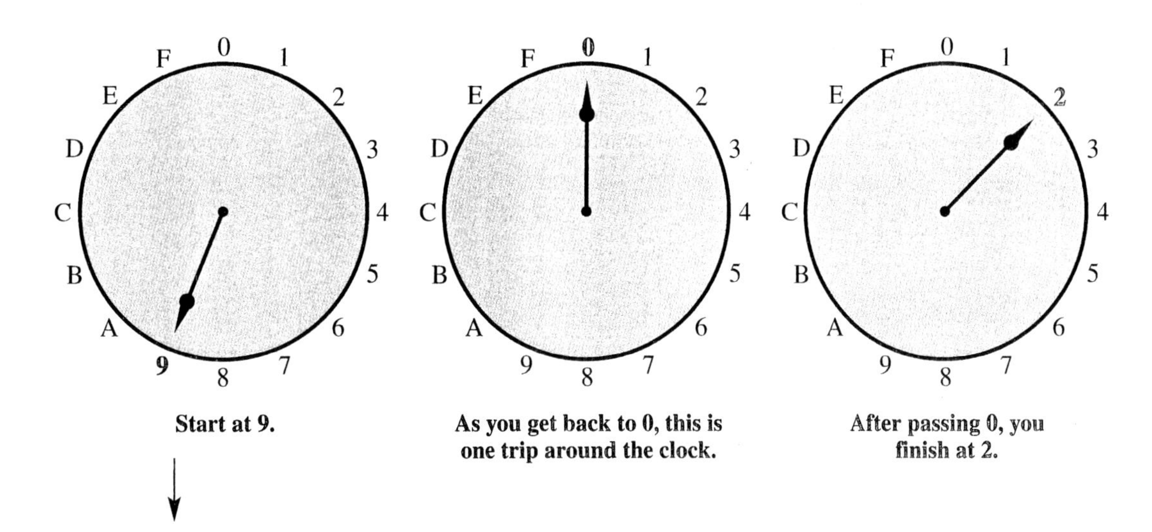

Start at 9.

As you get back to 0, this is one trip around the clock.

After passing 0, you finish at 2.

(b) $D_{16} + F_{16}$ — Using the following figure, start at the D, count F (15) digits. Again, we pass the 0 as we are counting. The result is $D_{16} + F_{16} = 1C_{16}$.

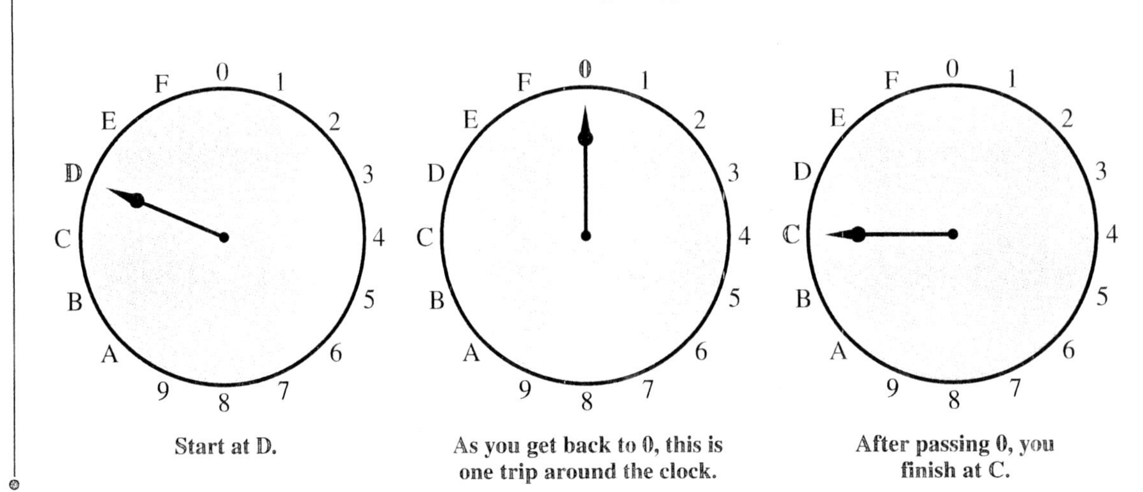

Start at D. | As you get back to 0, this is one trip around the clock. | After passing 0, you finish at C.

PRACTICE PROBLEMS 2 Use the base sixteen clock face to add each pair of hexadecimal numbers.

(a) $F_{16} + 3_{16}$

(b) $D_{16} + D_{16}$

Now that you can add single digits, the transition to multiple-digit addition requires only that you are prepared to carry.

EXAMPLE 3 Add the following pair of hexadecimal numbers.

$$\begin{array}{r} 23DE3_{16} \\ +\,38152_{16} \\ \hline \end{array}$$

$$\begin{array}{r} 23DE3_{16} \\ +\,38152_{16} \\ \hline 5_{16} \end{array}$$

$3_{16} + 2_{16} = 5_{16}$.

$$\begin{array}{r} \overset{1}{} \\ 23\overset{1}{D}E3_{16} \\ +\,38152_{16} \\ \hline 35_{16} \end{array}$$

$E_{16} + 5_{16} = 13_{16}$.

$$\begin{array}{r} 23\overset{1}{D}E3_{16} \\ +\,38152_{16} \\ \hline F35_{16} \end{array}$$

$1_{16} + D_{16} + 1_{16} = F_{16}$.

PRACTICE PROBLEM Answers on page 140

$$\begin{array}{r} \overset{1}{2}3DE3_{16} \\ +\ 38152_{16} \\ \hline 5BF35_{16} \end{array}$$

$3_{16} + 8_{16} = B_{16}$ and finally $2_{16} + 3_{16} = 5_{16}$.

PRACTICE PROBLEM 3 Add the following pair of hexadecimal numbers.

$$\begin{array}{r} 47AA9_{16} \\ +\ 583AE_{16} \\ \hline \end{array}$$

Let's look at a couple of hexadecimal subtraction problems.

EXAMPLE 4 **Subtract each pair of hexadecimal numbers.**

(a)
$$\begin{array}{r} 3E5_{16} \\ -\ 1B4_{16} \\ \hline 231_{16} \end{array}$$

From the right, we see that $5_{16} - 4_{16} = 1_{16}$, $E_{16} - B_{16} = 3_{16}$, and $3_{16} - 1_{16} = 2_{16}$.

(b)
$$\begin{array}{r} 56A2_{16} \\ -\ 45E0_{16} \\ \hline 2_{16} \end{array}$$

$2_{16} - 0_{16} = 2_{16}$

$$\begin{array}{r} 5\ _1 \\ 5\not{6}A2_{16} \\ -\ 45E0_{16} \\ \hline C2_{16} \end{array}$$

E_{16} is larger than A_{16}, so we borrow one group of 16 from the 6. The 6 becomes a 5 and the A becomes 1A. We can also use the base sixteen clock face for subtraction. If we start at A and count backward E places, we pass the 0 as we go in a counterclockwise direction, which indicates borrowing. Subtracting E from 1A can be rephrased as, "What do we add to E_{16} to get $1A_{16}$?" Counting forward, we find the answer is C_{16}.

$$\begin{array}{r} 5\ _1 \\ 5\not{6}A2_{16} \\ -\ 45E0_{16} \\ \hline 10C2_{16} \end{array}$$

$5_{16} - 5_{16} = 0_{16}$ and $5_{16} - 4_{16} = 1_{16}$

PRACTICE PROBLEM Answers
on page 140

PRACTICE PROBLEMS 4 **Subtract each pair of hexadecimal numbers.**

(a) $\begin{array}{r} B28_{16} \\ -\,837_{16} \\ \hline \end{array}$ (b) $\begin{array}{r} 3A8B_{16} \\ -\,1FF9_{16} \\ \hline \end{array}$

When you first learned to multiply decimals, you memorized the following one-digit multiplication table.

Decimal Multiplication Table

×	0	1	2	3	4	5	6	7	8	9
0	0	0	0	0	0	0	0	0	0	0
1	0	1	2	3	4	5	6	7	8	9
2	0	2	4	6	8	10	12	14	16	18
3	0	3	6	9	12	15	18	21	24	27
4	0	4	8	12	16	20	24	28	32	36
5	0	5	10	15	20	25	30	35	40	45
6	0	6	12	18	24	30	36	42	48	54
7	0	7	14	21	28	35	42	49	56	63
8	0	8	16	24	32	40	48	56	64	72
9	0	9	18	27	36	45	54	63	72	81

To easily multiply hexadecimal numbers, you should memorize the following one-digit multiplication table. This text will not assume that you have done so, but you will occasionally be reminded of this table.

PRACTICE PROBLEM Answers
on page 140

Hexadecimal Multiplication Table

×	0	1	2	3	4	5	6	7	8	9	A	B	C	D	E	F
0	0	0	0	0	0	0	0	0	0	0	0	0	0	0	0	0
1	0	1	2	3	4	5	6	7	8	9	A	B	C	D	E	F
2	0	2	4	6	8	A	C	E	10	12	14	16	18	1A	1C	1E
3	0	3	6	9	C	F	12	15	18	1B	1E	21	24	27	2A	2D
4	0	4	8	C	10	14	18	1C	20	24	28	2C	30	34	38	3C
5	0	5	A	F	14	19	1E	23	28	2D	32	37	3C	41	46	4B
6	0	6	C	12	18	1E	24	2A	30	36	3C	42	48	4E	54	5A
7	0	7	E	15	1C	23	2A	31	38	3F	46	4D	54	5B	62	69
8	0	8	10	18	20	28	30	38	40	48	50	58	60	68	70	78
9	0	9	12	1B	24	2D	36	3F	48	51	5A	64	6C	75	7E	87
A	0	A	14	1E	28	32	3C	46	50	5A	64	6E	78	82	8C	96
B	0	B	16	21	2C	37	42	4D	58	64	6E	79	84	8F	9A	A5
C	0	C	18	24	30	3C	48	54	60	6C	78	84	90	9C	A8	B4
D	0	D	1A	27	34	41	4E	5B	68	75	82	8F	9C	A9	B6	C3
E	0	E	1C	2A	38	46	54	62	70	7E	8C	9A	A8	B6	C4	D2
F	0	F	1E	2D	3C	4B	5A	69	78	87	96	A5	B4	C3	D2	E1

EXAMPLE 5 **Use the table above to find the product $A_{16} \times B_{16}$.**

The shading indicates that the intersection of column A and row B gives the result 6E. We can say that

$$A_{16} \times B_{16} = 6E_{16}.$$

PRACTICE PROBLEMS 5 **Use the hexadecimal multiplication table to find the product $5_{16} \times D_{16}$.**

Multiplying hexadecimal numbers with more than one digit follows an algorithm that parallels that of multiplying decimal numbers with more than one digit.

PRACTICE PROBLEM Answers on page 140

EXAMPLE 6 **Find the product.**

$3A_{16} \times B_{16}$

We will set up the problem in the vertical multiplication format.

$$\begin{array}{r} 3A_{16} \\ \times B_{16} \\ \hline \end{array}$$

From the table (and Example 5) we know that $A_{16} \times B_{16} = 6E_{16}$.

$$\begin{array}{r} {}^{6} \\ 3A_{16} \\ \times \ B_{16} \\ \hline E \end{array}$$

We write the E in the 1s' column and put the 6 over the next (16s') column to the left.

Again, using the table, we find that $B_{16} \times 3_{16} = 21_{16}$.

We need to add the 6 that we carried. $21_{16} + 6_{16} = 27_{16}$.

$$\begin{array}{r} 3A_{16} \\ \times \quad B_{16} \\ \hline 27E_{16} \end{array}$$

We see that

$3A_{10} \times B_{16} = 27E_{16}$

If we convert these numbers into decimal equivalents, we would have

$58 \times 11 = 638$

PRACTICE PROBLEMS 6 **Find the product.**

$4C_{16} \times 9_{16}$

In our final example, we demonstrate the multiplication of hexadecimal numbers with several digits.

EXAMPLE 7 **Find the product of the two hexadecimal numbers.**

(a)
$$\begin{array}{r} 15_{16} \\ \times \ 13_{16} \\ \hline \end{array}$$

Using the multiplication table, we first multiply by the 3.

$$\begin{array}{r} 15_{16} \\ \times \ 13_{16} \\ \hline 3F \end{array}$$

PRACTICE PROBLEM Answers on page 140

As in decimal multiplication, place a 0 in the 1s' column before multiplying by the 1 in the 16s' column.

$$\begin{array}{r} 15_{16} \\ \times\ 13_{16} \\ \hline 3F \\ 0\ \end{array}$$

Now, multiply by the 1.

$$\begin{array}{r} 15_{16} \\ \times\ 13_{16} \\ \hline 3F \\ 150 \\ \hline \end{array}$$

Add to get the result.

$$\begin{array}{r} 15_{16} \\ \times\ 13_{16} \\ \hline 3F \\ 150 \\ \hline 18F_{16} \end{array}$$

$15_{16} \times 13_{16} = 18F_{16}$

(b)

$$\begin{array}{r} 2AD3_{16} \\ \times\quad 12_{16} \\ \hline \end{array}$$

First, multiply by the 2.

$$\begin{array}{r} 2AD3_{16} \\ \times\quad 12_{16} \\ \hline 55A6 \end{array}$$

$$\begin{array}{r} 2AD3_{16} \\ \times\quad 12_{16} \\ \hline 55A6 \\ 2AD30 \\ \hline \end{array}$$

Use a 0 as a placeholder under the 6, and then multiply by 1.

$$\begin{array}{r} 2AD3_{16} \\ \times\quad 12_{16} \\ \hline 55A6 \\ 2AD30 \\ \hline 302D6_{16} \end{array}$$

Add the two results.

$2AD3_{16} \times 12_{16} = 302D6_{16}$

- Using a calculator, confirm the result.

PRACTICE PROBLEM Answers on page 140

PRACTICE PROBLEMS 7 Find the product of the two hexadecimal numbers.

(a) $1F_{16} \times 21_{16}$

(b) $3BB7_{16} \times 12_{16}$

ANSWERS TO PRACTICE PROBLEMS

1. (a) A_{16} (b) E_{16} (c) F_{16}
2. (a) 12_{16} (b) $1A_{16}$
3. $9FE57_{16}$
4. (a) $2F1_{16}$ (b) $1A92_{16}$
5. 41_{16}
6. $2AC_{16}$
7. (a) $3FF_{16}$ (b) $432DE_{16}$

SECTION 3.3

Exercises

Use the hexadecimal clock shown at the right to add the following hexadecimal numbers.

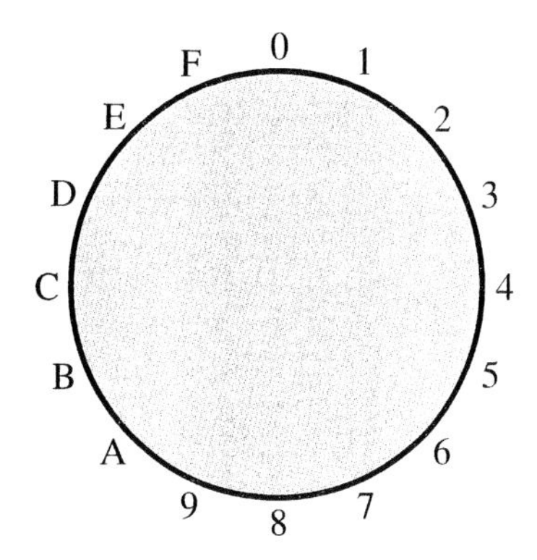

1. $\begin{array}{r} 4_{16} \\ +\ 5_{16} \\ \hline \end{array}$

2. $\begin{array}{r} 2_{16} \\ +\ 9_{16} \\ \hline \end{array}$

3. $\begin{array}{r} A_{16} \\ +\ 7_{16} \\ \hline \end{array}$

4. $\begin{array}{r} B2_{16} \\ +\ 1F_{16} \\ \hline \end{array}$

5. $\begin{array}{r} E9_{16} \\ +\ 18_{16} \\ \hline \end{array}$

6. $\begin{array}{r} C4_{16} \\ +\ A4_{16} \\ \hline \end{array}$

7. $\begin{array}{r} 9E2_{16} \\ +\ 3D0_{16} \\ \hline \end{array}$

8. $\begin{array}{r} 17F_{16} \\ +\ C15_{16} \\ \hline \end{array}$

9. $\begin{array}{r} AD2_{16} \\ +\ FCC_{16} \\ \hline \end{array}$

10. $\begin{array}{r} 2435_{16} \\ +\ 9E7B_{16} \\ \hline \end{array}$

11. $\begin{array}{r} 2039_{16} \\ +\ 8747_{16} \\ \hline \end{array}$

12. $\begin{array}{r} ABBA_{16} \\ +\ ACE_{16} \\ \hline \end{array}$

Represents additionally challenging problems.

Use the hexadecimal clock at the right to subtract the following hexadecimal numbers.

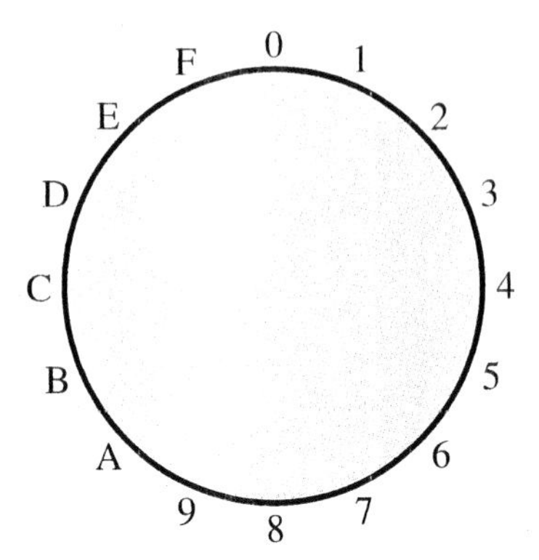

13. $9_{16} - 3_{16}$

14. $A_{16} - 9_{16}$

15. $12_{16} - 7_{16}$

16. $23_{16} - 4_{16}$

17. $E1_{16} - 1E_{16}$

18. $C4_{16} - 7F_{16}$

19. $912_{16} - 3D0_{16}$

20. $FF0_{16} - B14_{16}$

21. $62C_{16} - 1DC_{16}$

22. $FACE_{16} - EC2D_{16}$

23. $1601_{16} - 1342_{16}$

24. $420A_{16} - 1FF3_{16}$

Use the following table to help you multiply the following hexadecimal numbers.

×	0	1	2	3	4	5	6	7	8	9	A	B	C	D	E	F
0	0	0	0	0	0	0	0	0	0	0	0	0	0	0	0	0
1	0	1	2	3	4	5	6	7	8	9	A	B	C	D	E	F
2	0	2	4	6	8	A	C	E	10	12	14	16	18	1A	1C	1E
3	0	3	6	9	C	F	12	15	18	1B	1E	21	24	27	2A	2D
4	0	4	8	C	10	14	18	1C	20	24	28	2C	30	34	38	3C
5	0	5	A	F	14	19	1E	23	28	2D	32	37	3C	41	46	4B
6	0	6	C	12	18	1E	24	2A	30	36	3C	42	48	4E	54	5A
7	0	7	E	15	1C	23	2A	31	38	3F	46	4D	54	5B	62	69
8	0	8	10	18	20	28	30	38	40	48	50	58	60	68	70	78
9	0	9	12	1B	24	2D	36	3F	48	51	5A	64	6C	75	7E	87
A	0	A	14	1E	28	32	3C	46	50	5A	64	6E	78	82	8C	96
B	0	B	16	21	2C	37	42	4D	58	64	6E	79	84	8F	9A	A5
C	0	C	18	24	30	3C	48	54	60	6C	78	84	90	9C	A8	B4
D	0	D	1A	27	34	41	4E	5B	68	75	82	8F	9C	A9	B6	C3
E	0	E	1C	2A	38	46	54	62	70	7E	8C	9A	A8	B6	C4	D2
F	0	F	1E	2D	3C	4B	5A	69	78	87	96	A5	B4	C3	D2	E1

25. $8_{16} \times 3_{16}$

26. $A_{16} \times 4_{16}$

27. $5E_{16} \times 2_{16}$

28. $BED_{16} \times 12_{16}$

29. $59F_{16} \times 13_{16}$

30. $D8C_{16} \times 41_{16}$

Add the following octal numbers.

31. $4_8 + 2_8$

32. $6_8 + 3_8$

33. $12_8 + 6_8$

34. $32_8 + 17_8$

35. $50_8 + 47_8$

36. $54_8 + 36_8$

The ancient Maya of Central America used a number system partially based on twenty. A base twenty system is called a *vigesimal system* and requires twenty different symbols to represent numbers. The Maya often used elaborate symbols (many were human faces), but in modern notation the digits used are 0, 1, 2, 3, 4, 5, 6, 7, 8, 9, A, B, C, D, E, F, G, H, I, and J.

Find the decimal equivalent for each of the following vigesimal (base twenty) numbers.

37. 5_{20}

38. H_{20}

39. 10_{20}

40. 20_{20}

41. AI_{20}

42. $G0B_{20}$

Find the vigesimal (base twenty) equivalent for each of the following decimal numbers.

43. 17

44. 50

45. 192

46. 512

47. 971

48. 2004

Add the following vigesimal (base twenty) numbers.

49. $14_{20} + 5_{20}$

50. $21_{20} + 9_{20}$

51. $GI_{20} + E_{20}$

52. $3J_{20} + 4F_{20}$

53. $HE_{20} + HA_{20}$

54. $JJ_{20} + H0_{20}$

55. Why do you think the ancient Maya used a system based on twenty?

56. What other number bases make sense to you? Why?

Elements of Coding

BITS of HISTORY

A.D. 1600–1700 Gottfried Leibniz was a man of extraordinary talents and diverse interests. He is often listed with Newton as the cofather of calculus, and much like his contemporary, Leibniz dabbled in other areas of science and mathematics. He was possibly the first person to publish a paper about binary numbers. Along with a skilled craftsman, he built a calculating device called the stepped reckoner, which was more sophisticated than the Pascaline. It is speculated that Leibniz had contemplated designing a binary calculator, but no such device has ever been found.

A computer stores all information using only 1s and 0s. This includes all of the words that are typed into a word processor. To make the storage of information as consistent as possible, the computer industry has agreed on several standard codes. One of the first such major codes was the **American Standard Code for Information Interchange**, abbreviated as **ASCII** (or, redundantly, the ASCII code).

ASCII assigns a unique integer between 0 and 127 (decimal) for each symbol that can be entered into the computer, including numbers, letters, keyboard symbols, punctuation, and some control characters. The following table shows which integer (and its hexadecimal equivalent) is assigned to which character.

DECIMAL RANGE	HEXADECIMAL RANGE	CHARACTERS
0–31	0_{16}–$1F_{16}$	control characters
32	20_{16}	SPACE
33–47	21_{16}–$2F_{16}$	nonnumeric symbols on the top row of the keyboard
48–57	30_{16}–39_{16}	0, 1, 2, . . ., 9
58–63	$3A_{16}$–$3F_{16}$	:, ;, <, =, >, ?
64	40_{16}	@
65–90	41_{16}–$5A_{16}$	A, B, C, . . ., Z
91–96	$5B_{16}$–60_{16}	[, \,], ^, _, ‘
97–122	61_{16}–$7A_{16}$	a, b, c, . . ., z
123–127	$7B_{16}$–$7F_{16}$	{, |, }, -, DEL

There are many patterns that emerge from examining this table. Look at the hexadecimal and bit patterns for A and a.

CHARACTER	HEX	BIT PATTERN
A	41_{16}	$0100\ 0001_2$
a	61_{16}	$0110\ 0001_2$

Note: We often split ASCII bytes into nibbles for conversion purposes.

Can you see why these were chosen as the starting points for upper- and lowercase letters? You may see the justification more clearly when you are asked to find the bit pattern for a given letter!

EXAMPLE 1 **Find the ASCII bit pattern associated with each of the following: T, E, s, t, 2.**

"T" is the twentieth letter of the alphabet, so its decimal ASCII is 64 + 20 = 84.

The ASCII bit pattern for "T" is $0101\ 0100_2$ (5 sixteens and 4 ones is 84).

"E" is the fifth letter, so the decimal ASCII is 64 + 5 = 69.

The ASCII bit pattern for "E" is $0100\ 0101_2$ (4 sixteens and 5 ones).

"s" is the nineteenth lowercase letter. Its decimal ASCII is 96 + 19 = 115.

The ASCII bit pattern for "s" is $0111\ 0011_2$ (7 sixteens and 3 ones).

The ASCII bit pattern for "t" is right after the s, so it is $0111\ 0100_2$.

"2" is one more than 1 (which has decimal ASCII of 49), so its decimal ASCII is 50.

The ASCII bit pattern for "2" is $0011\ 0010_2$ (3 sixteens and 2 ones).

PRACTICE PROBLEMS 1 **Find the ASCII bit pattern associated with each of the following:**

(a) S **(b)** G **(c)** b **(d)** g **(e)** 5

In the next example, you are asked to interpret a bit pattern.

EXAMPLE 2 **What symbol is generated by each of the following bit patterns?**

(a) $0110\ 0011_2$ — **This is the third character after 96. It is the letter "c."**

(b) $0011\ 0111_2$ — **This is the seventh character after 48. It is the character "7."**

PRACTICE PROBLEMS 2 **What symbol is generated by each of the following bit patterns?**

(a) $0110\ 1110_2$

(b) $0011\ 1001_2$

Two extremely important elements of coding are **error detection** (finding that there is an error in transmission) and **error correction** (finding and fixing the error). Because bytes are sent through electronic channels, it is common to have a bit accidentally turned on or turned off in the

PRACTICE PROBLEM Answers on page 149

transmission process. The simplest method of error detection is the use of a **parity bit**. A parity bit (which we will assume to be the left-most bit) allows the computer that is receiving information to determine whether data was corrupted during transmission. There are many kinds of parity, but here we study two kinds, **even-parity** and **odd-parity**. The following example uses odd-parity.

EXAMPLE 3 **Assuming odd-parity, determine whether each of the following bytes was most likely correctly transmitted.**

(a) $0111\ 1001_2$ — **Because there is an odd number of 1s (in this case, 5) this byte has odd-parity and was most likely correctly transmitted.**

(b) $1101\ 0100_2$ — **This byte has an even number of 1s; it was not correctly transmitted.**

(c) $0111\ 1111_2$ — **This byte has odd-parity and was most likely correctly transmitted.**

PRACTICE PROBLEMS 3 **Assuming odd-parity, determine whether each of the following bytes was most likely correctly transmitted.**

(a) $1010\ 1001_2$ (b) $1111\ 0100_2$ (c) $0001\ 1010_2$

We mentioned that parity allows the computer to determine whether a byte is reasonably correct, but not whether it is absolutely correct. The difficulty comes when an even number of errors occurs in the transmission. An example follows.

Assume that we want to send the ASCII character "c" to another computer.

The ASCII bit pattern is

$\_110\ 0011_2$.

(The eighth bit is reserved as the parity bit.)

To encode our character using odd-parity, we set the parity bit to one and then transmit.

$1110\ 0011_2$.

Now, assume that the second and third bits are altered in the transmission, so that what is received is

$1110\ 0101_2$.

PRACTICE PROBLEM Answers on page 149

The computer receiving the message checks the parity and okays the byte because there is an odd number of 1s. The computer then translates the byte as the character "e." Is this reasonably correct? According to the parity, yes. We know there was a mistake only because we knew which character we were transmitting.

EXAMPLE 4 **Assuming odd-parity, translate each of the following bytes.**

(a) $1101\ 0101_2$

Because there is an odd number of 1s (in this case, 5) this byte has odd-parity. Eliminating the parity bit, we find the ASCII for the character "U."

(b) $1011\ 0100_2$

This byte has an even number of 1s; it was not correctly transmitted.

(c) $0011\ 0111_2$

This byte has odd-parity. It represents the character "7."

PRACTICE PROBLEMS 4 **Assuming odd-parity, translate each of the following bytes.**

(a) $1101\ 1001_2$

(b) $1011\ 1001_2$

(c) $1011\ 0010_2$

Although ASCII is still the most commonly used code, its limitations have encouraged the development and implementation of more extensive codes. What are the limitations of ASCII? Obviously, its 7-bit restriction (with the eighth bit acting as a parity bit) means that only 128 different characters can be represented. By freeing up the eighth bit, an additional 128 characters were added. Many of these characters appear in Latin-based alphabets and include "ñ" (Spanish), "ç" (French), and "ö" (German). This modified ASCII is commonly known as **Extended ASCII**, which consists of 256 different characters.

Given the increased input (including Cyrillic, Arabic, and Chinese characters) used on computers, still more flexibility is needed.

PRACTICE PROBLEM Answers on page 149

A code currently in use in many environments is Unicode. Unicode uses a 16-bit representation, so it can recognize up to 65,536 different characters. It also incorporates ASCII and Extended ASCII by setting the last 8 bits to 0 and keeping the original code. Here are the bit patterns for the letter "S" in all three codes.

ASCII	EXTENDED ASCII	UNICODE
01010011	01010011	00000000 01010011

A code being developed by ISO has 32-bit representation. This code will be large enough to encode languages that are no longer spoken, such as ancient Egyptian. By encoding the hieroglyphs into a modern script, Egyptologists will be able to communicate more efficiently and produce documents much more easily. How many different characters can this 32-bit code recognize?

ANSWERS TO PRACTICE PROBLEMS

1. (a) The ASCII for "S" is 01010011_2.
   (b) The ASCII for "G" is 01000111_2.
   (c) The ASCII for "b" is 01100010_2.
   (d) The ASCII for "g" is 01100111_2.
   (e) The ASCII for "5" is 00110101_2.

2. (a) n (b) 9

3. (a) $1010\ 1001_2$ does not have odd-parity.
   (b) $1111\ 0100_2$ has odd-parity.
   (c) $0001\ 1010_2$ has odd-parity.

4. (a) $1101\ 1001_2$ represents the character "Y."
   (b) $1011\ 1001_2$ represents the character "9."
   (c) $1011\ 0010_2$ does not have odd-parity.

SECTION 3.4

Exercises

Find the decimal ASCII code for each of the following characters.

1. X
2. F
3. G
4. q
5. 9
6. ?

Find the ASCII bit pattern associated with each character.

7. Y
8. 5
9. B
10. b
11. 0
12. z

Find the hexadecimal ASCII code for each of the following characters.

13. O
14. t
15. K
16. 7
17. q
18. R

Translate the following ASCII bit patterns into names of science fiction characters.

19. 01010010_2 00110010_2 01000100_2 00110010_2

20. 01010011_2 01100101_2 01110110_2 01100101_2 01101110_2

21. 01001110_2 01100101_2 01101111_2

22. 01011001_2 01101111_2 01100100_2 01100001_2

23. 01010011_2 01110000_2 01101111_2 01100011_2 01101011_2

24. 01010001_2 01110101_2 01101001_2 01100111_2 01101111_2 01101110_2

25. How many different characters could be encoded in a 12-bit code?

26. How many different characters could be encoded in a 24-bit code?

Determine whether each of the following bit patterns has an error (assuming odd-parity).

27. 01110010_2

28. 11001101_2

29. 01111010_2

30. 10010100_2

31. 00100101_2

32. 11101110_2

33. Assuming even-parity, translate the following ASCII bytes.

(a) 1100 0101 (b) 0111 0001 (c) 0111 1010

34. Does a parity bit allow you to identify the bit is in error (that is, a bit is on that should be off or a bit is off that should be on) if you know that an error has been encoded? Explain your answer.

35. The following block is coded in even-parity. The shaded bits in each column and row are the parity bits. Assuming the parity bits are correct and only one error is detected, circle the bit that most likely causes the error. Justify your answer.

	A	B	C	PARITY
1	1	1	1	1
2	1	0	1	0
3	0	1	1	1
PARITY	0	1	1	0

36. The following ASCII block is coded in even-parity. The shaded eighth bit is the parity bit, and the bottom row is the even block parity byte. Decode the message.

0	1	0	1	0	1	0	1
1	0	1	1	0	0	1	0
1	0	1	0	0	0	0	0
1	1	0	1	0	0	1	0
1	1	0	0	1	1	1	1
1	1	0	0	0	0	1	1
0	1	0	0	1	0	1	1
0	1	0	1	0	0	1	1
1	0	0	0	0	0	0	1

37. Assume an error-detecting code repeats every bit sent three times. For example, if 101 is the original message, then 111 000 111 should be sent. How would you correct an errant message? How certain would you be about your correction?

SUMMARY

3.1 The Octal System

- The base ten system is a natural choice for human beings because most people have ten fingers.

- The base two system is a natural model for computers because machine information is stored in a manner that accepts one of two states.

- Base eight is called the **octal number system**. The place value pattern is indicated below.

OCTAL NUMBER	1000_8	100_8	10_8	1_8
BASE TEN VALUE	512	64	8	1
POWER OF EIGHT	8^3	8^2	8^1	8^0

- To convert a binary number to its octal equivalent, break the binary number into a group of three bits, starting on the right. We will call these groups of three bits **tribbles**.

$$
\begin{aligned}
1010011011_2 &= 001 \quad 010 \quad 011 \quad 011_2 \\
&= 1 \quad 2 \quad 3 \quad 3_8 \\
1010011011_2 &= 1233_8
\end{aligned}
$$

3.2 Hexadecimal Representation

- Hexadecimal numbers have a base of sixteen.

- To convert a string of bits to hexadecimal, group the bits in sets of four (sometimes called a nibble) starting at the right. Then, convert each nibble by using the table.

- Use the division algorithm to convert a base ten number into its hexadecimal equivalent.

DECIMAL	BINARY	OCTAL	HEXADECIMAL
0	0000_2	0_8	0_{16}
1	0001_2	1_8	1_{16}
2	0010_2	2_8	2_{16}
3	0011_2	3_8	3_{16}
4	0100_2	4_8	4_{16}
5	0101_2	5_8	5_{16}
6	0110_2	6_8	6_{16}
7	0111_2	7_8	7_{16}
8	1000_2	10_8	8_{16}
9	1001_2	11_8	9_{16}
10	1010_2	12_8	A_{16}
11	1011_2	13_8	B_{16}
12	1100_2	14_8	C_{16}
13	1101_2	15_8	D_{16}
14	1110_2	16_8	E_{16}
15	1111_2	17_8	F_{16}
16	10000_2	20_8	10_{16}

3.3 Base 16 Arithmetic

- To add two hexadecimal digits, pick one of them as the starting point, then add the second number. If the total is beyond F_{16} (15), carry to the left. The hexadecimal clock can be used to facilitate addition.

- When subtracting two hexadecimal digits, use addition to check your work.

- To multiply two hexadecimal digits, use the hexadecimal multiplication table.

×	0	1	2	3	4	5	6	7	8	9	A	B	C	D	E	F
0	0	0	0	0	0	0	0	0	0	0	0	0	0	0	0	0
1	0	1	2	3	4	5	6	7	8	9	A	B	C	D	E	F
2	0	2	4	6	8	A	C	E	10	12	14	16	18	1A	1C	1E
3	0	3	6	9	C	F	12	15	18	1B	1E	21	24	27	2A	2D
4	0	4	8	C	10	14	18	1C	20	24	28	2C	30	34	38	3C
5	0	5	A	F	14	19	1E	23	28	2D	32	37	3C	41	46	4B
6	0	6	C	12	18	1E	24	2A	30	36	3C	42	48	4E	54	5A
7	0	7	E	15	1C	23	2A	31	38	3F	46	4D	54	5B	62	69
8	0	8	10	18	20	28	30	38	40	48	50	58	60	68	70	78
9	0	9	12	1B	24	2D	36	3F	48	51	5A	64	6C	75	7E	87
A	0	A	14	1E	28	32	3C	46	50	5A	64	6E	78	82	8C	96
B	0	B	16	21	2C	37	42	4D	58	64	6E	79	84	8F	9A	A5
C	0	C	18	24	30	3C	48	54	60	6C	78	84	90	9C	A8	B4
D	0	D	1A	27	34	41	4E	5B	68	75	82	8F	9C	A9	B6	C3
E	0	E	1C	2A	38	46	54	62	70	7E	8C	9A	A8	B6	C4	D2
F	0	F	1E	2D	3C	4B	5A	69	78	87	96	A5	B4	C3	D2	E1

- The shading indicates that $C_{16} \times D_{16} = 9C_{16}$

- Multiplying hexadecimal numbers with more than one digit follows an algorithm that parallels that of multiplying decimal numbers with more than one digit.

3.4 Elements of Coding

- One of the first major digital codes was the **American Standard Code for Information Interchange**. This is abbreviated as **ASCII** (or, redundantly, the ASCII code.)

- ASCII assigns a unique integer between 0 and 127 (decimal) for each symbol that can be entered into the computer. This includes numbers, letters, keyboard symbols, punctuation, and some control characters.

- Two extremely important elements of coding are **error detection** and **error correction**. Because bytes are sent through electronic channels, it is common to have a bit accidentally turned on or turned off in the transmission process.

- The simplest method of error detection is the use of a **parity bit**. The most common kinds of parity are **even-parity** and **odd-parity**.

- Although ASCII is still the most commonly used code, its limitations have encouraged the development and implementation of more extensive codes. By freeing up the eighth bit, an additional 128 characters can be added. Many of these characters appear in Latin-based alphabets and include "ñ" (Spanish), "ç" (French), and "ö" (German). This modified ASCII is commonly known as **Extended ASCII**, which consists of 256 different characters.

- Unicode uses a 16-bit representation, so it can recognize up to 65,536 different characters. It also incorporates ASCII and Extended ASCII by setting the last eight bits to zero and keeping the original code.

- A code being developed by ISO has 32-bit representation. This code will be large enough to encode languages that are no longer spoken, such as ancient Egyptian. By encoding the hieroglyphs into a modern script, Egyptologists will be able to communicate more efficiently and produce documents more easily.

GLOSSARY QUIZ

The following terms were introduced in Chapter 3 of the text. Match each with one of the definitions that follows.

hexadecimal number ______ ASCII ______ octal number ______

algorithm ______ parity bit ______ tribble ______

(a) a number written in base eight
(b) a group of three bits
(c) a series of steps that solves a problem
(d) a number written in base sixteen
(e) acronym for the American Standard Code for Information Interchange
(f) the simplest method of error detection

CHAPTER 3 REVIEW EXERCISES

[3.1] Find the decimal equivalent for each of the following octal numbers.

1. 6_8
2. 17_8
3. 65_8
4. 130_8
5. 241_8
6. 1503_8

[3.1] Find the octal equivalent for each of the following decimal numbers.

7. 3
8. 16
9. 49
10. 102
11. 357
12. 1492

[3.1] Find the octal equivalent for each of the following binary numbers.

13. 00001110_2
14. 00111101_2
15. 11000100_2
16. 110010011111_2

[3.1] **Find the binary equivalent for each of the following octal numbers.**

17. 15_8

18. 74_8

19. 236_8

20. 1043_8

[3.2] **Find the decimal equivalent for each of the following hexadecimal numbers.**

21. 6_{16}

22. C_{16}

23. $B2_{16}$

24. ADD_{16}

25. BEE_{16}

26. $F179_{16}$

[3.2] **Find the hexadecimal equivalent for each of the following decimal numbers.**

27. 3

28. 15

29. 43

30. 97

31. 608

32. 1812

[3.2] **Find the hexadecimal equivalent for each of the following binary numbers.**

33. 00001001_2 **34.** 00110011_2 **35.** 10011001_2

36. 10000001_2 **37.** 01011111011_2 **38.** 101011010111_2

[3.2] **Find the binary equivalent for each of the following hexadecimal numbers.**

39. 41_{16} **40.** $F8_{16}$ **41.** $23E_{16}$

42. CAB_{16} **43.** $52B9_{16}$ **44.** $F07D_{16}$

[3.2] **Convert each binary fraction into both its octal and hexadecimal equivalent by grouping into tribbles and nibbles working away from the binary point.**

45. 0.1100_2 **46.** 0.0110_2 **47.** 0.10100011_2

[3.3] Add the following hexadecimal numbers.

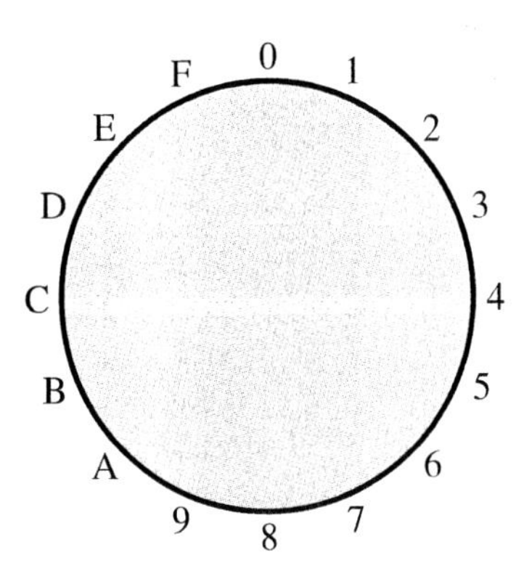

48. $3_{16} + 5_{16}$

49. $15_{16} + 9_{16}$

50. $F7_{16} + 2A_{16}$

51. $4C9_{16} + 2B6_{16}$

52. $BEDE_{16} + D1B2_{16}$

53. $DEC0_{16} + FAC_{16}$

[3.3] Subtract the following hexadecimal numbers.

54. $7_{16} - 2_{16}$

55. $7B_{16} - 44_{16}$

56. $F92_{16} - CC_{16}$

57. $E0B_{16} - 34D_{16}$

58. $D81E_{16} - E7F_{16}$

[3.3] **Use the table on page 137 to help you multiply the following hexadecimal numbers.**

59. $\begin{array}{r} D_{16} \\ \times 5_{16} \\ \hline \end{array}$

60. $\begin{array}{r} 18_{16} \\ \times C2_{16} \\ \hline \end{array}$

61. $\begin{array}{r} ACE_{16} \\ \times \quad 12_{16} \\ \hline \end{array}$

[3.4] **Find the decimal ASCII code for each of the following characters.**

62. 6

63. i

64. P

[3.4] **Find the ASCII bit pattern associated with each character.**

65. e

66. 7

67. Q

[3.4] **Find the character associated with each ASCII bit pattern.**

68. 01011001_2

69. 00110100_2

70. 01101101_2

[3.4] **Find the hexadecimal ASCII code for each of the following characters.**

71. I

72. 9

73. r

[3.4] Translate the following ASCII bit patterns into the names of musical artists or groups.

74. 01010011_2 01110100_2 01101001_2 01101110_2 01100111_2

75. 01000010_2 01101100_2 01101001_2 01101110_2 01101011_2 00110001_2 00111000_2 00110010_2

76. 01000101_2 01101110_2 01101001_2 01100111_2 01101101_2 01100001_2

77. 01010000_2 00101110_2 01000100_2 01101001_2 01100100_2 01100100_2 01111001_2

[3.4] Assuming even-parity, determine whether each of the following bit patterns has an error.

78. 00010011_2 **79.** 10100101_2 **80.** 10111111_2

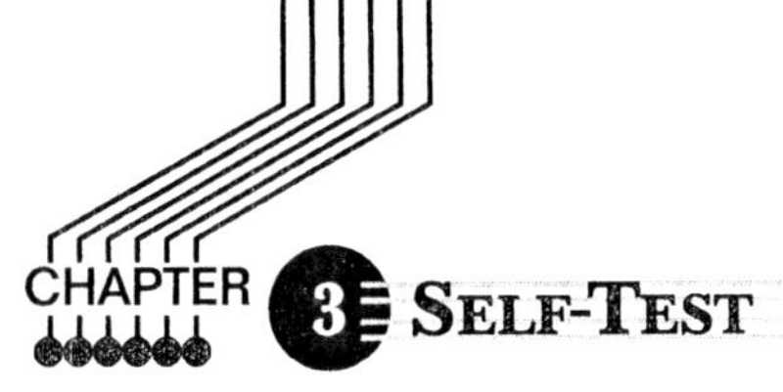

SELF-TEST

1. **Find the decimal equivalent for each of the following octal numbers.**

   (a) 13_8 (b) 46_8 (c) 257_8

2. **Find the octal equivalent for each of the following decimal numbers.**

   (a) 29 (b) 78 (c) 165

3. **Find the octal equivalent for each of the following binary numbers.**

   (a) 00001011_2 (b) 00011101_2 (c) 10111110_2

4. **Find the decimal equivalent for each of the following hexadecimal numbers.**

   (a) $3D_{16}$ (b) $B4_{16}$ (c) $A1A_{16}$

5. **Assuming even-parity, determine whether each of the following bit patterns contains an error.**

   (a) 10000000_2 (b) 01001001_2 (c) 11111111_2

6. **Find the hexadecimal equivalent for each of the following decimal numbers.**

   (a) 62 (b) 105 (c) 237

7. **Find the hexadecimal equivalent for each of the following binary numbers.**

   (a) 10110011_2 (b) 00101000_2

   (c) 0.1100_2 (d) 10000011.0011_2

8. **Find the binary equivalent for each of the following hexadecimal numbers.**

   (a) $A4_{16}$ (b) $B0E_{16}$ (c) $FC2_{16}$

9. **Find the decimal ASCII code for each of the following characters.**

   (a) T (b) 4

10. Perform the operations on the following hexadecimal numbers.

(a) $42_{16} + 53_{16}$

(b) $6E_{16} + 9D_{16}$

(c) $1C_{16} - 7_{16}$

(d) $A3_{16} - 5F_{16}$

(e) $A2_{16} \times 5_{16}$

(f) $D0_{16} \times 76_{16}$

11. Find the ASCII bit pattern associated with each character.

(a) 8

(b) d

12. Translate the following ASCII bit patterns into the names of U.S. cities.

(a) 01010000_2 01101000_2 01101111_2 01100101_2 01101110_2 01101001_2 01111000_2

(b) 01010011_2 01100101_2 01100001_2 01110100_2 01110100_2 01101100_2 01100101_2

(c) 01000100_2 01100101_2 01110100_2 01110010_2 01101111_2 01101001_2 01110100_2

Sets and Algebra

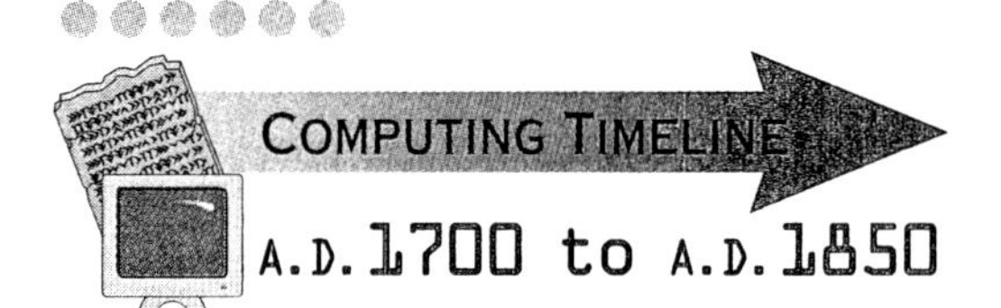

This era produced tremendous innovation in both the development of computing machines and the development of the mathematics behind the modern-day computer. In 1820, Charles-Xavier Thomas invented a mechanical calculator that became the first commercially available calculator. In 1847, George Boole published *Mathematical Analysis of Logic*. This became the foundation for what we now call Boolean algebra.

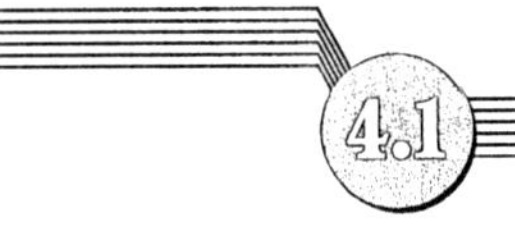

4.1 The Language of Sets

BITS of HISTORY

A.D. 1800–1900 Much of set theory is attributed to the brilliant, but tragic mathematical genius, Georg Cantor. In 1897, he published the first formal paper on set theory, introducing such concepts as set and subset as well as establishing rules for operating on sets. Cantor, with some aid of his contemporaries, was also the first to prove there were different orders or magnitude of infinity. In an attempt to free all of mathematics using sets, Cantor found himself in various stages of mental instability throughout his life.

DEFINITION: A **set** is a collection of distinct objects. Those objects can be people, books, numbers, letters, or anything else that is described. An object must be clearly defined as part of a set, or not part of a set. An **element** is an object contained in a particular set.

We use braces, "{" and "}", to designate the contents of a set. We usually name a set with a capital letter, so that we can refer to it without listing all of its elements. Each of the following is a set.

$A = \{a, b, c, d\}$
$C =$ {Segur, Wiechec, Cook, Nunn}
$D = \{0, 1\}$
$N = \{1, 2, 3, \ldots\}$

Note that the fourth set ends with three dots (called an ellipsis). Ellipses indicate that the remaining elements of the set can be found by continuing the same pattern. This can be used only when the pattern is clear.

$S =$ {Madison, Thomas, Mare, . . .}

and

$T = \{6, 24, 13, 7, \ldots\}$

do not indicate patterns that can be continued.

EXAMPLE 1 **Use set notation to list the elements for each of the following.**

(a) the U.S. presidents who served in the eighteenth century

(b) the natural numbers that are divisible by 3

Solutions

(a) {George Washington, John Adams}

(b) $\{3, 6, 9, 12, \ldots\}$

PRACTICE PROBLEMS 1 **Use set notation to list the elements for each of the following.**

(a) the U.S. presidents whose fathers were U.S. presidents

(b) negative integers that are divisible by 5

PRACTICE PROBLEM Answers on page 171

DEFINITION: A **subset** is a set of elements that are all in a specified set.

For example, the set

$A = \{2, 4, 6, 8\}$ is a subset of the set

$N = \{1, 2, 3, \ldots\}$.

EXAMPLE 2 **Which of the following are subsets of the set $S = \{a, b, c, d, e, f, g, h\}$?**

(a) $A = \{b, c, d\}$ is a subset, as all three elements are in S.

(b) $B = \{f, a, c, e\}$ is a subset (note that the order of elements in a set is unimportant).

(c) $C = \{a, c, t, e\}$ is not a subset because the element t is not in set S.

PRACTICE PROBLEMS 2 **Which of the following are subsets of the set $S = \{0, 1, 2, 3, 4, 5, 6, 7, 8, 9, \text{A}, \text{B}, \text{C}, \text{D}, \text{E}, \text{F}\}$?**

(a) $A = \{6, 9, \text{A}, 10\}$

(b) $B = \{F, A, C, E\}$

(c) $C = \{0, F\}$

Following are two important definitions that are part of the language of sets.

DEFINITION: The **universal set** (sometimes referred to as the **universe**) is the collection of all elements under consideration. The universal set is usually designated with the symbol U.

DEFINITION: A set that contains no elements is called the **empty set**. The empty set is designated with either a pair of empty brackets, **{ }** or the symbol Ø.

PRACTICE PROBLEM Answers on page 171

EXAMPLE 3 **Use set notation to designate each of the following.**

(a) the universal set of Greek letters

(b) the set of all even prime numbers greater than 2

Solutions

(a) $U = \{\alpha, \beta, \gamma, \ldots, \omega\}$, (the ellipsis indicates all of the letters between γ and ω).

(b) $\varnothing$

PRACTICE PROBLEMS 3 **Use set notation to designate each of the following.**

(a) the universal set of English letters

(b) the set of all twentieth century women U.S. presidents

It is sometimes important to know how many elements are in a set.

DEFINITION: The **cardinality** of a set is the number of elements in the set. Cardinality is designated by placing the set name between two vertical lines. $|A|$ is the number of elements in set A.

Note the similarity between cardinality and absolute value. In mathematics, we say that both cardinality and absolute value represent magnitude.

What about sets with an infinite number of elements? In subsequent courses you may encounter two infinite sets with different cardinalities, but in this text, we do not distinguish between sizes of infinity and just say that such sets have a cardinality that is infinite.

EXAMPLE 4 **Given the following sets, find the cardinality for each.**

(a) $A = \{2, 4, 6, 8, 10\}$ **(b)** $B = \{$the letters of the English alphabet$\}$

(c) $C = \{2, 4, 6, 8, \ldots, 48\}$ **(d)** $D = \{1, 2, 3, 4, \ldots\}$ **(e)** $E = \varnothing$

Solutions

(a) $|A| = 5$ **(b)** $|B| = 26$ **(c)** $|C| = 24$ **(d)** $|D|$ is infinite **(e)** $|E| = 0$

PRACTICE PROBLEM Answers
on page 171

PRACTICE PROBLEMS 4 **Given the following sets, find the cardinality for each.**

(a) $A = \{a, b, c, d, e, f\}$

(b) $B = \{$the letters in the word *Mississippi*$\}$

(c) $C = \{3, 4, 5, 6, \ldots, 44\}$

(d) $D = \{\ \}$

ANSWERS TO PRACTICE PROBLEMS

1. **(a)** {John Q. Adams, George W. Bush} **(b)** $\{\ldots, -20, -15, -10, -5\}$
2. **(a)** not a subset **(b)** a subset **(c)** a subset
3. **(a)** $U = \{a, b, c, \ldots, z\}$, (the ellipsis indicates all of the letters between c and z).

   (b) $\varnothing$
4. **(a)** $|A| = 6$ **(b)** $|B| = 4$ (Note that we do not repeat elements in a set, so $B = \{m, i, s, p\}$.)

   (c) $|C| = 42$ **(d)** $|D| = 0$

SECTION 4.1

Exercises

1. Use set notation to list the elements for each of the following.
   (a) the U.S. states whose names start with the letter *A*

   (b) the days of the week

   (c) the natural numbers that are divisible by 10

   (d) the prime numbers that are less than 14

2. Use set notation to list the elements for each of the following.
   (a) the months of the year whose names start with the letter *J*

   (b) the natural numbers that are divisible by 7

   (c) the first five digits of the number π

   (d) the negative integers that are divisible by 6

3. Which of the following are subsets of the set F, where F = {Taurus, Ranger, Explorer, Escort, Windstar, Contour, Mustang}?
   (a) {Explorer, Escort, Mustang}

   (b) {Escort, Ranger, Expedition, Windstar}

   (c) {Mustang, Taurus, Explorer, Ranger}

   (d) {Explorer, Blazer, Escort}

4. Which of the following are subsets of the set C, where C = {black, blue, cyan, green, magenta, red, white, yellow}?
   (a) {red, green, blue}

(b) {black, white, gray, blue}

(c) {white}

(d) {cyan, magenta, black, yellow}

(e) {red, orange, yellow}

5. Which of the following are subsets of the set N, where $N = \{1, 2, 3, \ldots\}$?
(a) $\{0, 1\}$

(b) $\{1, 3, 5, 7, 9\}$

(c) $\{2, 4, 6, \ldots\}$

6. Which of the following are subsets of the set Z, where $Z = \{\ldots, -2, -1, 0, 1, 2, \ldots\}$?
(a) $\{6, 28\}$

(b) $\{0, 1, 2, \ldots, E, F\}$

(c) $\{-5, -6, -7, \ldots\}$

(d) N (from Exercise 5)

(e) $\{1, 1.1, 1.11, \ldots\}$

7. Use set notation to represent each of the following.
(a) the universal set of vowels in the English alphabet (excluding y)

(b) the set of all U.S. states in Europe

(c) the universal set of octal digits

(d) the set of all prime numbers less than 2

8. Use set notation to represent each of the following.
   (a) the universal set of English letters on the fourth row of a standard keyboard (the row starting with the Shift key)

   (b) the set of oceans that border the U.S. state of Missouri

   (c) the set of all U.S. states that are islands

   (d) the universal set of hexadecimal digits

9. Given the following sets, find the cardinality of each.
   (a) $A = \{3, 5, 7, 9\}$

   (b) B = {the letters in the word *Oregon*}

   (c) $C = \{10, 20, 30, 40, \ldots 100\}$

   (d) $D = \{1, 3, 5, \ldots\}$

   (e) $E = \{\emptyset\}$

   (f) F = {the set of U.S. capitols}

10. Given the following sets, find the cardinality of each.
   (a) A = {digits on a standard clock}

   (b) B = {the letters in the word *coffee*}

   (c) $C = \emptyset$

   (d) D = {all real numbers}

   (e) E = {months that have 28 days}

4.2 Set Operators

BITS of HISTORY

A.D. 1800–1900 A bank clerk named William S. Burroughs invented an adding and listing machine in the late nineteenth century. This was the first mechanical calculating device with keys and a printer that was both reliable and affordable for commerce. The Burrough's machine continued its popularity until World War I, when it was replaced by more sophisticated devices.

In arithmetic, we can add two numbers and get a third number. For example, we say

$$4 + 3 = 7.$$

In the expression 4 + 3, we call the plus sign an **operator**. In particular, it is a binary operator, because it calls for two numbers to be added. (Note that this is a different use of the word binary than we encountered in Chapter 2.)

There are also different operators that can be used with sets. The first we will look at is called the union.

DEFINITION: If A and B are sets, their **union**, which we write as $A \cup B$, is the set consisting of all elements that are in A or B or in both A and B.

EXAMPLE 1 **Let A = {e, g, b, d, f}, and let B = {f, a, c, e}. Find $A \cup B$.**

Since $A \cup B$ consists of all the elements from either set A or set B, then

$$A \cup B = \{a, b, c, d, e, f, g\}.$$

PRACTICE PROBLEM 1 **Let $A = \{1, 3, 5, 7, 9\}$, and let $B = \{2, 3, 5, 7, 11\}$. Find $A \cup B$.**

In arithmetic, multiplication is also a binary operator. There is also a second operator for sets.

DEFINITION: If A and B are sets, their **intersection**, written as $A \cap B$, is the set consisting of all elements that are common to **both** A and B.

EXAMPLE 2 **Let $A = \{e, g, b, d, f\}$, and let $B = \{f, a, c, e\}$. Find $A \cap B$.**

Since $A \cap B$ consists of all the elements that belong in common to both A and B, then

$$A \cap B = \{e, f\}.$$

PRACTICE PROBLEM Answers on page 179

PRACTICE PROBLEM 2 **Let $A = \{1, 3, 5, 7, 9\}$, and let $B = \{2, 3, 5, 7, 11\}$. Find $A \cap B$.**

> **DEFINITION:** If two sets have nothing in common, they are said to be **disjoint**. The next example illustrates such a case.

EXAMPLE 3 **Find $A \cap B$ if $A = \{\text{Stan, Kyle, Kenny, Cartmen}\}$ and $B = \{\text{Simon, Alvin, Theodore}\}$.**

In this case, the sets are disjoint, so they do not share any common elements. Therefore, the intersection is empty, and we write

$A \cap B = \varnothing$ or $A \cap B = \{\,\}$.

PRACTICE PROBLEMS 3 **Let A = {チ, ヌ, ム, ヨ}, and let $B = \{\pi, \phi, \psi, \alpha\}$. Find $A \cap B$.**

In arithmetic, a negative sign is used to make a positive number negative (or a negative number positive). Negation is called a **unary operator** because it requires only one number. There is also a unary operator for sets.

> **DEFINITION:** Sometimes we are interested in finding the items that are not in a set. If A is a set, then we define the **complement** of A, which we write as A' to be the set of elements that are in the universal set that are not in A.

EXAMPLE 4 **Let $U = \{0, 1, 2, 3, 4, 5, 6, 7, 8, 9\}$ and $A = \{0, 2, 4, 6, 8\}$. Find A'.**

Because our universal set is the set of digits for base ten and A is the set of even digits,

$$A' = \{1, 3, 5, 7, 9\}, \text{ which is also the set of odd digits.}$$

PRACTICE PROBLEM 4 **Let $U = \{0, 1, 2, 3, 4, 5, 6, 7, 8, 9, A, B, C, D, E, F\}$ and $A = \{0, 2, 4, 6, 8, A, C, E\}$. Find A'.**

PRACTICE PROBLEM Answers
on page 179

In Section 4.1, we looked at the cardinality of a set. The following equation relates to the cardinality of the union of two sets, which we will call the **cardinality principle for two sets**:

$$|A \cup B| = |A| + |B| - |A \cap B|.$$

EXAMPLE 5 **Given that $|A \cup B| = 132$, $|B| = 55$, and $|A \cap B| = 33$, find $|A|$.**

From the previous equation, we know that

$$132 = |A| + 55 - 32.$$

Solving the equation for $|A|$, we find

$$|A| = 132 - 23 = 109.$$

There are 109 elements in set A.

PRACTICE PROBLEM 5 **Given that $|A \cup B| = 72$, $|A| = 44$, and $|B| = 55$, find $|A \cap B|$.**

In the next example, the same equation is used to solve an application.

EXAMPLE 6 **Let A = {VSU students who like math} and B = {VSU students who like computer science}.**

Assume that there are 50 VSU students who like computer science, 30 who like math, and 65 students who like at least one of them. How many students like both?

Solution

From the sentence above, we have $|A| = 30$, $|B| = 50$, and $|A \cup B| = 65$. We wish to find $|A \cap B|$.

We know that

$|A \cup B| = |A| + |B| - |A \cap B|$, so
$65 = 30 + 50 - |A \cap B|$. Solving for $|A \cap B|$,
$|A \cap B| = 80 - 65 = 15.$

There are 15 VSU students who like both math and computer science.

PRACTICE PROBLEM Answers on page 179

PRACTICE PROBLEM  **In a random sample of MS patients, it was found that 25 had allergies to milk products, 32 had allergies to peanuts, and 15 had allergies to both. How many had allergies to at least one of the two?**

The final example uses the **cardinality principle for three sets:**

$$|A \cup B \cup C| = |A| + |B| + |C| - |A \cap B| - |A \cap C| - |B \cap C| + |A \cap B \cap C|$$

EXAMPLE 7 A = {students taking a foreign language class}, with $|A| = 125$,
B = {students who own the DVD *Star Trek VI*}, with $|B| = 40$, and
C = {students who have their own apartment}, with $|C| = 140$.

Also let $|A \cap B| = 2$, $|A \cap C| = 88$, $|B \cap C| = 16$, and $|A \cup B \cup C| = 200$.

(a) Describe an element of the set $A \cap B \cap C$.

(b) Find $|A \cap B \cap C|$.

Solutions

(a) $A \cap B \cap C$ represents the set of students taking a foreign language who have their own apartment and own the DVD *Star Trek VI.*

(b) Because $|A \cup B \cup C| = |A| + |B| + |C| - |A \cap B| - |A \cap C| - |B \cap C| + |A \cap B \cap C|$,

we have
$$\begin{aligned} 200 &= 125 + 40 + 140 - 2 - 88 - 16 + |A \cap B \cap C| \\ 200 &= 305 - 106 + |A \cap B \cap C| \\ 200 &= 199 + |A \cap B \cap C| \\ |A \cap B \cap C| &= 1. \end{aligned}$$

There is only one foreign-language student who has an apartment and owns the DVD *Star Trek VI.*

PRACTICE PROBLEM Answers on page 179

PRACTICE PROBLEMS **7** A = {**CDs that cost more than $15**}, **with** $|A| = 20$,
B = {**CDs that are played on radio station WILD in their entirety**}, **with** $|B| = 6$, **and**
C = {**CDs that contain at least one single from a hit movie**}, **with** $|C| = 40$.

Also let $|A \cap B| = 4$, $|A \cap C| = 10$, $|B \cap C| = 3$, and $|A \cap B \cap C| = 2$.

(a) Describe an element of the set $A \cup B \cup C$.

(b) Find $|A \cup B \cup C|$.

ANSWERS TO PRACTICE PROBLEMS

1. $A \cup B = \{1, 2, 3, 5, 7, 9, 11\}$
2. $A \cap B = \{3, 5, 7\}$
3. $A \cap B = \varnothing$ or $A \cap B = \{\,\}$
4. $A' = \{1, 3, 5, 7, 9, B, D, F\}$, which is also the set of odd hexadecimal digits.
5. $|A \cap B| = 27$
6. 42
7. **(a)** a CD that costs more than $15, is played on WILD in its entirety, or that contains at least one single from a hit movie

   (b) $|A \cup B \cup C| = 51$

SECTION

4.2 Exercises

1. Let $U = \{0, 1, 2, 3, 4, 5, 6, 7, 8, 9\}$, $A = \{2, 4, 5, 7, 8, 9\}$, and $B = \{1, 2, 5, 6, 8\}$.

Find the following:

(a) $A \cup B$ **(b)** $A \cap B$ **(c)** A'

(d) B' **(e)** $|A \cup B|$ **(f)** $|A \cap B|$

(g) $|A|$ **(h)** $|A'|$ **(i)** $|U|$

2. Let $U = \{0, 1, 2, 3, 4, 5, 6, 7, 8, 9\}$, $A = \{1, 3, 5, 7, 9\}$, $B = \{0, 2, 4, 6, 8\}$, and $C = \{3, 6, 9\}$.

Find the following:

(a) $A \cup C$ **(b)** $A \cap C$ **(c)** $A \cup B$

(d) U' **(e)** $|A \cup B|$ **(f)** $|A \cap B|$

(g) $|C|$ **(h)** $|C'|$ **(i)** $|U'|$

3. Let $U = \{0, 1, 2, 3, 4, 5, 6, 7, 8, 9\}$, $A = \{0, 2, 5, 7, 8\}$, and $B = \{1, 2, 6, 7\}$.

Find the following:

(a) A' **(b)** B' **(c)** $A' \cap B'$

(d) $A \cup B$ **(e)** $(A \cup B)'$

(f) What is the relationship of the answers to parts c and e?

Represents additionally challenging problems.

4. Let $U = \{0, 1, 2, 3, 4, 5, 6, 7, 8, 9, 10, 11\}$, $P = \{2, 3, 5, 7, 11\}$, $O = \{1, 3, 5, 7, 9, 11\}$, and $A = \{2, 5, 8, 11\}$.

Find the following:

(a) $P \cup A$ **(b)** $O \cap A$ **(c)** O'

(d) A' **(e)** $O' \cup A'$ **(f)** $(O \cap A)'$

(g) What is the relationship of the answers to parts e and f?

5. Let U = {films listed below},
D = {*Taxi Driver, Cape Fear, Raging Bull, Casino, Goodfellas*},
P = {*Raging Bull, Casino, Goodfellas, Home Alone, Lethal Weapon 2*}, and
G = {*Braveheart, Road Warrior, Lethal Weapon 2, Hamlet*}.

Find the following:

(a) $D \cap P$ **(b)** $P \cup G$ **(c)** $D \cap G$

(d) D' **(e)** P' **(f)** $P \cap G$

6. Let U = {Disney films listed below}
A = {*Aladdin, Fantasia, The Lion King, The Little Mermaid, Snow White*},
N = {*Flubber, Herbie, The Parent Trap, Escape to Witch Mountain, Mary Poppins*}, and
Y = {*Aladdin, Flubber, The Lion King, The Little Mermaid, The Parent Trap*}.

Find the following:

(a) $A \cap Y$ **(b)** $N \cap A$ **(c)** $A \cap N$

(d) Y' **(e)** $A \cup N \cup Y$ **(f)** $N' \cap A$

7. Given $|A \cap B| = 8$, $|A \cup B| = 37$ and $|A| = 30$, find $|B|$.

8. Given $|A \cup B| = 44$, $|A| = 13$, and $|B| = 39$, find $|A \cap B|$.

9. Let A = {Community College students in Math 92} and
   B= {Community College students in CS 160}.

   Assume there are 31 Community College students in Math 92, 45 in CS 160, and 62 in at least one of the two courses. How many Community College students are in both courses?

10. Let N = {students who use Netscape} and
    E = {students who use Internet Explorer}.

    Assume there are 15 students who use Netscape, 38 who use Internet Explorer, and 7 who use both. How many students use at least one of the two Web browsers?

11. Let P = {computers with a Pentium 4 processor}, with $|P| = 125$,
    R = {computers with 256 MB of RAM}, with $|R| = 45$, and
    D = {computers with a DVD-ROM drive}, with $|D| = 35$.

    Also, let $|P \cap R| = 40$, $|R \cap D| = 27$, $|P \cap D| = 32$, and $|P \cap R \cap D| = 25$. Answer the following:

    (a) Describe an element of the set $P \cup R \cup D$.

    (b) Find $|P \cup R \cup D|$.

12. Let W = {computers with Winamp}, with $|W| = 143$,
    R = {computers with RealPlayer}, with $|R| = 70$, and
    C = {computers with a CD writer}, with $|C| = 33$.

    Also, let $|W \cap C| = 20$, $|R \cap C| = 7$, and $|W \cap R| = 28$, and let 193 machines have at least one of the three.

    How many computers have Winamp, RealPlayer, and a CD writer?

4.3 Venn Diagrams

In Section 4.2, we looked at the set operators **union**, **intersection**, and **complementation**. In this section, we use a common set theory tool, a **Venn diagram**, to illustrate these operators.

BITS of HISTORY

In 1881, the English mathematician John Venn published the book, *Symbolic Logic*, which made lasting contributions to the fields of logic and probability. Venn provided a visual model of representing the union and intersection of sets by producing three circles inside a box that generated eight nonoverlapping regions. Each region could be filled or unfilled, thus creating 256 different combinations or outcomes.

DEFINITION: A **Venn diagram** consist of a rectangle that represents the universal set, together with inner circles, each representing a specific subset of the universal set.

The first systematic use of these diagrams was made by the German philosopher and mathematician Gottfried W. von Leibniz. The diagrams used today are sometimes called Euler diagrams, after the Swiss mathematician Leonhard Euler who devised them. The English logician John Venn greatly improved and popularized the use of these diagrams.

EXAMPLE 1 **Shade $A \cup B$ on a Venn diagram.**

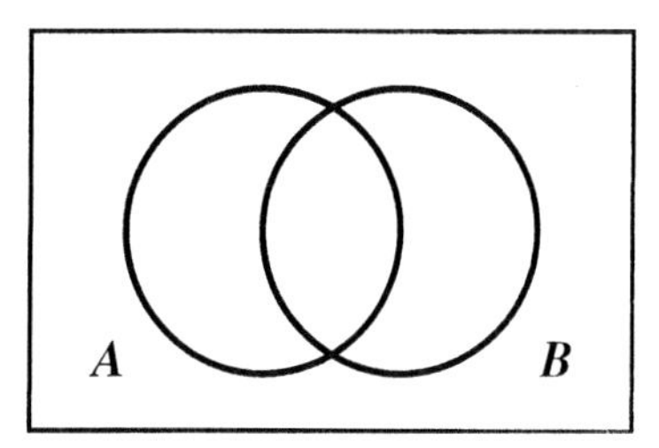

Solution

$A \cup B$

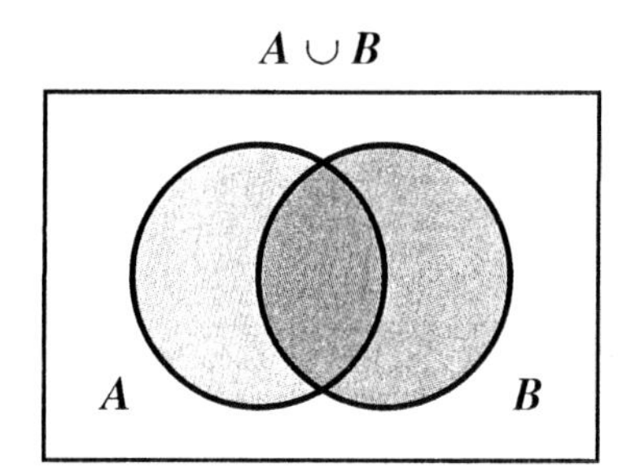

PRACTICE PROBLEM Answers on page 190

PRACTICE PROBLEM  **Shade $A \cap B$ on a Venn diagram.**

A Venn diagram can be an excellent tool for solving some logic problems. The first step requires that elements be properly placed within the diagram. The next example illustrates.

EXAMPLE 2 **Let $A = \{e, g, b, d, f\}$, and let $B = \{f, a, c, e\}$. Use a Venn diagram to illustrate the sets.**

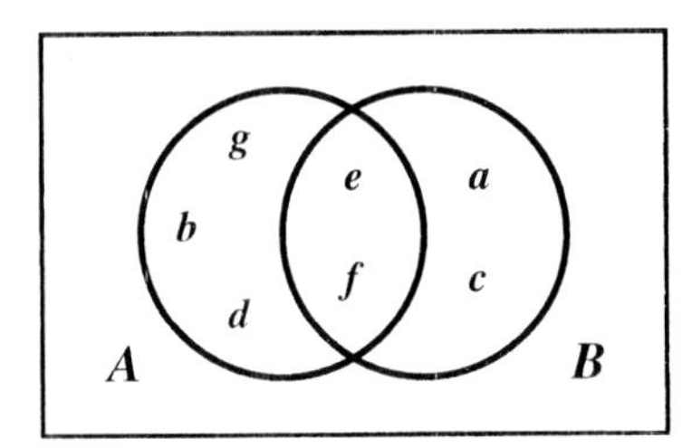

PRACTICE PROBLEM 2 **Let $A = \{1, 3, 5, 7, 9\}$, and let $B = \{2, 3, 5, 7, 11\}$. Use a Venn diagram to illustrate the sets.**

In most applications of Venn diagrams, there are important elements of the universal set that are not in one of the specified subsets.

EXAMPLE 3 **Let $A = \{e, g, b, d, f\}$, $B = \{f, a, c, e\}$, and $U = \{a, b, c, d, e, f, g, h, i, j\}$. Use a Venn diagram to illustrate the sets.**

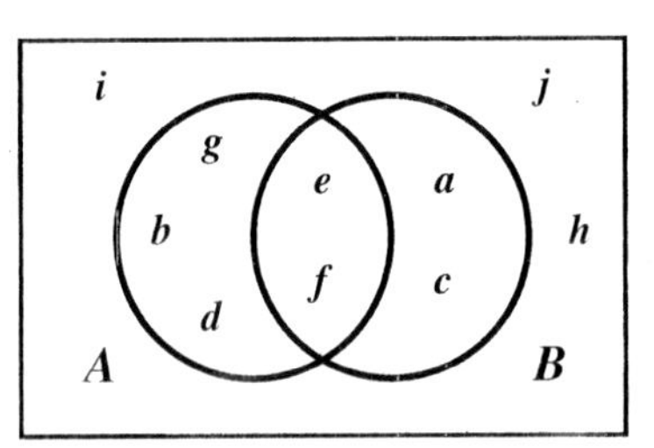

PRACTICE PROBLEM Answers
on page 190

PRACTICE PROBLEM 3 **Let $A = \{1, 3, 5, 7, 9\}$, $B = \{2, 3, 5, 7, 11\}$, and $U = \{1, 2, 3, 4, 5, 6, 7, 8, 9, 10, 11, 12\}$. Use a Venn diagram to illustrate the sets.**

EXAMPLE 4 **Let $A = \{e, g, b, d, f\}$, $B = \{f, a, c, e\}$, and $U = \{a, b, c, d, e, f, g, h, i, j\}$. Use a Venn diagram to illustrate A'.**

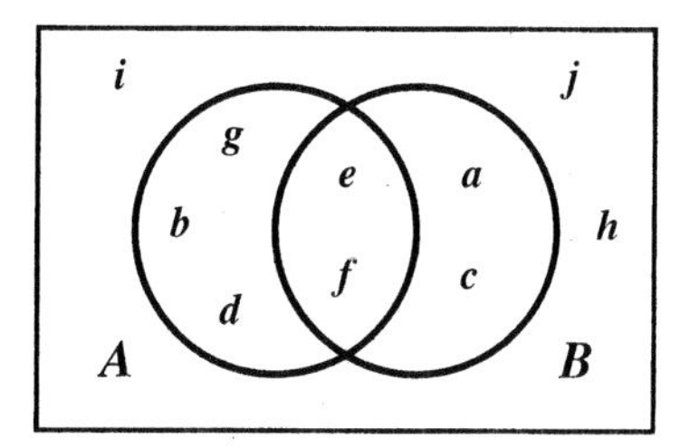

Solution

A'

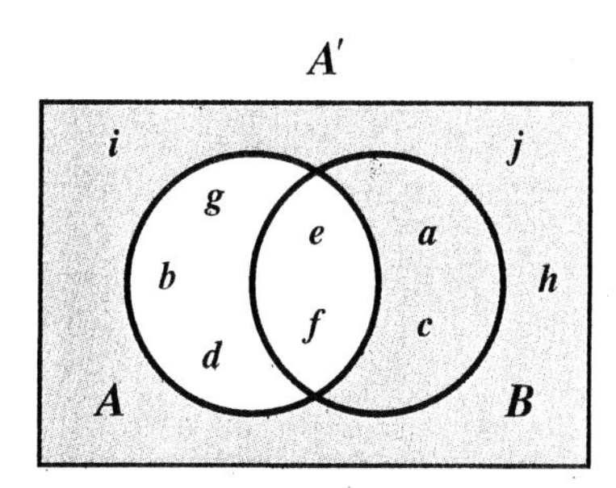

PRACTICE PROBLEM 4 **Let $A = \{1, 3, 5, 7, 9\}$, $B = \{2, 3, 5, 7, 11\}$, and $U = \{1, 2, 3, 4, 5, 6, 7, 8, 9, 10, 11, 12\}$. Use a Venn diagram to illustrate A'.**

In the next example, Venn diagrams are used to illustrate an important mathematical idea credited to Augustus De Morgan, an early nineteenth-century British mathematician.

EXAMPLE 5 **Use Venn diagrams to show that $(A \cup B)' = A' \cap B'$ (This is one form of De Morgan's law.)**

PRACTICE PROBLEM Answers
on page 190

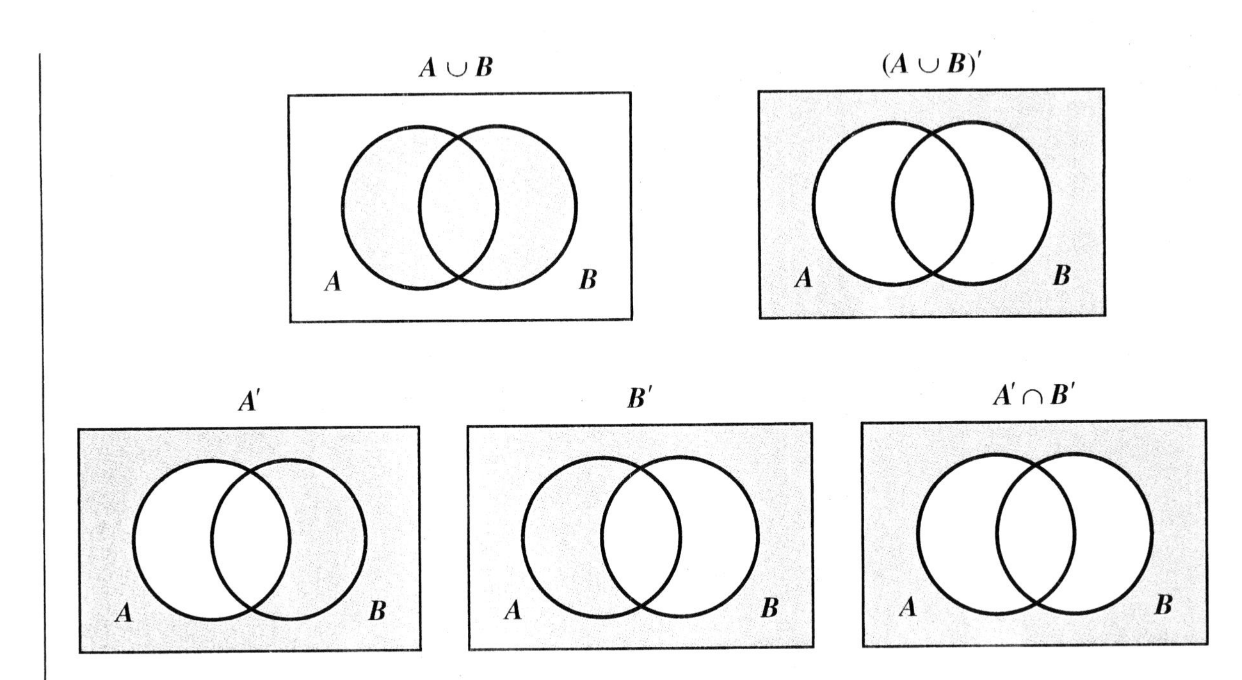

Note that while the diagrams at the end are obtained in different ways, they are the same, which means that the expressions $(A \cup B)'$ and $A' \cap B'$ are equivalent.

PRACTICE PROBLEM 5 **Use a Venn diagram to show that $(A \cap B)' = A' \cup B'$. (This is the second form of De Morgan's law.)**

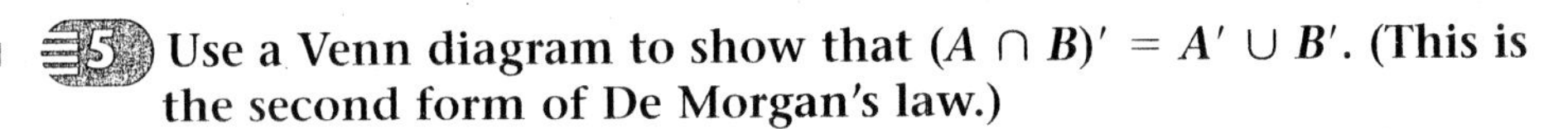

Venn diagrams can also be used to indicate cardinality, as in the following example.

EXAMPLE 6 **Draw the Venn diagram for which the following cardinalities apply:**

$$|A| = 10, \ |B| = 20, \text{ and } |A \cap B| = 7.$$

Solution

To draw such a diagram, we must start with the intersection. Note that we write the cardinality, 7, inside the intersection.

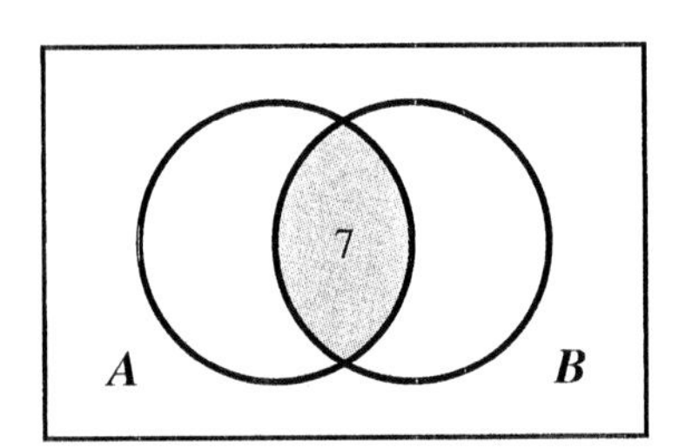

PRACTICE PROBLEM Answers on page 190

Now, we must make certain that the total inside the circle for set A is 10. That means that there are $10 - 7 = 3$ elements of A that are not in the intersection of A and B. Likewise, there are $20 - 7 = 13$ elements of B that are not in the intersection of A and B. Here is the Venn diagram.

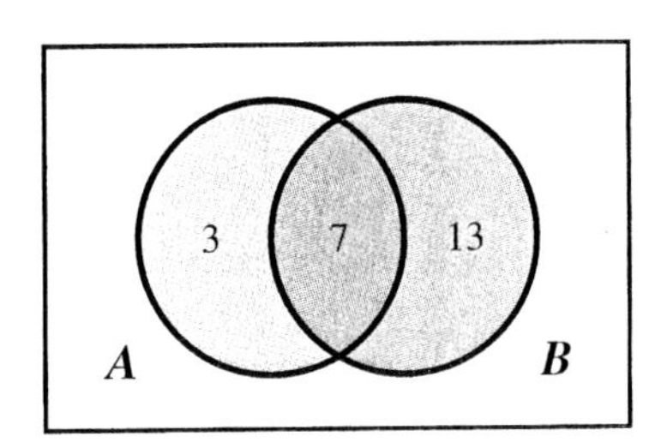

Example 6 should remind you of the cardinality principle for two sets from Section 4.2,

$$|A \cup B| = |A| + |B| - |A \cap B|$$

← (Again, we subtract the intersection once because it is counted twice, once in $|A|$ and once in $|B|$).

PRACTICE PROBLEM 6 **Draw the Venn diagram for which the following cardinalities apply:**

$|A| = 200$, $|B| = 150$, and $|A \cup B| = 325$.

EXAMPLE 7 **Draw the Venn diagram for which the following cardinalities apply:**

$|A| = 20$, $|B| = 19$, $|C| = 24$, $|A \cap B| = 6$, $|A \cap C| = 10$,
$|B \cap C| = 7$, and $|A \cup B \cup C| = 46$.

Solution

To draw such a diagram, we first determine the number of elements in the intersection and then work outward. This is similar to the previous example, but now we are dealing with the intersection of three sets. We can use an algorithm to help solve this problem.

First, we identify the eight pieces of information needed to solve this problem.

These are the eight parts of the cardinality principle for three sets in Section 4.2. We then substitute the known quantities into the equation:

$$\begin{aligned} |A \cup B \cup C| &= |A| + |B| + |C| - |A \cap B| - |A \cap C| - |B \cap C| + |A \cap B \cap C| \\ 46 &= 20 + 19 + 24 - 6 - 10 - 7 + |A \cap B \cap C|. \end{aligned}$$

This is just like solving a linear equation in a beginning algebra class; our unknown quantity, or "x," is $|A \cap B \cap C|$.

PRACTICE PROBLEM Answers on page 190

Combining like terms on the right, we obtain

$$46 = 40 + |A \cap B \cap C|.$$

Subtracting 40 from both sides of the equation, we find

$$6 = |A \cap B \cap C|.$$

Now that we know the cardinality of the intersection of all three sets, we place that value on the Venn diagram and work outward, determining the cardinality of the intersecting regions just outside the center.

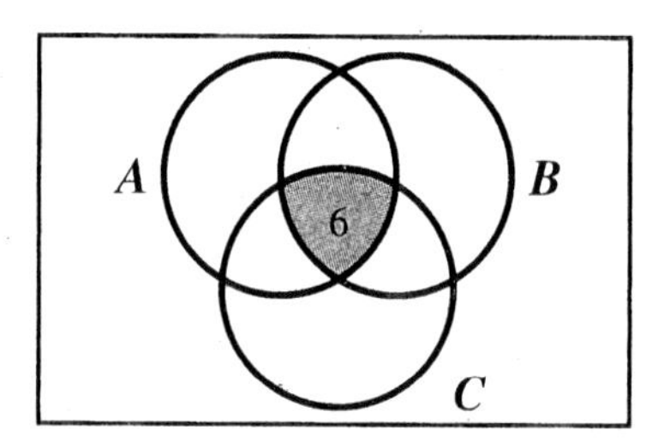

Note that $A \cap B$ is made up of two parts, $A \cap B \cap C$ and $A \cap B \cap C'$. We know that $|A \cap B| = |A \cap B \cap C| + |A \cap B \cap C'|$

$$6 = 6 + |A \cap B \cap C'|.$$

So, $|A \cap B \cap C'| = \mathbf{0}$, which means we put a 0 in that region on the Venn diagram.

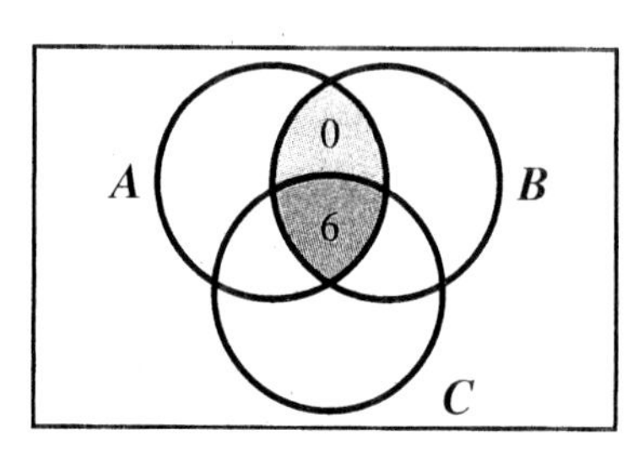

Continuing the process for the other two regions, we obtain

$$|A \cap C| = |A \cap B \cap C| + |A \cap B' \cap C| = 10 = 6 + \mathbf{4}$$

$$|B \cap C| = |A \cap B \cap C| + |A' \cap B \cap C| = 7 = 6 + \mathbf{1}.$$

To find the cardinality of the final three regions, we use a similar process. To determine the cardinality of the region inside A, we find the cardinality of the four parts that make up region A.

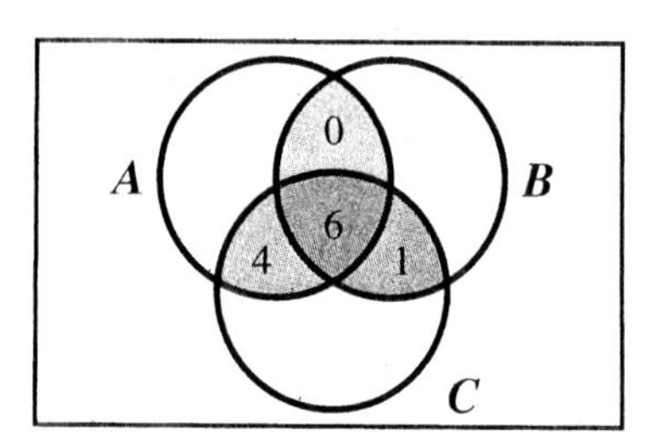

$$|A| = |A \cap B \cap C| + |A \cap B \cap C'| + |A \cap B' \cap C| + |A \cap B' \cap C'|$$

$$20 = 6 + 0 + 4 + |A \cap B' \cap C'|$$

so, $|A \cap B' \cap C'| = \mathbf{10}$,

$$\begin{aligned} |B| &= |A \cap B \cap C| + |A \cap B \cap C'| + |A' \cap B \cap C| + |A' \cap B \cap C'| \\ 19 &= \quad 6 \quad + \quad 0 \quad + \quad 1 \quad + \quad \mathbf{12} \end{aligned},$$

and

$$\begin{array}{ccccccccc} |C| & = |A \cap B \cap C| & + & |A' \cap B \cap C| & + & |A \cap B' \cap C| & + & |A' \cap B' \cap C| \\ 24 & = \quad 6 & + & 1 & + & 4 & + & \mathbf{13} \end{array}.$$

Finally, we fill in the last of the Venn diagram.

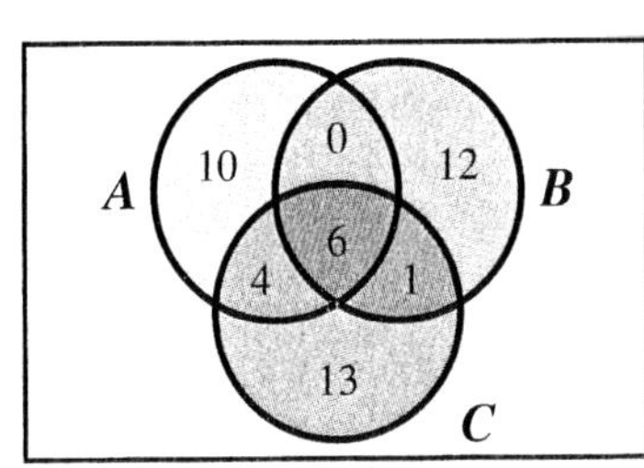

Practice Problem 7 **Draw the Venn diagram for which the following cardinalities apply:**

$|A| = 29$, $|B| = 30$, $|C| = 45$, $|A \cap B| = 24$, $|A \cap C| = 16$, $|B \cap C| = 20$, and $|A \cup B \cup C| = 60$.

Practice Problem Answers on page 190

ANSWERS TO PRACTICE PROBLEMS

1.

$A \cap B$

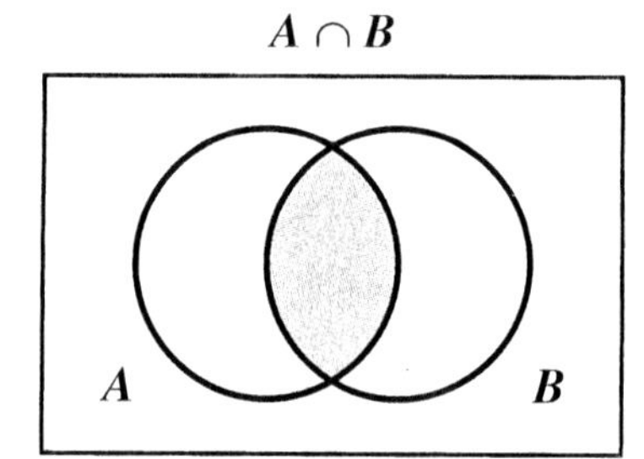

2.

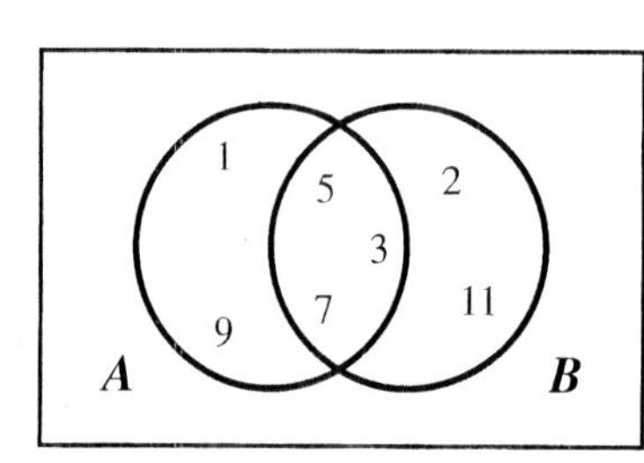

3.

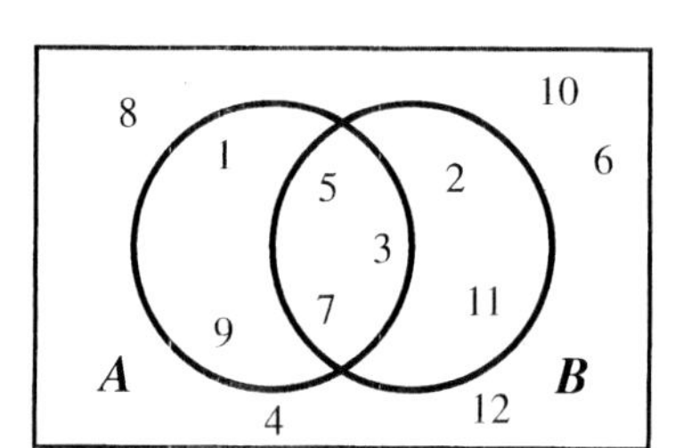

4.

A'

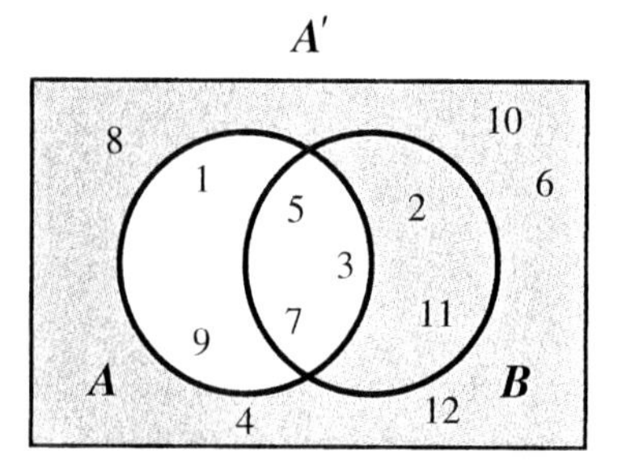

5.

$(A \cap B)' = A' \cup B'$

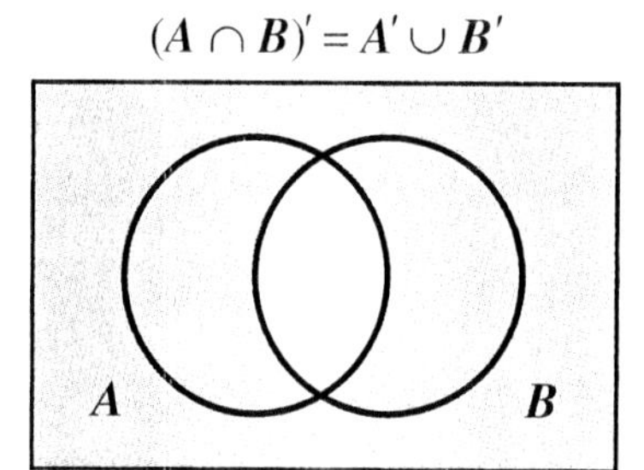

6. Recall that $|A \cup B| = |A| + |B| - |A \cap B|$, $|A \cap B| = 25$.

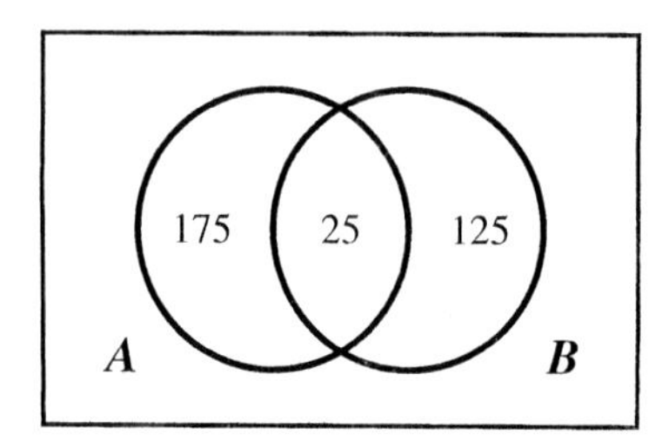

7.

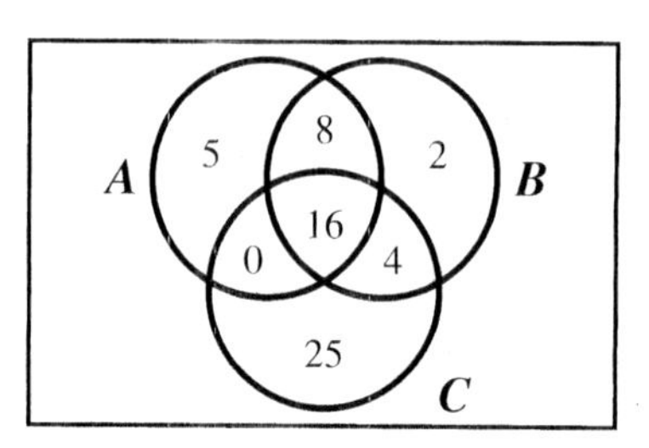

SECTION

4.3 Exercises

1. Shade the specified regions on the following Venn diagrams:

$A \cup B$

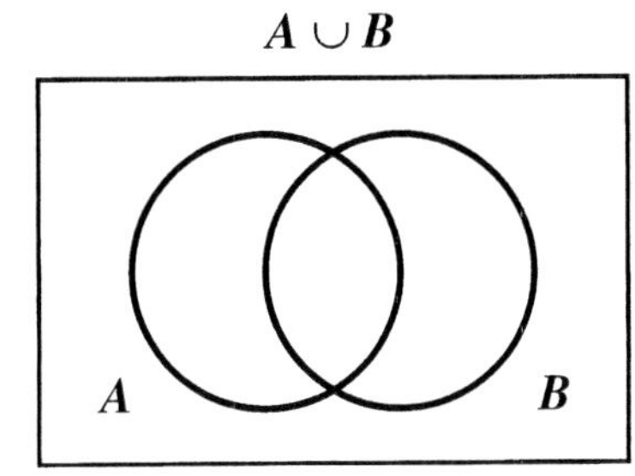

B'

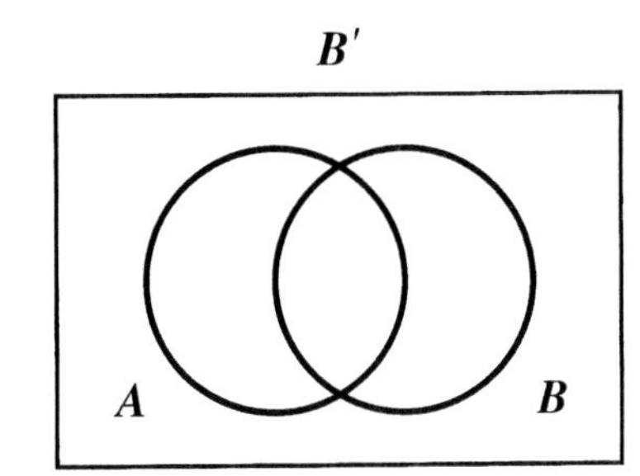

2. Shade the specified regions on the following Venn diagrams:

A'

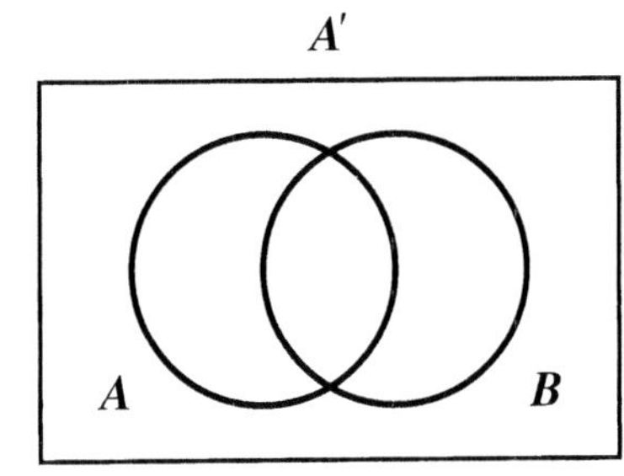

U

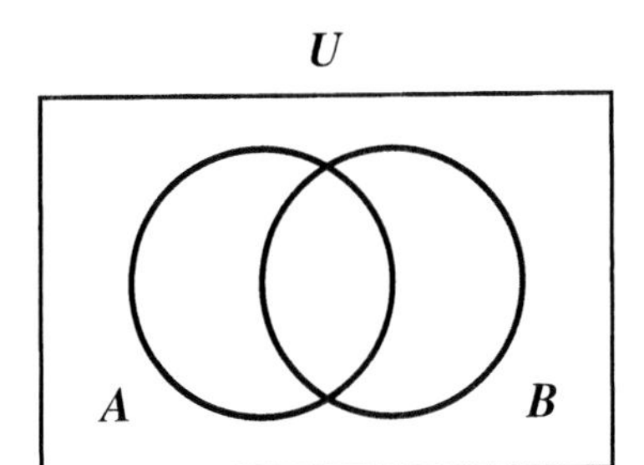

3. Label the following sets on the Venn diagram below.
   Let $A = \{0, 3, 6, 9\}$
   $B = \{0, 2, 4, 6, 8\}$
   $U = \{0, 1, 2, 3, 4, 5, 6, 7, 8, 9\}$

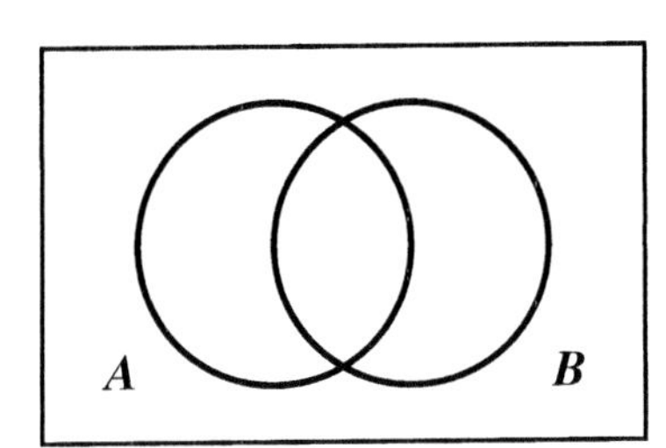

 **Represents additionally challenging problems.**

4. Label the following sets on the Venn diagram below.
Let $A = \{0, 2, 4, 7\}$
$B = \{1, 2, 4, 8, 9\}$
$U = \{0, 1, 2, 3, 4, 5, 6, 7, 8, 9\}$

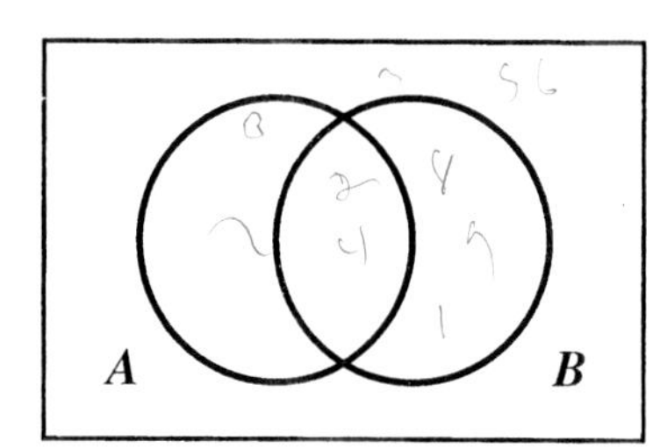

5. Draw a Venn diagram to illustrate the set A'.
Let $A = \{0, 1, 6, 7\}$
$B = \{1, 2, 4, 7, 8, 9\}$
$U = \{0, 1, 2, 3, 4, 5, 6, 7, 8, 9\}$

6. Draw a Venn diagram to illustrate the set B'.
Let $A = \{0, 1, 6, 7\}$
$B = \{1, 2, 4, 7, 8, 9\}$
$U = \{0, 1, 2, 3, 4, 5, 6, 7, 8, 9\}$.

7. Label the following sets on the Venn diagram below.
Let $A' = \{1, 2, 3, 4\}$
$B' = \{1, 2, 4, 8, 9\}$
$U = \{0, 1, 2, 3, 4, 5, 6, 7, 8, 9\}$

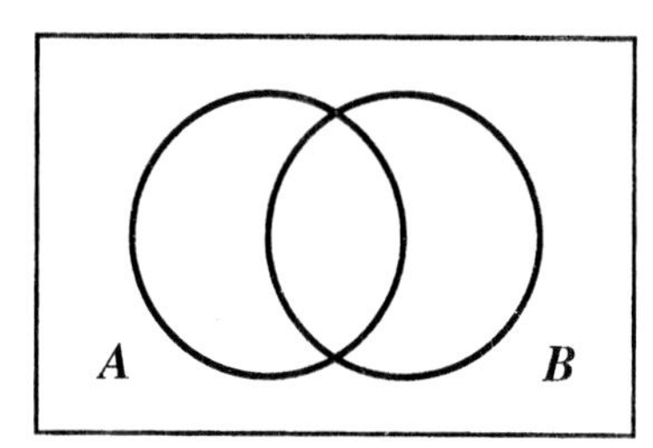

8. Shade the specified regions on the following Venn diagrams.

$A \cup B'$

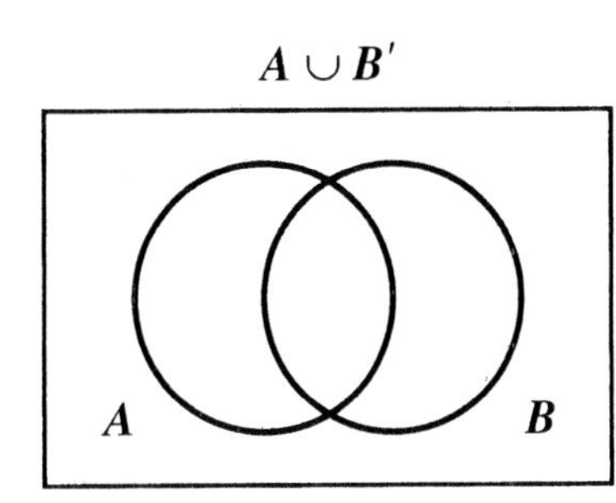

$A \cap B'$

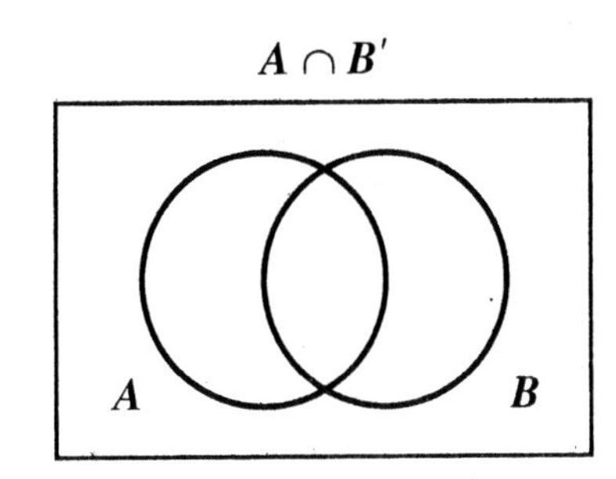

9. Draw a Venn diagram that indicates $|A \cap B| = 13$, $|A| = 52$, and $|B| = 25$. What is $|A \cup B|$?

10. Draw a Venn diagram that indicates $|A \cup B| = 40$, $|A| = 11$, and $|B| = 35$. What is $|A \cap B|$?

11. Let A = {Community College students in Math 92} and
B = {Community College students in CS 160}.

Assume there are 37 Community College students in Math 92, 43 in CS 160, and 65 in at least one of the two courses. How many Community College students are in both courses? Illustrate your solution on the following Venn diagram.

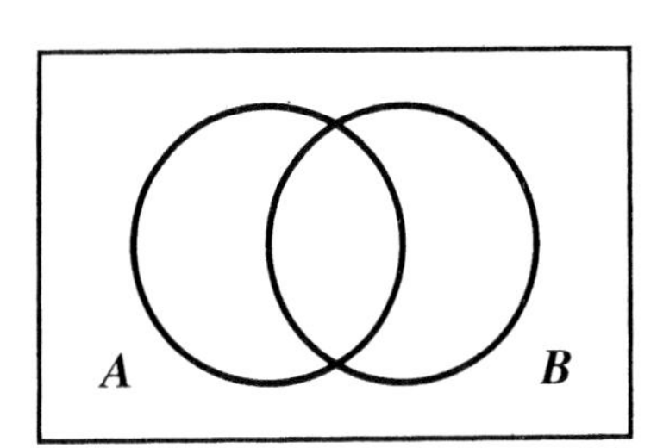

12. Let N = {students who use Netscape} and
E = {students who use Internet Explorer}.

Assume there are 14 students who use Netscape, 39 who use Internet Explorer, and 9 who use both. How many students use at least one of the two Web browsers? Illustrate your solution on the following Venn diagram.

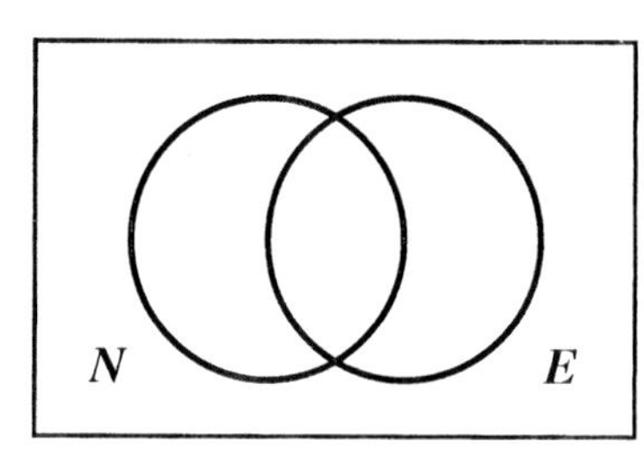

13. Shade the specified regions on the following Venn diagrams:

$A \cup C$

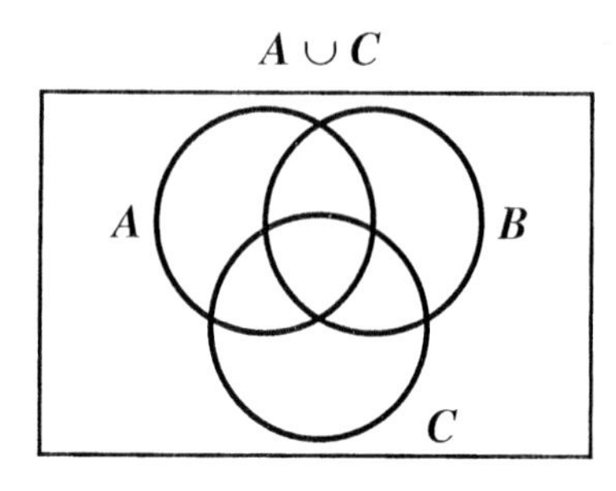

$(A \cap B)'$

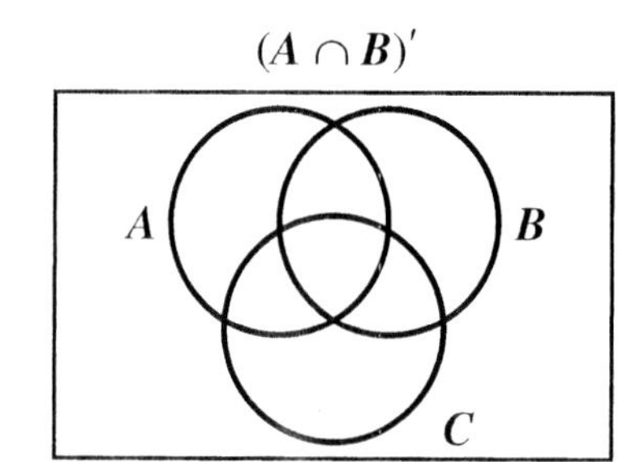

14. Shade the specified regions on the following Venn diagrams:

C'

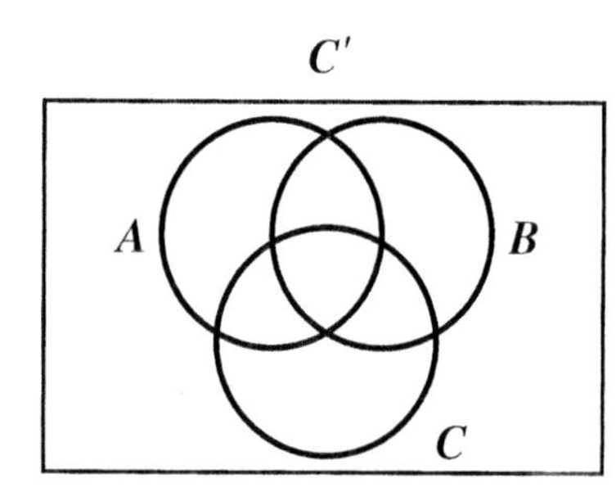

$A' \cap (B \cup C)$

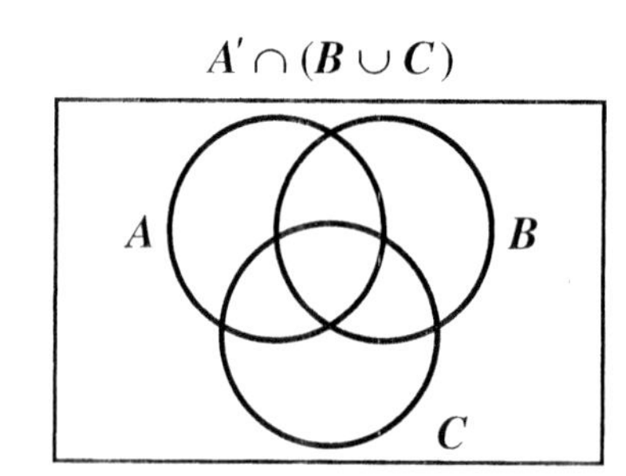

15. Draw the Venn diagram for which the following cardinalities apply:

$|A \cap B| = 8$, $|A| = 41$, and $|B| = 42$, $|A \cap C| = 24$, $|B \cap C| = 11$, $|C| = 42$, and $|A \cup B \cup C| = 85$. What is $|A \cap B \cap C|$?

16. Draw the Venn diagram for which the following cardinalities apply:
$|A \cap B| = 36$, $|A| = 216$, and $|B| = 41$, $|A \cap C| = 123$, $|B \cap C| = 23$, $|C| = 126$, and $|A \cap B \cap C| = 21$. What is $|A \cup B \cup C|$?

17. Let P = {computers with a Pentium 4 processor}, with $|P| = 125$,
R = {computers with 512 MB of RAM}, with $|R| = 45$, and
D = {computers with a DVD writer drive}, with $|D| = 35$.

Also, let $|P \cap R| = 40$, $|R \cap D| = 27$, $|P \cap D| = 32$, and $|P \cap R \cap D| = 25$.

(a) Draw the Venn diagram for which the given cardinalities apply.

(b) How many computers had only a DVD writer drive?

(c) How many computers had only a Pentium 4 processor?

18. Let W = {computers with Winamp}, with $|W| = 143$,
R = {computers with RealPlayer}, with $|R| = 70$, and
C = {computers with a CD writer}, with $|C| = 33$.

Also, let $|W \cap C| = 20$, $|R \cap C| = 7$, and $|W \cap R| = 28$, and let 193 machines have at least one of the three.

(a) Draw the Venn diagram for which the given cardinalities apply.

(b) How many computers have only Winamp?

(c) How many computers have only a CD writer?

(d) How many computers have only RealPlayer?

4.4 Propositions and Truth Tables

BITS of HISTORY

William Thomson, who would later be known as Lord Kelvin, studied mathematics as well as several branches of science. The absolute temperature scale, or Kelvin scale, was named for the title he was later awarded by the British government. Lord Kelvin also studied tidal flows and invented a 20-foot-long analog machine known as the tidal analyzer, which could predict tides based on previously recorded data.

In Chapter 3, we discussed the basic elements of coding. We know how a computer stores a letter or a number, but we do not yet know how a computer makes a decision.

The mathematics behind decision-making is called **logic**. All of our work in this chapter is based on mathematical logic.

In arithmetic, we create expressions using numbers and operators. Examples of such expressions include

$27 + 38$ and $457 - (2 \times 3)^3$.

In algebra, we use variables to represent numbers. Those same expressions can be written as algebraic expressions

$x + y$ and $a - (bc)^3$

where $x = 27$, $y = 38$, $a = 457$, $b = 2$, and $c = 3$.

In logic, instead of numbers, we work with propositions.

A **proposition** is a declarative statement that is either true or false.

Here are a few examples of propositions.

It rained in West Virginia yesterday.
Tex ate barbecued tofu for dinner last night.
$12_{16} + 9_{16} = 21_{16}$. **Do you recognize this as a false statement?**

EXAMPLE 1 **Determine whether each of the following is a proposition.**

(a) Is it raining? — **No, this is a question and not a declarative statement.**

(b) $4 + 2 = 6$ — **Yes, this is a proposition that is always true.**

(c) The temperature in Dallas is over 90°F. — **Yes, this is a proposition that is sometimes true and sometimes false.**

(d) Turn in your homework on Friday. — **No, this statement cannot be labeled as true or false.**

(e) I did all of the homework last night. — **Yes, this statement is either true or false.**

PRACTICE PROBLEM Answers on page 200

PRACTICE PROBLEMS 1 **Determine whether each of the following is a proposition.**

(a) $8 - 2 = 5$

(b) Is this difficult for you?

(c) The final will be 12 pages long.

(d) The surface is slippery when wet.

(e) Turn left at the next light.

As in algebra, we could assign a variable name to each statement. We put braces around each statement to make it clear where the statement begins and ends.

p = {It rained in West Virginia yesterday.}
q = {Tex ate barbecued tofu for dinner last night.}
$r = \{12_{16} + 9_{16} = 21_{16}\}$

Also, as in algebra, each proposition has a negative. We write the negative of p as $\sim p$, which we read as "not p" or "the **negation** of p."

EXAMPLE 2 Given p = {It rained in West Virginia yesterday.}
q = {Tex ate barbecued tofu for dinner last night.}
$r = \{12_{16} + 9_{16} = 21_{16}\}$
$s = \{3x + 2 < 7\}$

write each of the following propositions in words or symbols.

Solutions

(a) $\sim p$ We could write, "It did not rain in West Virginia last night."
(b) $\sim q$ "Tex did not eat barbecued tofu for dinner last night."
(c) $\sim r$ $12_{16} + 9_{16} \neq 21_{16}$
(d) $\sim s$ $3x + 2 \geq 7$

PRACTICE PROBLEMS 2 Given p = {Trinh is going to law school.}
q = {Maria got an A on her math test.}
$r = \{12_{16} + 9_{16} = 1B_{16}\}$
$s = \{x - 5 \leq -2\}$,

write each of the following propositions in words or symbols. There may be more than one correct answer.

(a) $\sim p$ (b) $\sim q$ (c) $\sim r$ (d) $\sim s$

PRACTICE PROBLEM Answers on page 200

The relationship of the truth value of a statement and its negation can be shown in a **truth table**.

p	$\sim p$
True	False
False	True

Each row of a truth table is independent. The first row indicates that when p is true, its negation is false. The second row indicates that when p is false, its negation is true. We could abbreviate truth value with the letters T and F, as below.

p	$\sim p$
T	F
F	T

Because a computer computes the value of a statement, true is given a value of 1 and false is given a value of 0. When we are working with computer logic, our truth tables will consist of 1s and 0s rather than Ts and Fs. The table for negation becomes:

p	$\sim p$
1	0
0	1

EXAMPLE 3 **Complete the following truth table.**

p	q	$\sim p$	$\sim q$
1	1		
1	0		
0	1		
0	0		

PRACTICE PROBLEM Answers
on page 200

Before proceeding, note that, with two statements, there are four possible combinations, 11, 10, 01, and 00.

To complete the table, we examine each row, assigning the opposite value for the negation of a statement.

p	q	$\sim p$	$\sim q$
1	1	0	0
1	0	0	1
0	1	1	0
0	0	1	1

PRACTICE PROBLEMS 3 **Complete the following truth table.**

r	s	$\sim s$	$\sim r$
1	1		
1	0		
0	1		
0	0		

ANSWERS TO PRACTICE PROBLEMS

1. (a) "8 − 2 = 5" is a proposition. (It is never true.)
   (b) "Is this difficult for you?" is not a proposition.
   (c) "The final will be 12 pages long." is a proposition.
   (d) "The surface is slippery when wet." is a proposition.
   (e) "Turn left at the next light." is not a proposition.
2. *Note*: There may be more than one correct answer to these problems.
   (a) Trinh is not going to law school.
   (b) Maria did not get an A on her math test.
   (c) $12_{16} + 9_{16} \neq 1B_{16}$
   (d) $x - 5 > -2$
3.

r	s	$\sim s$	$\sim r$
1	1	0	0
1	0	1	0
0	1	0	1
0	0	1	1

SECTION 4.4

Exercises

1. Which of the following are propositions?
   (a) Is the sky blue?

   (b) $4 + 5 = 9$

   (c) $x = 3$

   (d) Do your homework.

   (e) 7 is less than 3.

   (f) Eat your vegetables.

   (g) $24(-36) < 0$

   (h) Is this a proposition?

2. For each of the following propositions, find a proposition that is equivalent.
   (a) $x + 7 = -3$

   (b) 4 is less than 7.

   (c) It is Tuesday.

   (d) $4 > -6$

   (e) $25 < -26$

3. For each of the following propositions, find a negation.
   (a) $x - 7 = 2$

   (b) 5 is more than 3.

   (c) It is 97°F outside.

   (d) $12 > -4$

   (e) Math is fun.

   (f) $9 < 2$

▲ **Represents additionally challenging problems.**

4. For each of the following statements, find a negation.

(a) $x + 7 \neq 10$

(b) It is not raining today.

(c) He never takes out the trash.

(d) $12 \not\leq 13$

(e) $4 \not\geq -6$

(f) Mark did not fail his test.

5. The following is a truth table containing truth values for two propositions, p and q. Complete the table.

p	q	$\sim p$	$\sim q$	$\sim(\sim p)$	$\sim(\sim q)$	$\sim(\sim(\sim p))$	$\sim(\sim(\sim q))$
1	1						
1	0						
0	1						
0	0						

6. The following is a truth table containing truth values for two propositions, p and $\sim q$. Complete the table.

p	$\sim q$	$\sim p$	$\sim(\sim q)$	$\sim(\sim p)$	$\sim(\sim(\sim q))$	$\sim(\sim(\sim p))$	$\sim(\sim(\sim(\sim q)))$
1	0						
1	1						
0	0						
0	1						

7. Find a negation of the following propositions.

(a) Pat never has none of the answers.

(b) John ain't got no friends.

8. In logic, if two negations occur successively, what happens? In English, what happens? What property is similar in mathematics?

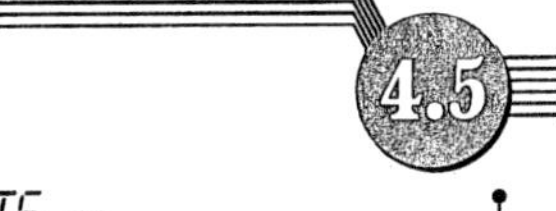

4.5 Logical Operators and Internet Searches

BITS of HISTORY

Not formally educated in advanced mathematics, George Boole studied the works of some of the great mathematicians on his own. Perhaps this independent research led to his new approach of bridging logic and mathematics, which he published in 1854. By creating analogs between logic and mathematical symbols, a branch of mathematics later known as Boolean algebra resulted that would eventually form the basis of computer logic.

Propositions are like numbers and variables in that we can operate on them. The operations we use are not addition and multiplication. Instead they are what we call **logical operators**. These include AND, OR, and NOT.

If p is a true statement and q is a false statement, then the statement p AND q is false, so it has a value of 0. In fact, the statement p AND q is true only when both p and q are true. We summarize all of the possibilities for a logical operator in a truth table. The following is the truth table for the proposition ***p* AND *q*.**

p	q	p AND q
1	1	1
1	0	0
0	1	0
0	0	0

Just as we use the symbol "·" to represent the operator "times," we can use the symbol "∧" to represent the operator "AND."

p	q	$p \wedge q$
1	1	1
1	0	0
0	1	0
0	0	0

Examine the following truth table for the proposition p OR q.

p	q	p OR q
1	1	1
1	0	1
0	1	1
0	0	0

Notice that p OR q is true when at least one of the two statements is true. The symbol "$\vee$" represents "OR."

p	q	$p \vee q$
1	1	1
1	0	1
0	1	1
0	0	0

EXAMPLE 1 **Use the tables to find the value of each statement.**

(a) $1 \wedge 1$

(b) $1 \vee 0$

(c) $0 \wedge 1$

(d) ~ 1

Solution

(a) $1 \wedge 1 = 1$

(b) $1 \vee 0 = 1$

(c) $0 \wedge 1 = 0$

(d) $\sim 1 = 0$

PRACTICE PROBLEMS 1 **Use the tables to find the value of each statement.**

(a) $1 \wedge 0$

(b) $1 \vee 1$

(c) $0 \wedge 0$

(d) ~ 0

Now go back and compare the three tables to multiplication, addition, and negation. How do they compare?

In the next example, we look at the compound proposition $\sim(p \wedge q)$. To do this, we find the truth table for p AND q, and change every result.

PRACTICE PROBLEM Answers on page 209

EXAMPLE 2 **Find the truth table for the statement $\sim(p \wedge q)$.**

p	q	$p \wedge q$	$\sim(p \wedge q)$
1	1	1	0
1	0	0	1
0	1	0	1
0	0	0	1

PRACTICE PROBLEM 2 **Find the truth table for the statement $\sim(p \vee q)$.**

One of the most important mathematical concepts that is relevant to the computer field is that of the function.

DEFINITION: A **function** is a rule that guarantees that a given input always results in the same output.

Think about what difficulty you would have if you put the same expression into your calculator twice and got two different answers. If you contend that this has happened, then either you or your calculator did not function properly!

Example 2 could be written as:

Find the truth table for the function $f(p, q) = \sim(p \wedge q)$.

p	q	$p \wedge q$	$f(p, q)$
1	1	1	0
1	0	0	1
0	1	0	1
0	0	0	1

The statement $f(p, q) = \sim(p \wedge q)$ simply says that the value of $\sim(p \wedge q)$ is a function of the values of p and q. The next example illustrates this concept.

EXAMPLE 3 **Given the function $f(p, q) = \sim(p \wedge q)$, find $f(1, 0)$.**

$f(1, 0)$ is asking for the value in the final column of the table above when $p = 1$ and $q = 0$.

In the second row, p has a value of 1 and q has a value of 0. The function value (from the last column) is 1. We can say

$$f(1, 0) = 1.$$

PRACTICE PROBLEM Answers on page 209

PRACTICE PROBLEM 3 **Given the function $f(p, q) = \sim(p \wedge q)$, find $f(0, 0)$.**

DEFINITION: De Morgan's laws comprise two important rules in logic.

Version 1 **not (*p* and *q*)** is equivalent to the statement **not *p* or not *q***.

In symbols, we write $\sim(p \wedge q) \Leftrightarrow \sim p \vee \sim q$.

Version 2 **not (*p* or *q*)** is equivalent to the statement **not *p* and not *q***.

In symbols, we write $\sim(p \vee q) \Leftrightarrow \sim p \wedge \sim q$.

EXAMPLE 4 **Show that $\sim(p \wedge q)$ is equivalent to $\sim p \vee \sim q$.**

We can let $f(p, q) = \sim p \vee \sim q$ and $g(p, q) = \sim(p \wedge q)$.

p	q	$\sim p$	$\sim q$	$f(p, q)$	$p \wedge q$	$g(p, q)$
1	1	0	0	0	1	0
1	0	0	1	1	0	1
0	1	1	0	1	0	1
0	0	1	1	1	0	1

Note that the columns labeled $f(p, q)$ and $g(p, q)$ are identical. The two statements are equivalent.

PRACTICE PROBLEM 4 **Show that $\sim(p \vee q)$ is equivalent to $\sim p \wedge \sim q$. The truth table from this practice problem should demonstrate the other law of De Morgan.**

PRACTICE PROBLEM Answers on page 209

With the ever-increasing amount of information being added to the Internet, understanding the logic behind search engines will help you find results more quickly. It is important to know what types of input particular search engines accept. Some require quotations around the terms being searched, and others may require the logical operator to be in ALL CAPS. For our exercises, we put the logical operators in ALL CAPS but omit the quotation marks.

EXAMPLE 5 **Suppose we are doing a search on the musical artist Don Henley. Because Don Henley was also in a group called the Eagles, we may want to broaden our search to include the Eagles. We would accomplish this by doing an "OR" search. The "OR" command allows any of the specified search terms to be present on the Web pages listed in results. The "OR" search can also be described as a "Match Any" search.**

The command typed into the search engine could be: Don Henley OR the Eagles.

Logically represented, this equates to $p \vee q$.

PRACTICE PROBLEMS 5 **What command could you type into a search engine if you wanted information about**

(a) the baseball player Babe Ruth or one of his famous teams, the 1927 Yankees?

(b) the state of Missouri?

EXAMPLE 6 **Suppose you are doing a more specific search about the relationship between mathematics and art. These two topics are very broad, so doing an OR search would give you more information than you need. The "AND" command requires that all search terms be present on the Web pages listed in results. It can also be described as a "Match All" search.**

The command typed into the search engine could be: mathematics AND art.

Logically represented, this equates to $p \wedge q$.

PRACTICE PROBLEMS 6 **What command could you type into a search engine if you wanted information about**

(a) the mathematics associated with M. C. Escher's artwork?

(b) car seat safety for children?

PRACTICE PROBLEM Answers on page 209

EXAMPLE 7 **Suppose you want some information on disc golf, a sport in which one throws a Frisbee or disc into a basket. Because the word "golf" is probably more recognizable, an AND search may still yield too much unwanted information about golf. In this case, a combination search using both the AND and NOT operators would probably work better.**

The command typed into the search engine could be: disc golf AND NOT golf.

The search engine would look for Web pages that contain the words "disc" and "golf" but exclude pages that just contain "golf."

Logically represented, this equates to $p \wedge (\sim q)$.

Most search engines now use the negative sign instead of the word "NOT." To get Frisbee and not golf, you could type, "Frisbee –golf

PRACTICE PROBLEMS 7 **What command could you type into a search engine if you wanted information about**

(a) the comedians, the Marx Brothers?

(b) a python (a type of snake)?

(c) pollution in rivers but not in oceans?

PRACTICE PROBLEM Answers
on page 209

Answers to Practice Problems

1. **(a)** $1 \wedge 0 = 0$ **(b)** $1 \vee 1 = 1$ **(c)** $0 \wedge 0 = 0$ **(d)** $\sim 0 = 1$

p	q	$p \vee q$	$\sim(p \vee q)$
1	1	1	0
1	0	1	0
0	1	1	0
0	0	0	1

3. $f(0, 0) = 1$ 4. We can let $f(p, q) = \sim p \wedge \sim q$ and $g(p, q) = \sim(p \vee q)$.

p	q	$\sim p$	$\sim q$	$f(p, q)$	$p \vee q$	$g(p, q)$
1	1	0	0	0	1	0
1	0	0	1	0	1	0
0	1	1	0	0	1	0
0	0	1	1	1	0	1

$f(p, q) = g(p, q)$

5. **(a)** Babe Ruth OR 1927 Yankees
   (b) Missouri OR MO (many Web sites will use the postal abbreviation)
6. **(a)** mathematics AND Escher **(b)** car seat AND child safety
7. **(a)** Marx Brothers NOT Karl **(b)** snake AND python NOT Monty
   (c) pollution AND rivers NOT oceans

SECTION

4.5 Exercises

1. Consider the two propositions:

   p: Austin can have coffee with cream.
   q: Austin can have coffee with sugar.

   (a) Write the proposition $p \wedge q$ in sentence form.

   (b) Write the proposition $p \vee q$ in sentence form.

2. Consider the two propositions:

   p: Jane went to the movies.
   q: Jane went to the beach.

   (a) Write the proposition $p \wedge q$ in sentence form.

   (b) Write the proposition $p \vee q$ in sentence form.

3. Compare the propositions in Exercises 1b and 2b. How are they the same? How are they different?

4. Complete the following truth table.

p	q	$\sim p$	$q \vee \sim p$
1	1		
1	0		
0	1		
0	0		

 **Represents additionally challenging problems.**

5. Using Exercise 1, write the proposition $q \vee \sim p$ in sentence form.

6. Using Exercise 2, write the proposition $\sim q \vee p$ in sentence form.

7. Complete the following truth table.

p	q	$\sim q$	$\sim q \vee p$
1	1		
1	0		
0	1		
0	0		

8. Complete the following truth table.

p	q	$\sim p$	$\sim p \vee p$
1	1	0	1
1	0	0	1
0	1	1	1
0	0	1	1

9. Complete the following truth table.

p	q	$\sim q$	$\sim q \wedge q$
1	1		
1	0		
0	1		
0	0		

10. Look at the last column of your truth tables in Exercises 8 and 9. What can you say about the proposition $\sim p \vee p$? What can you say about the proposition $\sim q \wedge q$?

11. Complete the following truth table for $f(A, B) = \sim A \wedge B$.

A	B	~A	f(A, B)
1	1		
1	0		
0	1		
0	0		

12. Complete the following truth table for $g(A, B) = A \vee \sim B$.

A	B	~B	g(A, B)
1	1	0	1
1	0	1	1
0	1	0	0
0	0	1	1

13. Use Exercise 11 to find the following.

(a) $f(1, 1)$ (b) $f(1, 0)$ (c) $f(0, 0)$

14. Use Exercise 12 to find the following.

(a) $g(1, 1)$ (b) $g(1, 0)$ (c) $g(0, 0)$

1 1 0

15. Compare the last column of your truth tables in Exercises 11 and 12. What can you say about the two columns? What rule does this demonstrate?

16. Consider the proposition: "He is not going to the store, or he is not going to the bank." Use De Morgan's law to write an equivalent proposition.

17. Consider the proposition: "Abby is not playing softball and is playing soccer." Use De Morgan's law to write an equivalent proposition.

18. Choose a sport and a famous person associated with that particular sport, and then describe how you would do an Internet search using logical operators.

19. Choose a film and a famous actor or actress in that particular film, and then describe how you would do an Internet search using logical operators.

20. What command could you type into a search engine if you wanted information about the state of Minnesota?

21. What command could you type into a search engine if you wanted information about Portland, Maine?

22. Find three specific propositions (or strings) that would match a logical search of the following form: $\sim p \wedge (q \vee r)$.

CHAPTER 4 SUMMARY

4.1 The Language of Sets

- A **set** is a well-defined collection of distinct objects. Those objects can be people, books, numbers, letters, or anything else that is described. An object must be clearly defined as part of a set or not part of a set.
- An **element** is an object contained in a particular set. Use braces, "{" and "}", to designate the contents of a set. Name a set with a capital letter, so that you can refer to it without listing all of its elements.
- A **subset** is a set of elements that are all in a specified set.
- The **universal set** (sometimes referred to as the **universe**) is the collection of all elements under consideration. The universal set is usually designated with the symbol U.
- A set that contains no elements is called the **empty set**. The empty set is designated with either a pair of empty brackets, { }, or with the symbol Ø.
- The **cardinality** of a set is the number of elements in the set. Cardinality is designated by placing the set name between two vertical lines. $|A|$ is the number of elements in set A.

4.2 Set Operators

- A **binary operator** connects two elements and maps them to one. Addition is an example of a binary operator.
- If A and B are sets, their **union**, which we write as $A \cup B$, is the set consisting of all elements that belong to either A or B or both.
- If A and B are sets, their **intersection**, written as $A \cap B$, is the set consisting of all elements that belong to **both** A and B.
- If two sets have nothing in common, they are said to be **disjoint**.
- A **unary operator** maps one number to another. Negation is an example of a unary operator.
- If A is a set, then we define the **complement** of A to be the set of elements that are in the universal set that are not in A and denote it by A'.
- The following equation relates to the cardinality of the union of two sets, which we will call the **cardinality principle for two sets**.

$$|A \cup B| = |A| + |B| - |A \cap B|$$

- The equation below is the cardinality principle for three sets:

$$|A \cup B \cup C| = |A| + |B| + |C| - |A \cap B| - |A \cap C| - |B \cap C| + |A \cap B \cap C|$$

4.3 Venn Diagrams

- A **Venn diagram** is a rectangle that represents the universal set together with inner circles, each representing a specific subset of the universal set.
- $(A \cup B)' = A' \cap B'$ (This is one form of De Morgan's law.)
- $(A \cap B)' = A' \cup B'$ (This is the second form of De Morgan's law.)
- Venn diagrams can be used to indicate cardinality.

4.4 Propositions and Truth Tables

- The mathematics behind decision-making is called **logic**.
- A **proposition** is a declarative statement that is either true or false.
- We write the negative of p as $\sim p$, which we read as "not p," or "the **negation** of p."
- The relationship of the truth value of a statement and its negation can be shown in a **truth table**.

p	$\sim p$
True	False
False	True

or

p	$\sim p$
1	0
0	1

4.5 Logical Operators and Internet Searches

- **Logical operators** include AND, OR, and NOT.
- The following is the truth table for the proposition p **AND** q.

p	q	$p \wedge q$
1	1	1
1	0	0
0	1	0
0	0	0

- The following is the truth table for the proposition ***p* OR *q***.

p	q	$p \vee q$
1	1	1
1	0	1
0	1	1
0	0	0

- A **function** is a rule that guarantees that a given input always results in the same output.
- We can evaluate the function $f(p, q)$ for given values of p and q.
- Two important rules in logic are called **De Morgan's laws**.

Version 1 **not (*p* and *q*)** is equivalent to the statement **not *p* or not *q***.

In symbols, we write $\sim(p \wedge q) \Leftrightarrow \sim p \vee \sim q$.

Version 2 **not (*p* or *q*)** is equivalent to the statement **not *p* and not *q***.

In symbols, we write $\sim(p \vee q) \Leftrightarrow \sim p \wedge \sim q$.

4 Glossary Quiz

The following terms were introduced in Chapter 4 of the text. Match each with one of the definitions that follow.

universal set ______	element ______	set ______
subset ______	union ______	empty set ______
cardinality ______	complement ______	Venn diagram ______
intersection ______	$(A \cup B)' = A' \cap B'$ ______	logic ______
unary operator ______	proposition ______	

(a) a collection of distinct objects
(b) an object contained in a particular set
(c) a set of elements that are all in a specified set
(d) the collection of all elements under consideration
(e) a set that contains no elements
(f) the number of elements in a set
(g) the set consisting of all elements that belong to either of two sets
(h) the set consisting of all elements that belong in both of two sets
(i) a sign that changes the value of a number or variable
(j) the set of elements that are in the universal set that are not in a specified set
(k) a rectangle that represents the universal set together with inner circles, each representing a specific subset of the universal set
(l) one form of De Morgan's law
(m) the mathematics behind decision-making
(n) a declarative statement that is either true or false

CHAPTER 4 REVIEW EXERCISES

[4.1] Use set notation to list the elements for each of the following.

1. The countries that are located on the continent of North America.

2. The first names of the *Brady Bunch* kids.

3. The seven colors in the visible spectrum.

4. The prime numbers between 8 and 16.

5. The letters in the word *Windows*.

6. The natural numbers that are divisible by 5.

[4.1] **7.** Which of the following are subsets of the set O, where O = {Portland, Eugene, Salem, Corvallis, Bend, Klamath Falls, Medford}

(a) {Eugene, Medford, Salem}

(b) {Corvallis, Vancouver, Eugene, Portland}

(c) {Bend, Portland, Oregon}

(d) {Salem, Portland, Eugene, Bend}

(e) $\varnothing$

[4.1] **8.** Which of the following are subsets of E, where $E = \{2, 4, 6, 8, \ldots\}$

(a) $\{32, 16, 24\}$

(b) $\{6, 9, 12, 15, 18\}$

(c) $\{0, 2, 4\}$

(d) $\{2000, 4000, 8000, \ldots\}$

(e) $\{2, 4, 6, 8, \ldots\}$

[4.1] **Use set notation to represent each of the following:**

9. The set of all U.S. states that border California.

10. The universal set of hexadecimal digits.

11. The set of the seven castaways on the television show, *Gilligan's Island*.

12. The universal set of Zodiac signs.

13. The set of odd integers between 2 and 100.

[4.1] Given the following set, find the cardinality of each.

14. P = {pine nuts, basil, olive oil, parmesan cheese}

15. S = {current members of the U.S. Senate}

16. F = {Gandalf, Frodo, Sam, Aragorn, Merry, Pippin, Legolas, Gimli, Boromir}

17. O = {all odd integers}

18. B = {all natural numbers less than 100 that are divisible by 15}

[4.2] 19. Let $U = \{0, 1, 2, 3, 4, 5, 6, 7, 8, 9, A, B, C, D, E, F\}$
$A = \{0, 2, 4, 6, 8, A, C, E\}$
$B = \{2, 3, 5, 7, B, D\}$
$C = \{0, 3, 6, 9, C, F\}$

Find the following:

(a) $A \cup B$

(b) $A \cap B$

(c) $A \cap C$

(d) B'

(e) $A' \cap C'$

(f) $|A'|$

(g) $|A \cup C|$

(h) $|U|$

[4.2] 20. Let U = {films listed below},
P = {*Emma, Shakespeare in Love, Bounce, The Talented Mr. Ripley, Seven, Possession, Sliding Doors*},
A = {*Shakespeare in Love, Good Will Hunting, Dogma, Bounce, Pearl Harbor, Phantoms, Armageddon, Changing Lanes, Chasing Amy, Boiler Room*}, and
D = {*The Talented Mr. Ripley, Good Will Hunting, The Bourne Identity, School Ties, Dogma, All the Pretty Horses, Courage Under Fire, Chasing Amy, Mystic Pizza*}.

Find the following:

(a) $P \cap A$

(b) $P \cup A$

(c) $A \cap D$

(d) A'

(e) $A \cap D \cap P$

(f) $|D'|$

(g) $|A \cap D \cap P|$

(h) $|U|$

(i) $|P \cap A|$

[4.2] **21.** Let A = {students who use Frontpage}
B = {students who use Dreamweaver}

Assume there are 29 students who use Frontpage, 21 students who use Dreamweaver, and 7 students who use both. How many students use at least one of the two Web page design programs?

22. Let W = {computers running Windows XP}
R = {computers with 1 GB of RAM}
D = {computers with a DVD+RW drive}

A computer technician needs to get the computer lab ready for the new term. She starts by doing an inventory and finds that

- There are 39 computers in the lab.
- Nineteen are running Windows XP
- Sixteen have a DVD+RW drive, and 20 of them have 1 GB of RAM
- Seven of the machines running Windows XP have a DVD+RW drive
- Six of the machines running Windows XP have 1 GB of RAM
- Eight of the machines that have a DVD+RW drive also have 1 GB of RAM

How many of the computers running Windows XP have both 1 GB of RAM and a DVD+RW drive?

[4.3] 23. Shade the specified regions on the following Venn diagrams.

$B \cap A$

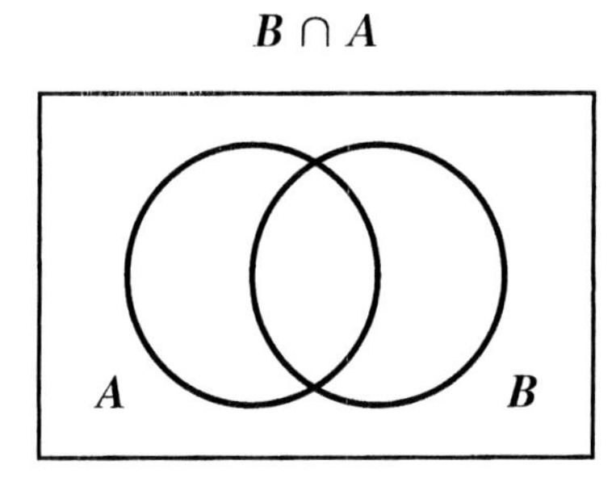

A'

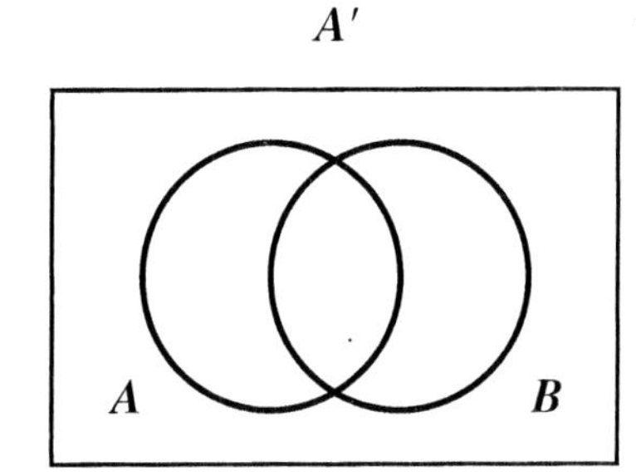

[4.3] **24.** Label the following sets on the Venn diagram below.
Let $U = \{0, 1, 2, 3, 4, 5, 6, 7, 8, 9, A, B, C, D, E, F\}$
$O = \{1, 3, 5, 7, 9, B, D, F\}$
$P = \{2, 3, 5, 7, B, D\}$

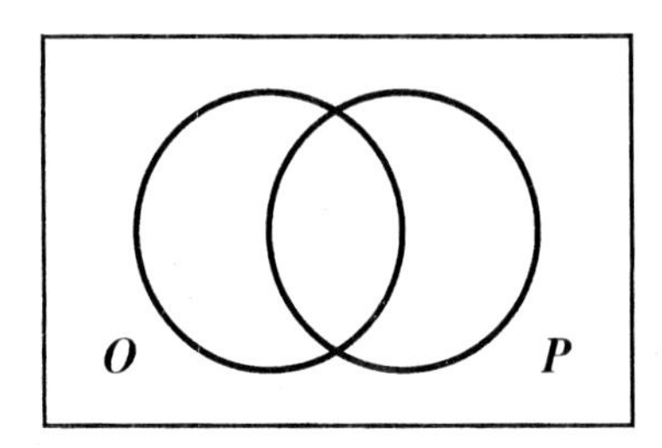

[4.3] **25.** Draw a Venn diagram to illustrate the set B'.
Let $U = \{0, 1, 2, 3, 4, 5, 6, 7, 8, 9\}$
$A = \{2, 4, 6, 8, 0\}$,
$B = \{8, 2, 5, 1\}$

[4.3] **26.** Shade the specified regions on the following Venn diagrams.

$A' \cup B$

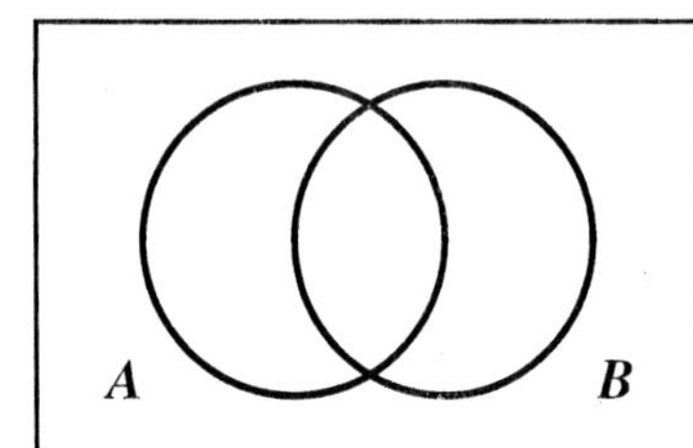

$A' \cap B'$

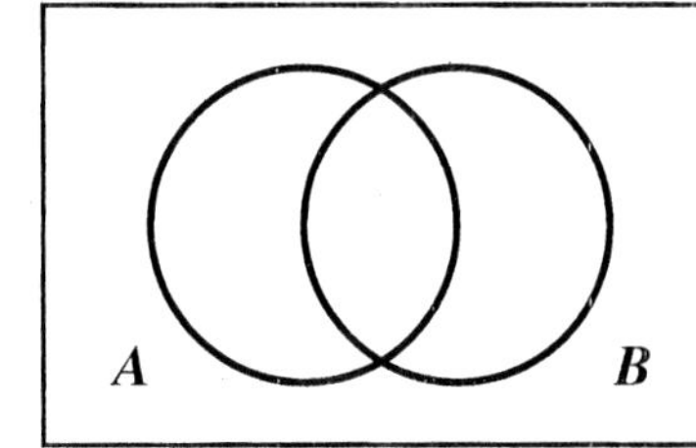

[4.3] 27. Draw a Venn diagram that indicates $|A \cap B| = 12$, $|A| = 30$, and $|B| = 19$. What is $|A \cup B|$?

28. Draw a Venn diagram that indicates $|A \cup B| = 95$, $|A| = 23$, and $|B| = 80$. What is $|A \cap B|$?

29. Let A = {students taking computer math}
B = {students taking PC repair}

Assume there are 22 students taking computer math, 17 taking PC repair, and 31 taking at least one of the two courses. How many students are taking both courses? Illustrate your solution on the following Venn diagram.

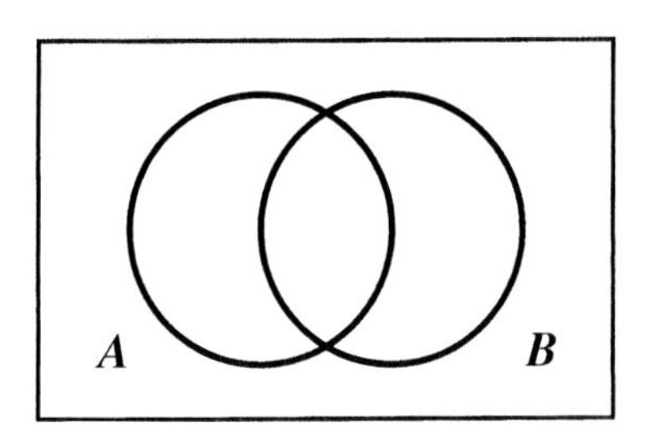

[4.3] **30.** Shade the specified regions on the following Venn diagrams.

$(A' \cap B)$

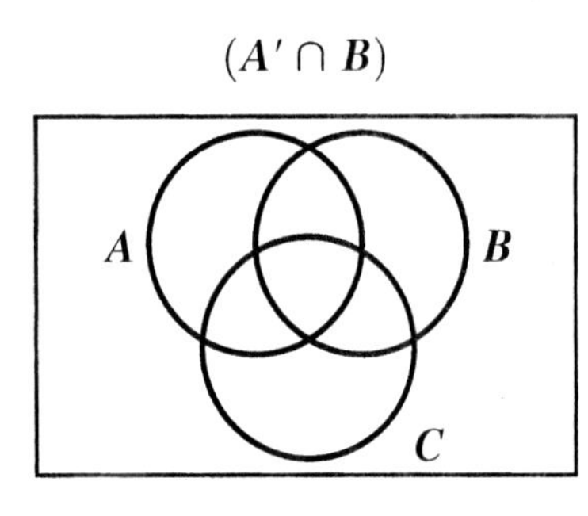

$(A \cup B') \cap C$

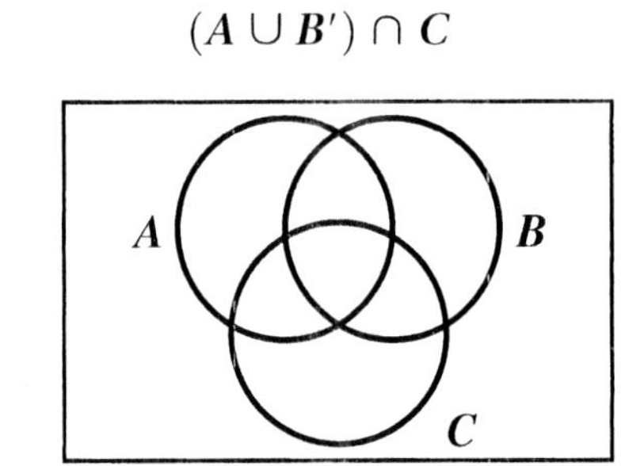

[4.3] **31.** Draw a Venn diagram that indicates $|A \cup B \cup C| = 63$, $|A| = 58$, $|B| = 12$, $|C| = 38$, $|A \cap B| = 11$, $|A \cap C| = 33$, and $|B \cap C| = 6$. What is $|A \cap B \cap C|$?

[4.3] **32.** Let N = {computers with Nero}, with $|N| = 21$,
T = {computers with two CD-ROM drives}, with $|T| = 30$, and
W= {computers with WinMX}, with $|W| = 21$

Also, let $|N \cap T| = 16$, $|N \cap W| = 9$, $|T \cap W| = 20$, and $|N \cap T \cap W| = 8$

(a) Draw a Venn diagram for which the given cardinalities apply.

(b) Describe an element in the set $N \cap T \cap W$.

(c) How many computers have only WinMX?

(d) How many computers have Nero and two CD-ROM drives, but not WinMX?

[4.4] **33.** Which of the following are propositions?

(a) $4 \times 16 > 50$

(b) Walk this way.

(c) $5 - 8 > -1$

(d) Does Kate know how to play the board game *Go*?

(e) $5 + 8 = 12$

(f) The pug is snoring.

(g) $x = 2$ or $x = -1$

(h) Is the computer working properly?

34. For each of the following propositions, find a negation.

(a) $x + 9 = 16$

(b) The Bulldogs won the swim meet.

(c) $6 + 2 \nleq 5$

(d) Sam didn't tell the truth.

(e) $x = 0$ or $x = 1$

(f) The coffee has neither cream nor sugar added.

(g) There is no way out.

[4.4] **35.** Complete the following truth table.

p	q	$\sim p$	$\sim(\sim p)$	$\sim(\sim(\sim p))$	$\sim q$
1	1				
1	0				
0	1				
0	0				

[4.5] **36.** Consider following the two propositions:

p: Chris can study for his math test.

q: Chris can watch football.

(a) Write the proposition $p \vee q$ in sentence form.

(b) Write the proposition $p \wedge q$ in sentence form.

(c) Write the proposition p in sentence form.

(d) Write the proposition $p \wedge \sim q$ in sentence form.

[4.5] **37.** Complete the following truth tables.

(a)

p	q	$\sim p$	$\sim p \vee q$
1	1		
1	0		
0	1		
0	0		

(b)

p	q	$\sim q$	$\sim q \vee q$
1	1		
1	0		
0	1		
0	0		

(c)

p	q	$\sim p$	$p \wedge \sim p$
1	1		
1	0		
0	1		
0	0		

[4.5] **38.** Complete the following truth table for $f(A, B) = A \wedge \sim B$.

A	B	~B	f(A, B)
1	1		
1	0		
0	1		
0	0		

[4.5] **39.** Use Exercise 38 to find the values of the following.

(a) $f(0, 1)$ (b) $f(0, 0)$ (c) $f(1, 1)$

[4.5] **40.** Consider the proposition and use De Morgan's law to write its negation. "Allison is not working in Chicago and is moving to St. Louis."

[4.5] **41.** What command could you type into a search engine if you wanted information about the NFL team, the Miami Dolphins? What command could you type into a search engine if you wanted information about swimming with dolphins in the city of Miami?

[4.5] **42.** What command could you type into a search engine if you wanted information about the town of California in the state of Missouri?

CHAPTER 4 SELF-TEST

1. **Use set notation to represent each of the following.**

   (a) The set of five oceans on the earth.

   (b) The set of prime numbers between 20 and 40.

   (c) The universal set of octal digits.

   (d) The set of letters in the word *spanakopita*.

   (e) The universal set of all natural numbers that are divisible by 11.

2. **Which of the following are subsets of the set S, where $S = \{1, 2, 4, 8, 16, 32, 64, 128, 256, ...\}$**

   (a) $\{1, 16, 4, 64\}$ **(b)** $\{2, 4, 6, 8\}$ **(c)** $\{8, 64, 512, ...\}$

   (d) $\{2^0, 2^1, 2^2, ...\}$ **(e)** $\{1^2, 2^2, 3^2, ...\}$

3. **Let $U = \{0, 1, 2, 3, 4, 5, 6, 7, 8, 9, a, b\}$,**
   $A = \{0, 3, 6, 9\}$,
   $B = \{2, 3, 5, 7, b\}$,
   $C = \{1, 3, 5, 7, 9, b\}$

   Find the following:

   (a) $A \cup B$ **(b)** $C \cap B$

(c) C'

(d) $B' \cup C$

(e) $A \cap C'$

(f) $|C'|$

(g) $|B' \cup C|$

(h) $|U|$

4. **Shade the specified regions on the following Venn diagrams.**

$(A \cup B)'$

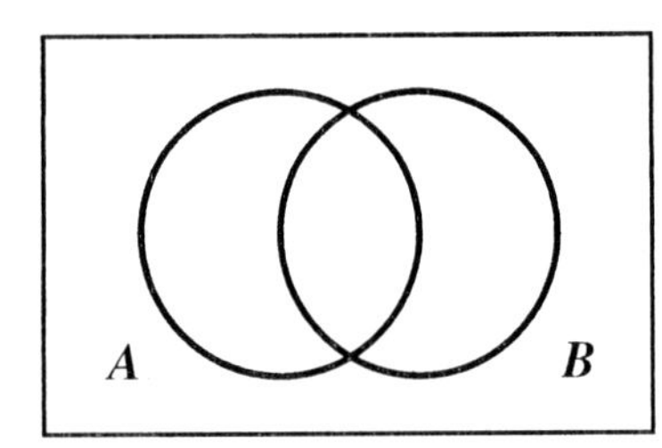

B'

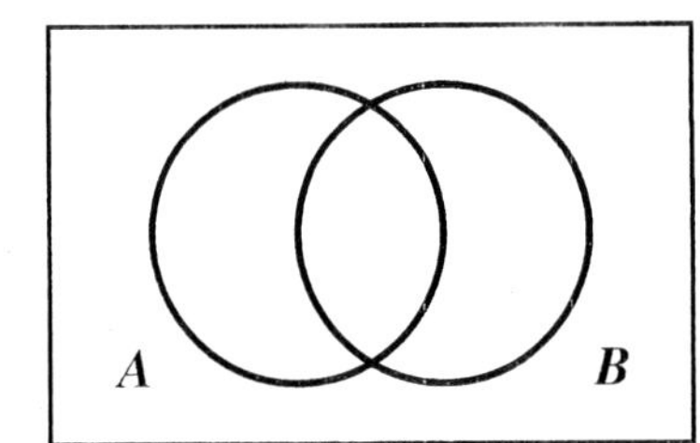

5. **Let U = {Bengals, Bills, Broncos, Browns, Chargers, Chiefs, Colts, Dolphins, Jaguars, Jets, Patriots, Raiders, Ravens, Steelers, Texans, Titans}**

A = {Dolphins, Ravens, Colts, Bengals, Browns, Jaguars, Broncos}
B = {Browns, Bills, Broncos, Bengals}
E = {Dolphins, Patriots, Jets, Bills}

Find the following:

(a) $E \cap A$

(b) $A \cup B$

(c) $E \cup B$

(d) A'

(e) $|E'|$

(f) $|A' \cap B|$

6. **Label the following regions on the Venn diagram below:**

   Let $U = \{0, 1, 2, 3, 4, 5, 6, 7, 8, 9, A, B\}$
   $O = \{1, 3, 5, 7, 9\}$
   $P = \{2, 3, 5, 7, B\}$

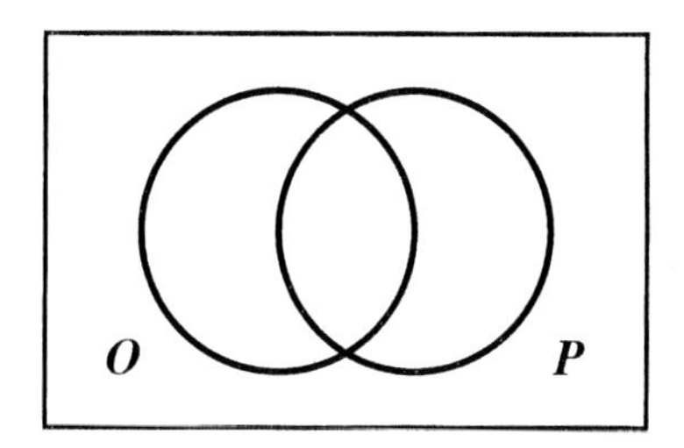

7. **Let A = {students taking a Web-design class}**
   B = {students taking a networking class}

   Assume there are 30 students taking a Web-design class, 22 students taking a networking class, and 39 students taking at least one of the two classes. How many students are taking both a Web-design class and a networking class? Draw a Venn diagram to illustrate your solution.

8. **Draw a Venn diagram for which the following cardinalities apply:**
$|A \cap B \cap C| = 26$, $|A| = 39$, $|B| = 45$, $|C| = 76$, $|A \cap B| = 27$,
$|A \cap C| = 35$, **and** $|B \cap C| = 26$.

What is $|A \cup B \cup C|$**?**

9. **For each of the following propositions, find a negation.**

(a) $x + 4 = 12$

(b) Yellow and blue make green.

(c) Stefan was late for class.

(d) $-3y < 15$

10. **Let** $F =$ **{computers with a flat screen monitor}, with** $|F| = 31$
$H =$ **{computers with a 100 GB hard drive}, with** $|H| = 24$
$D =$ **{computers with a DVD+RW drive}, with** $|D| = 40$

Also, let $|F \cap H| = 15$, $|F \cap D| = 16$, $|D \cap H| = 20$, and $|F \cup H \cup D| = 58$.

(a) Draw a Venn diagram for which the given cardinalities apply.

(b) Describe an element in the set $F \cup H \cup D$.

(c) How many computers have a flat screen monitor, a 100 GB hard drive, and a DVD+RW drive?

(d) How many computers have only a DVD+RW drive?

(e) How many computers have a 100 GB hard drive and a DVD+RW drive, but do not have a flat screen monitor?

11. Complete the following truth table for $f(A, B) = \sim A \vee B$, then use the truth table to evaluate the following:

(a) $f(0, 1)$ (b) $f(0, 0)$ (c) $f(1, 1)$

A	B	$\sim A$	$f(A, B)$
1	1		
1	0		
0	1		
0	0		

12. What command could you type into a search engine if you wanted information about the music group REM?

REM is also germane to the field of medicine. What command would you type into a search engine if you want information on this type of REM?

13. Which of the following are propositions?

(a) White is not a primary color.

(b) September comes before December.

(c) Is your birthday in September?

(d) $x + x = 3$

(e) Get your feet off the coffee table.

14. Complete the following truth tables.

(a)

p	q	$\sim p$	$\sim p \wedge p$
1	1		
1	0		
0	1		
0	0		

(b)

p	q	$\sim q$	$q \vee \sim q$
1	1		
1	0		
0	1		
0	0		

(c)

p	q	$\sim p$	$\sim q$	$\sim p \vee \sim q$
1	1			
1	0			
0	1			
0	0			

CHAPTERS 1–4 CUMULATIVE REVIEW EXERCISES

1. Complete the following table so that each row represents equivalent numbers with the base indicated in the column heading.

HEXADECIMAL	DECIMAL	BINARY
$AF7_{16}$		
	861	
		1010101.100101_2

2. Complete the following hexadecimal arithmetic problems:

(a) $E53_{16} + A97_{16}$

(b) $AD3_{16} - B9_{16}$

(c) $1B9_{16} \times 4F_{16}$

3. Complete the following octal arithmetic problems:

(a) $1763_8 + 5224_8$

(b) $763_8 - 274_8$

4. Put the following decimal numbers in 8-bit, two's complement notation.

(a) 38

(b) −10

(c) −121

5. You intercept an unidentified ASCII transmission where the last four bytes appear to be a signature. Decode this set of ASCII bit patterns to discover the identity of the sender (one character per blank).

Hexadecimal ASCII Key

[0–9]	30_{16} + Number (in hex)
[A–Z]	40_{16} + Letter (in hex)
[a–z]	60_{16} + Letter (in hex)

______	______	______	______
0101 0010	0011 0010	0100 0100	0011 0010

6. Let $S = \{0, 1, 2, 3, 4, 5, 6, 7, 8, 9\}$
$A = \{2, 4, 5, 7, 8, 9\}$
$B = \{1, 2, 6, 7\}$

Find the following:

(a) $A \cup B$ **(b)** $A \cap B$ **(c)** $(A \cup B)'$

(d) $|A|$ **(e)** $|A \cup B|$ **(f)** $|A' \cap B|$

7. The following block is coded in even parity. The coded bits are shaded. The unshaded 4th bit in each column and row is the parity bit. Assuming the parity bits are correct and only one error was detected, circle the bit that most likely caused the error. Justify your answer.

	A	*B*	*C*	*parity*
1	1	1	1	1
2	1	0	1	0
3	0	1	1	1
parity	0	1	1	

8. Given $A = \{C, M, Y, K\}$, list **all** the subsets of A.

9. Complete the following truth tables. Looking only at the final column, indicate whether each is a *tautology*, *fallacy*, or *conditional* proposition.

(a)

P	Q	P OR Q
1	1	
1	0	
0	1	
0	0	

(b)

P	Q	$P \wedge Q$	$\sim(P \wedge Q)$
1	1		
1	0		
0	1		
0	0		

(c)

P	Q	$\sim Q$	$P \vee \sim Q$	$(P \vee \sim Q) \vee Q$
1	1			
1	0			
0	1			
0	0			

10. Answer the following.

(a) What is one-half of 2^{2000}?

(b) What is one-half of 2^{-30}?

(c) What is one-half of 3^8?

(d) What is one-half of 10^2?

11. Shade the specified regions on the following Venn diagrams:

$A' \cap B'$

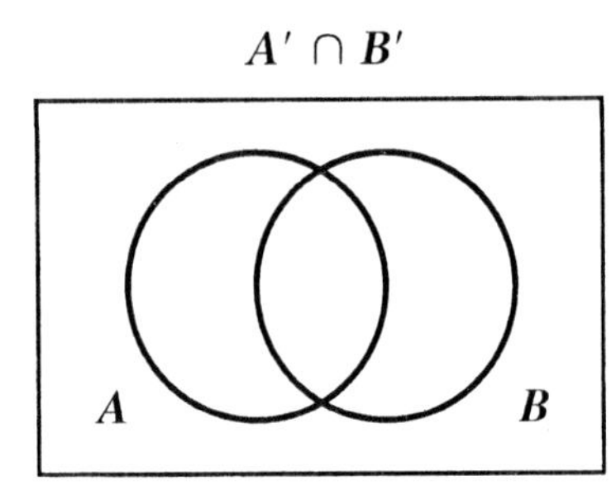

$A' \cup B$

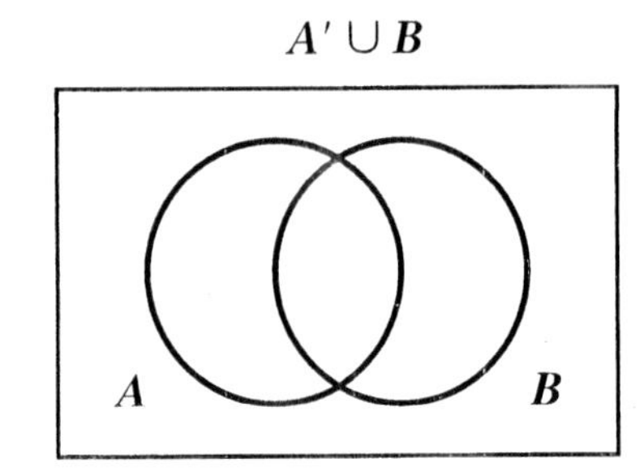

12. Let A = {students who have a Hotmail e-mail address}
B = {students who have a Yahoo e-mail address}

Assume there are 45 students who have a Yahoo e-mail address, 63 students who have a Hotmail e-mail address, and 27 students who have both. How many students have at least one of the two e-mail addresses? Illustrate your answer with a Venn diagram.

13. Draw a Venn diagram that indicates $|A \cup B \cup C| = 91$, $|A| = 54$, $|B| = 68$, $|C| = 16$, $|A \cap B| = 31$, $|A \cap C| = 12$, and $|B \cap C| = 15$. What is $|A \cap B \cap C|$?

14. What command could you type into a search engine if you wanted information about the HBO series, *The Sopranos*? What command could you type into a search engine if you wanted information about famous sopranos?

15. Use set notation to represent each of the following:

(a) The set of all vowels (excluding y) in the English language.

(b) The universal set of binary digits.

(c) The set of the months with 32 days.

(d) The universal set of months of the year.

(e) The set of even integers between 3 and 51.

16. Find the binary, octal and hexadecimal equivalents for each decimal number.

(a) 6 **(b)** 27 **(c)** 92 **(d)** 289

17. Simplify each expression, first as a base to a single power, then if possible as a decimal. You may assume the variables are nonzero.

(a) $2^5 \cdot y^8 \cdot y^{14}$ **(b)** $\frac{x^{19}}{x^{11}}$ **(c)** $\frac{2^{20}}{2}$

18. Approximate the following with your calculator. Report your answers to four significant digits if the value is not exact.

(a) $\frac{6.921 \times 3.125}{2.005}$ **(b)** $\sqrt{5}$ **(c)** $\frac{\pi}{6}$

19. Convert each number into decimal notation.

(a) 6.0913×10^{-7} **(b)** 5.0229×10^{11}

20. Simplify the following and write your answer in scientific notation.

$$\frac{(3.14 \times 10^{12})(7.546 \times 10^{-7})}{4.02 \times 10^{-3}}$$

21. Exactly how many *bytes* are in each of the following?

(a) 10 MB **(b)** 5600 K

22. Find the mean for the following set of data:

29 min, 32 min, 31 min, 39 min, 34 min, 33 min

Assume the data above is used to predict the average time to install network software on a hard drive. Use the mean from the data set to compute the absolute error and relative error if the actual installation time is 35 minutes.

23. Determine the number of significant digits.

(a) 0.002020 **(b)** 8,004,705,700,000,000

24. Given that the average American sleeps 7.25 hours a day, determine the total amount of sleep in minutes per lifetime for a person whose life expectancy is 80 years. You may assume 365.25 days per year. How many hours during that same person's life is he or she awake?

25. Perform the following binary subtractions. Rewrite each problem in decimal notation to check your work.

(a) $\begin{array}{r} 1111_2 \\ -\ 0101_2 \\ \hline \end{array}$ **(b)** $\begin{array}{r} 1001_2 \\ -\ 0110_2 \\ \hline \end{array}$

Boolean Circuits

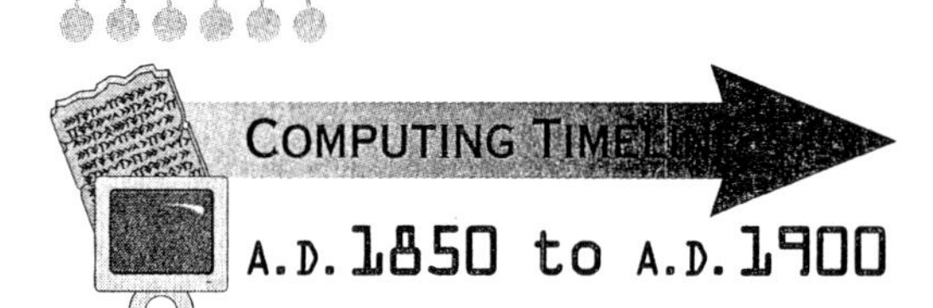

While both the typewriter and the adding machine with keyboard were invented during this period, perhaps the most significant development was Charles Babbage's analytic engine, which was a forerunner of our modern-day calculator. Although Babbage never quite developed a working model, his forty years on this project is generally considered to be the foundation from which all subsequent computer research grew.

5.1 Boolean Logic

The daughter of noted poet Lord Byron, Augusta Ada Lovelace was a skilled mathematician and a key player in the development of the computer. She translated a paper by the Italian Menabrea, who critiqued Babbage's analytical engines. Ada's annotated translation on the programming possibilities of the analytical engine was widely circulated, thus further advancing computer science.

Boolean algebra is concerned with any idea or object that has only two states. Those states could be

1, 0 yes, no **true, false** open, closed

or any of an infinite number of possibilities.

The branch of algebra related to variables that always have only 0 and 1 as possible values was devised in 1847 by the English logician George Boole. In his honor, we call this branch of mathematics Boolean algebra. It was not until the middle of the twentieth century that Boolean algebra was applied to the analysis of relay and switching circuits.

Symbols previously used in this text can be used in Boolean algebra, but the following symbols are customary and will be used here.

$'$ for negation (or complement) — **a' means not a or the complement of a.**

$\vee$ for "or" (also called "join")

$\wedge$ for "and" (also called "meet") — **Two variables together can also mean "meet," so xy is the same as $x \wedge y$.**

Negation ($'$) is a **unary operator** (it takes only one variable). Join ($\vee$) and meet ($\wedge$) are **binary operators** (they each take two variables).

The results of the symbols $\vee$ and $\wedge$ are referred to as the sum and product, respectively.

A Boolean set is usually referred to as ***B***, and two distinct elements of ***B*** are denoted by **0** and **1; 0** is the **zero element** and **1** is the **unit element**.

Order of Operation

If a Boolean expression has several operators, the parentheses take precedence. Parentheses are often used to minimize confusion. Then, in order, we perform the operations of negation, intersection, and then union.

Many of the properties that you learned in arithmetic and algebra have parallel properties in Boolean algebra. As you did in both arithmetic and algebra, you will learn to use some of these rules to simplify expressions. In this section, we examine only a couple of those laws. In Sections 5.3 and 5.5, we examine other ways to simplify Boolean expressions that we hope will be easier.

We first look at a couple of properties that should be familiar to you from your earlier work in mathematics.

Commutative Properties

BOOLEAN EXAMPLE	ARITHMETIC PARALLEL	ALGEBRA PARALLEL
$x \vee y = y \vee x$	$1 + 2 = 2 + 1$	$a + b = b + a$
$x \wedge y = y \wedge x$	$3 \times 2 = 2 \times 3$	$a \cdot b = b \cdot a$

Identity Properties

BOOLEAN EXAMPLE	ARITHMETIC PARALLEL	ALGEBRA PARALLEL
$x \vee 0 = x$	$3 + 0 = 3$	$a + 0 = a$
$x \wedge 1 = x$	$3 \times 1 = 3$	$a \cdot 1 = a$

EXAMPLE 1 **Name the property demonstrated in each Boolean expression.**

(a) $y \wedge 1 = y$

(b) $ab = ba$

(c) $z \vee 0 = z$

Solution

(a) $y \wedge 1 = y$ demonstrates the identity property of "and."

(b) $ab = ba$ demonstrates the commutative property of "and."

(c) $z \vee 0 = z$ demonstrates the identity property of "or."

PRACTICE PROBLEMS 1 **Name the property demonstrated in each Boolean expression.**

(a) $y \wedge x = x \wedge y$

(b) $a \vee b = b \vee a$

(c) $z \wedge 1 = z$

We now examine two more properties of Boolean algebra.

PRACTICE PROBLEM Answers on page 250

Distributive Properties

BOOLEAN EXAMPLE	ARITHMETIC PARALLEL	ALGEBRA PARALLEL
$x \wedge (y \vee z) = (x \wedge y) \vee (x \wedge z)$	$3(2 + 5) = 3 \times 2 + 3 \times 5$	$a(b + c) = a \cdot b + a \cdot c$
$x \vee (y \wedge z) = (x \vee y) \wedge (x \vee z)$	no parallel property	no parallel property

Note that, in Boolean algebra, we can distribute meet over join or join over meet. In arithmetic and algebra, there is only one distributive property, the distribution of multiplication over addition.

EXAMPLE 2 **Where possible, use the distributive properties to expand each Boolean expression.**

(a) $y \wedge (x \vee z)$

(b) $a \vee (b \wedge c)$

(c) $x \wedge (y \wedge z)$

Solution

(a) $y \wedge (x \vee z) = (y \wedge x) \vee (y \wedge z)$

Notice that the pattern is identical to the pattern that emerges when you distribute multiplication over addition.

(b) $a \vee (b \wedge c) = (a \vee b) \wedge (a \vee c)$

(c) $x \wedge (y \wedge z)$

The distributive law does not apply here. Both operations are the same.

We now look at the parts (a) and (b) from Example 2, but use xy instead of $x \wedge y$.

(a) $y \wedge (x \vee z) = y(x \vee z) = yx \vee yz$

(b) $a \vee (b \wedge c) = a \vee (bc) = (a \vee b)(a \vee c)$

PRACTICE PROBLEM Answers on page 250

PRACTICE PROBLEMS 2 **Where possible, use one of the distributive properties to expand each Boolean expression.**

(a) $y \vee (x \vee z)$ (b) $m \vee (n \wedge p)$ (c) $x \wedge (y \vee z)$

There are a couple more properties that we will use in this section. We present each of them with their arithmetic and algebra parallels in the following tables.

Complement Property

BOOLEAN EXAMPLE	ARITHMETIC PARALLEL	ALGEBRA PARALLEL
$x \wedge x' = 0$	$5 + (-5) = 0$	$x + (-x) = 0$
$x \vee x' = 1$	$3 \times \frac{1}{3} = 1$	$x \cdot \frac{1}{x} = 1$

Reduction

BOOLEAN EXAMPLE	ARITHMETIC PARALLEL	ALGEBRA PARALLEL
$x \wedge 0 = 0$	$5 \times 0 = 0$	$x \cdot 0 = 0$
$x \vee 1 = 1$	no parallel	no parallel

Simplifying Algebraic Expressions

In your algebra classes, you learned to use properties to simplify expressions such as

$$3x^2y^2 + 2xy - 2x^2y^2 + 6xy - 5xy - 2xy.$$

By collecting like terms and factoring, we can reduce that expression to its equivalent

$$x^2y^2 + xy$$

or even

$$xy(xy + 1).$$

Simplifying Boolean Expressions

In a similar way, the properties we have seen in this section can be used to simplify Boolean expressions. Simplifying Boolean expressions often requires patience, ingenuity, or even luck. Here is an example.

EXAMPLE 3 **Simplify $y(x \vee y')$.**

By the distributive law, we have

$$y(x \vee y') = yx \vee yy'.$$

By the complement law, $yy' = 0$, so we have

$$yx \vee yy' = yx \vee 0.$$

By the identity property, we have

$$yx \vee 0 = yx.$$

Putting all of this together, we find that

$$y(x \vee y') = yx.$$

PRACTICE PROBLEM 3 **Simplify $a(a' \vee b')$.**

ANSWERS TO PRACTICE PROBLEMS

1. (a) $y \wedge x = x \wedge y$ demonstrates the commutative property of "and."
   (b) $a \vee b = b \vee a$ demonstrates the commutative property of "or."
   (c) $z \wedge 1 = z$ demonstrates the identity property of "and."
2. (a) $y \vee (x \vee z)$ The distributive property does not apply.
   (b) $m \vee (n \wedge p) = (m \vee n) \wedge (m \vee p)$
   (c) $x \wedge (y \vee z) = (x \wedge y) \vee (x \vee z)$
3. $a(a' \vee b') = aa' \vee ab' = 0 \vee ab' = ab'$

SECTION

5.1 Exercises

1. Name the property demonstrated in each real variable expression.

(a) $x + (-x) = 0$

(b) $3(x + y) = 3x + 3y$

(c) $x + (z \cdot y) = (z \cdot y) + x$

(d) $x \cdot (z \cdot y) = (x \cdot z) \cdot y$

2. Name the property demonstrated in each real variable expression.

(a) $x + 0 = x$

(b) $yx = xy$

(c) $x \cdot (z + y) = (z + y) \cdot x$

(d) $x + (z + y) = (x + z) + y$

3. Name the property demonstrated in each Boolean expression.

(a) $x \wedge x' = 0$

(b) $x \wedge (z \vee y) = (x \wedge z) \vee (x \wedge y)$

(c) $x \vee (z \wedge y) = (z \wedge y) \vee x$

▲ **Represents additionally challenging problems.**

4. Name the property demonstrated in each Boolean expression.

(a) $x \wedge 1 = x$

(b) $x \wedge (zy) = (zy) \wedge x$

(c) $x \vee (z \wedge y) = (x \vee z) \wedge (x \vee y)$

5. Use real variable algebra to simplify the following expressions.

(a) $2(x + 7y)$

(b) $(x + y)(x - y)$

6. Use real variable algebra to simplify the following expressions.

(a) $5(x - 2) + 9x - 6$

(b) $(x - y) + 4(x - y) - 2x$

For Exercises 7–18, match each expression on the left with its equivalent on the right. Some answers may be used more than once or not at all.

7. $y \wedge 0$
8. $y \wedge 1$
9. y
10. $(x')'$
11. xx'
12. $(x \vee y)'$
13. $x \vee 0$
14. $x \vee 1$
15. $(x' \vee x')'$
16. $x \wedge (y')'$
17. $(y' \wedge x')'$
18. $(y' \vee x')'$

A. 1
B. 0
C. x
D. y
E. $x \wedge y$
F. $x \vee y$
G. $x' \wedge y'$
H. $x' \vee y'$

19. Use Boolean algebra to simplify the following expression, and then evaluate the expression for $x = 1$ and $z = 1$.

$$x \vee (z \wedge x')$$

20. Use Boolean algebra to simplify the following expression, and then evaluate the expression for $x = 1$ and $z = 1$.

$$xz \vee x'z$$

21. Use Boolean algebra to simplify the following expression.

$$xyz \vee xyz'$$

22. Use Boolean algebra to simplify the following expression.

$$(x \vee y)(x' \vee y)$$

5.2 Logic Circuits Part I: Switching Circuits

BITS of HISTORY

A.D. 1800–1900 Perhaps the most significant achievement in the development of the computer in the nineteenth century came from the British mathematician, Charles Babbage. Babbage's difference engine could calculate tables and record results on metal plates, thus eliminating typographical errors. With the success of this prototype, he was able to secure funding from the English government to build the first mechanical computer, which he called the analytical engine. Unfortunately, Babbage's analytical engine was never fully functional, although the theory behind it would serve as the foundation for the modern computer.

We have examined three Boolean operators:

$\wedge$	**AND**	(also called *meet*)
$\vee$	**OR**	(also called *join*)
$'$	**NOT**	(also *complement*)

Now we look at the connection between these operators and series-parallel circuits. First, we define a logic circuit.

A **logic circuit** is a set of symbols that relate to a Boolean expression. If the expression is true, the circuit is closed (complete), and if the statement is false, the circuit is open (incomplete).

The Boolean expression $A \vee B$ can be represented by the following circuit. This is called a **parallel circuit.**

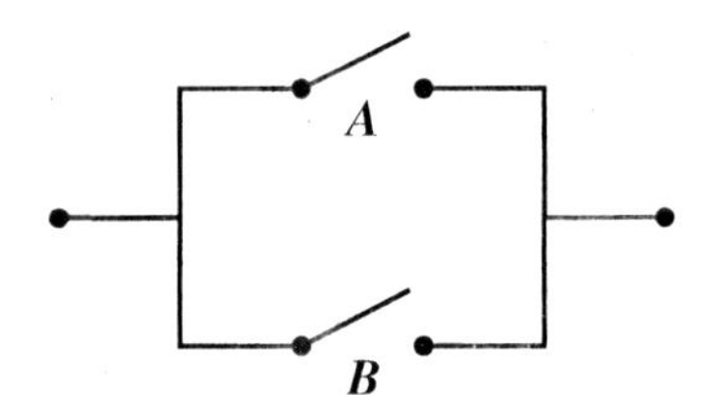

The A under the open switch indicates that the switch is closed when A is true, open when A is false. Note that, if A is true or B is true, there is a path from the beginning of the circuit on the left to the end of the circuit on the right.

The Boolean expression $A \wedge B$ can be represented by the following circuit. This is called a **series circuit**.

In this case, the circuit is closed (complete) only when both A is true and B is true.

A Boolean expression can also be represented by a set of switches, which we call a **switching circuit**. Each variable is represented by a two-position switch. The following graphic represents the expression $A \wedge B$. Note that both switches are in the up position pointing at the value 1.

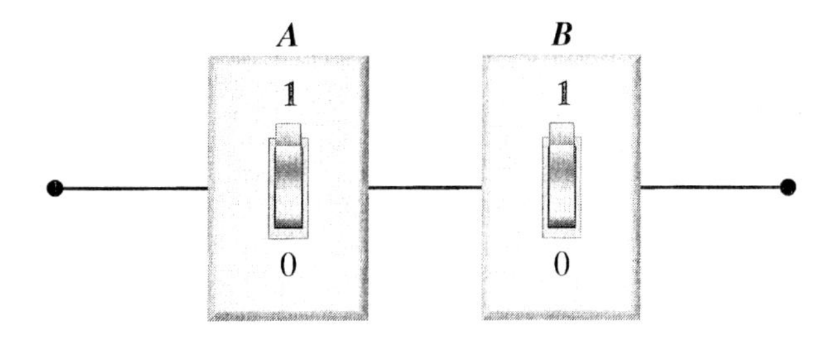

The Boolean expression $A \vee B$ can also be represented by a switching circuit.

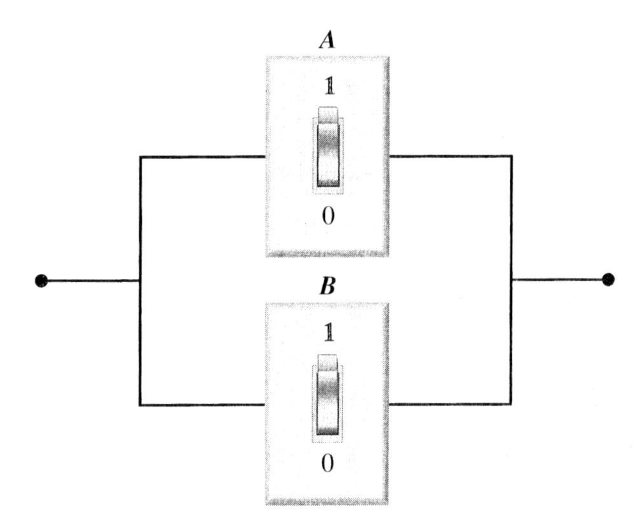

EXAMPLE 1 **Sketch the switching circuit associated with the Boolean expression $A \wedge B'$.**

Solution

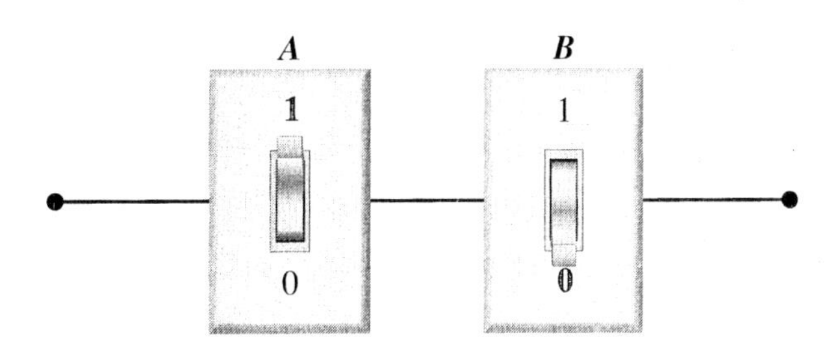

Note that variable A has a value of 1 while variable B has a value of 0.

PRACTICE PROBLEM 1 **Sketch the switching circuit associated with the Boolean expression $A \vee B'$.**

To better see the switching circuit related to a particular Boolean expression, we will henceforth imply the operator "AND" whenever it occurs. Such a grouping of Boolean variables without the operator "$\wedge$" is called a conjunctive. When these conjunctives are joined by an "OR," the expression becomes a disjunctive.

That means that in each of the following pairs of expressions, the expression on the left is equivalent to the one on the right.

BOOLEAN EXPRESSION	EQUIVALENT TERM
$A \wedge B$	AB
$A \wedge B' \wedge C$	$AB'C$

EXAMPLE 2 **Sketch the Boolean expression $B'C \vee ABC'$.**

Solution

Each term is sketched in series. The terms are then joined in parallel. The resulting circuit looks like this:

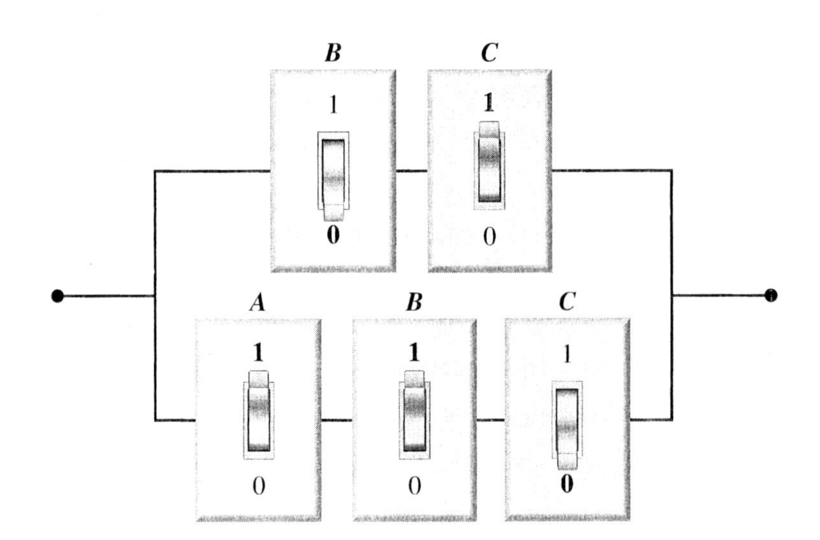

PRACTICE PROBLEMS Answers
on page 257

PRACTICE PROBLEM 2 Sketch the Boolean expression $A'B \vee A'BC$.

ANSWERS TO PRACTICE PROBLEMS

1. 2.

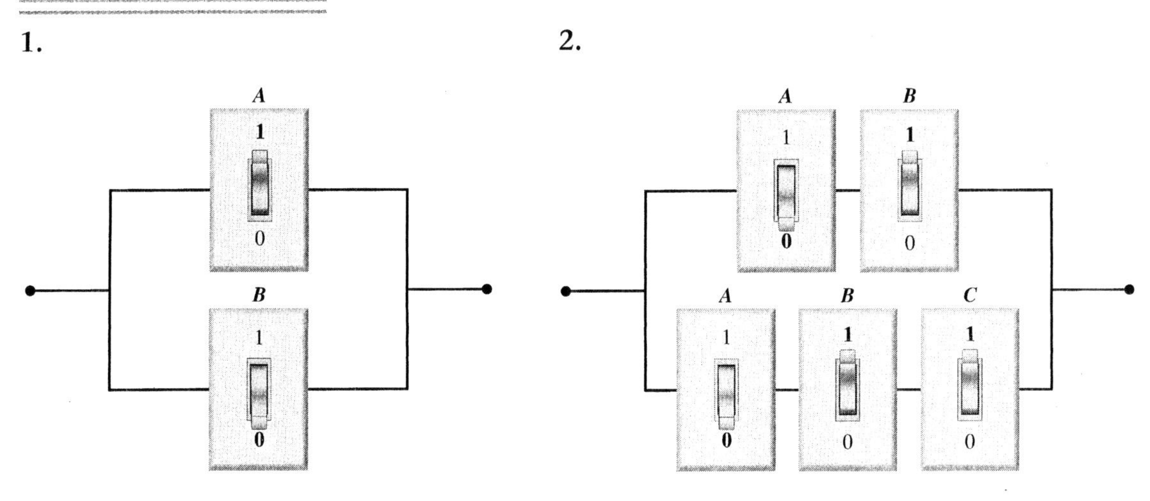

SECTION 5.2

Exercises

1. Sketch the switching circuit associated with the Boolean expression $A' \vee B$.

2. Sketch the switching circuit associated with the Boolean expression $A' \wedge B$.

3. Sketch the switching circuit associated with the Boolean expression $A'B \vee AB'$.

4. Sketch the switching circuit associated with the Boolean expression $B'C \vee ABC'$.

5. Give the Boolean expression associated with the following switching circuit.

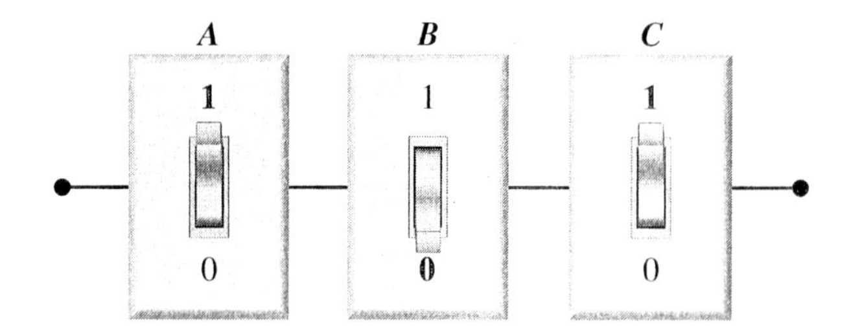

Represents additionally challenging problems.

6. Give the Boolean expression associated with the following switching circuit.

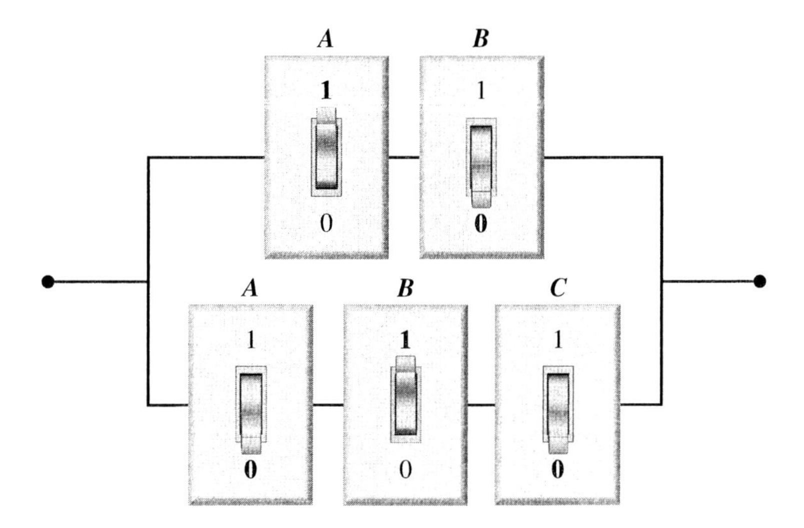

7. Give the Boolean expression associated with the following switching circuit.

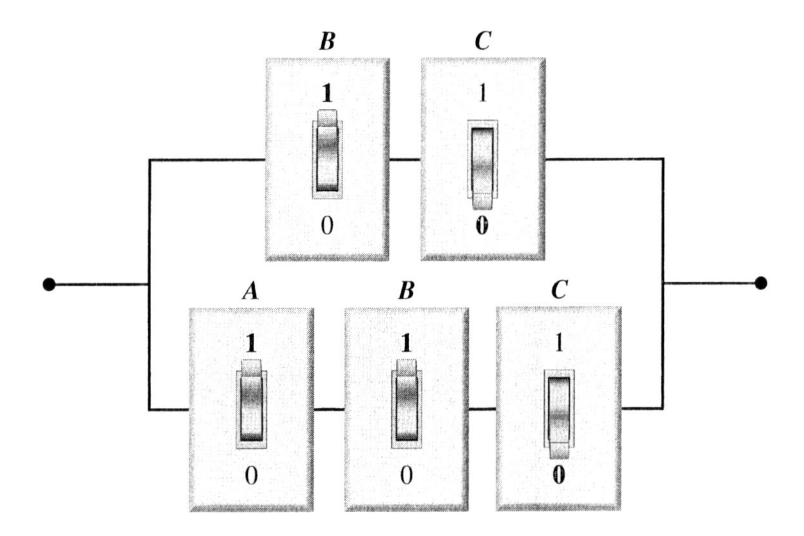

8. Find a Boolean expression that is equivalent but simpler than the one you found in Exercise 7.

5.3 Truth Tables and Disjunctive Normal Form

BITS of HISTORY

A.D. 1800–1900 William Stanley Jevons, who studied under De Morgan, unveiled his logical piano in 1870. This machine was capable of solving both algebraic and syllogistic problems—the first to do so. A contemporary who corresponded with George Boole, Jevons also made lasting contributions in the field of logic. His concept of including "both" when using the operator OR is still used today.

A Boolean expression that consists of a series of terms in which every variable appears in every term is said to be in **disjunctive normal form (dnf)**. Both of the following expressions are in dnf for the variables A, B, and C,

$$ABC' \vee AB'C$$
$$ABC \vee A'BC \vee A'B'C,$$

whereas the following three expressions are not in dnf:

$$AB \vee AB'C$$
$$A \vee B' \vee C$$
$$ABC \vee AB'C \vee AC$$

To this point, we have only implied that there is a relationship between truth tables and Boolean expressions. In fact, *every truth table represents a Boolean expression, and every Boolean expression can be rewritten as a truth table.*

The Boolean expression $AB \vee A'B$ can be represented by the following truth table.

A	B	$AB \vee A'B$
1	1	1
1	0	0
0	1	1
0	0	0

There are two 1s in the final column. The first is in the row AB and the second in the row $A'B$.

Because the values in the final column of a truth table are a function of the values of the variables, we sometimes use function notation to abbreviate that column. In the example above, we could say that

$$f(A, B) = AB \vee A'B.$$

The table could then be written as it is below.

A	B	$f(A, B)$
1	1	1
1	0	0
0	1	1
0	0	0

Note: We use function notation throughout this section.

EXAMPLE 1 **Write each of the following functions as a Boolean expression in disjunctive normal form (dnf).**

Rewrite $f(A, B)$ as a Boolean expression.

(a)

A	B	$f(A, B)$
1	1	1
1	0	1
0	1	1
0	0	0

The Boolean expression is made up of each variable combination that makes the function true (gives it a value of 1 in the final column). In this case, we have

$$f(A, B) = AB \vee AB' \vee A'B$$

Because A and B are the only variables and are represented in every term, this expression is written in dnf.

(b) Rewrite $f(A, B, C)$ as a Boolean expression.

A	B	C	$f(A, B, C)$
1	1	1	1
1	1	0	0
1	0	1	0
1	0	0	0
0	1	1	1
0	1	0	1
0	0	1	1
0	0	0	0

There are four sets of conditions that give this function a value of 1. That means the Boolean expression will have four terms:

$$f(A, B, C) = ABC \vee A'BC \vee A'BC' \vee A'B'C.$$

PRACTICE PROBLEMS 1 Write the following Boolean functions as Boolean expressions in dnf.

(a)

A	*B*	*f(A, B)*
1	1	1
1	0	0
0	1	1
0	0	1

(b)

A	*B*	*C*	*f(A, B, C)*
1	1	1	1
1	1	0	1
1	0	1	0
1	0	0	0
0	1	1	1
0	1	0	1
0	0	1	0
0	0	0	0

When we defined a logic circuit, we mentioned that the circuit was true whenever the expression was true. This definition makes it quite easy to find the circuit related to a truth table.

EXAMPLE 2 **Find the circuit associated with each truth table.**

(a)

A	*B*	*f(A, B)*
1	1	1
1	0	1
0	1	1
0	0	0

PRACTICE PROBLEM Answers on page 268

Solution

From Example 1(a), we know we have $f(A, B) = AB \vee AB' \vee A'B$.

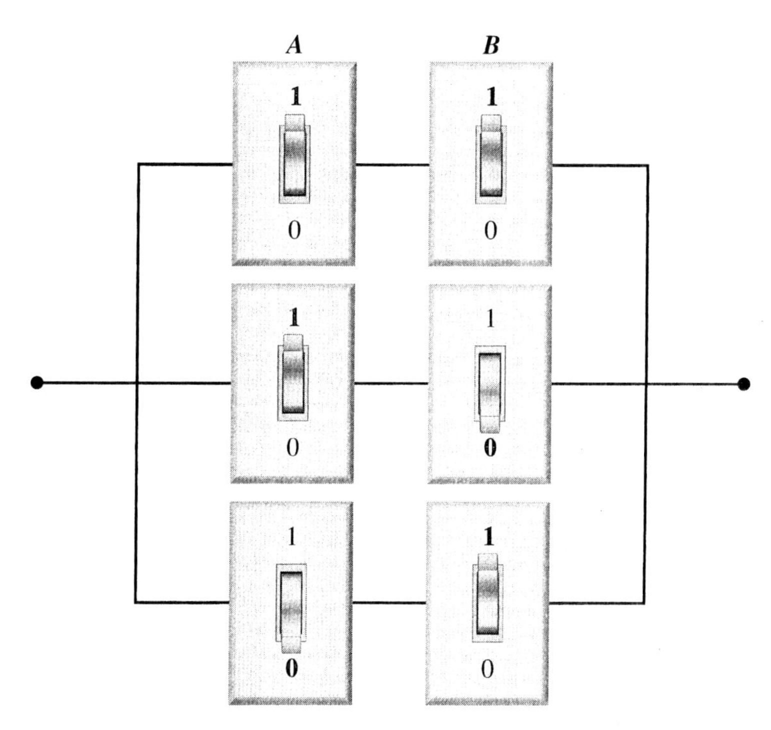

(b)

A	B	C	f (A, B, C)
1	1	1	1
1	1	0	0
1	0	1	1
1	0	0	0
0	1	1	1
0	1	0	1
0	0	1	0
0	0	0	0

Solution

The dnf is $ABC \vee AB'C \vee A'BC \vee A'BC'$.

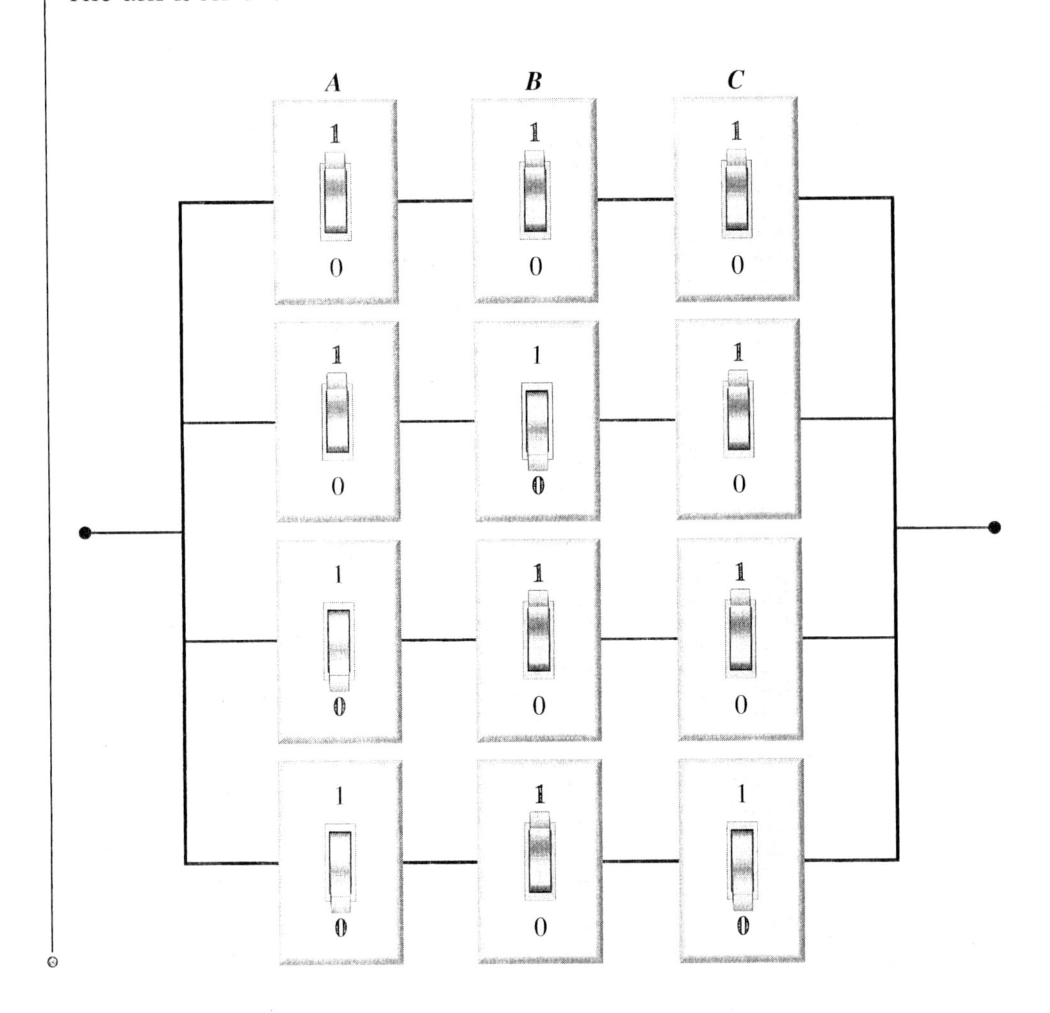

PRACTICE PROBLEMS 2 Find the circuit associated with each truth table.

(a)

A	B	$f(A, B)$
1	1	1
1	0	0
0	1	1
0	0	1

PRACTICE PROBLEM Answers
on page 268

(b)

A	B	C	$f(A, B, C)$
1	1	1	1
1	1	0	1
1	0	1	0
1	0	0	0
0	1	1	1
0	1	0	1
0	0	1	0
0	0	0	0

Every switching circuit can be represented by a unique truth table. Examples 3 and 4 illustrate a technique for finding the truth table for a given circuit.

EXAMPLE 3 **Find the truth table associated with the following circuit, circuit F.**

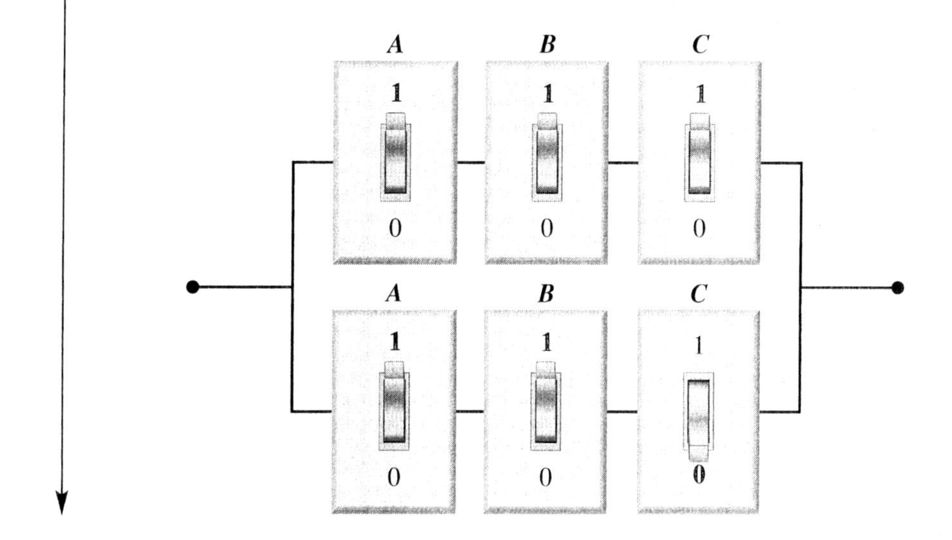

Solution

There are two different settings that allow the circuit to be complete. Either A, B, and C are all up (ABC), or A and B are up but C is down (ABC'). We place 1s in only those two rows in the truth table. All other rows get a 0.

A	B	C	CIRCUIT F
1	1	1	1
1	1	0	1
1	0	1	0
1	0	0	0
0	1	1	0
0	1	0	0
0	0	1	0
0	0	0	0

PRACTICE PROBLEM 3 **Find the truth table associated with the following circuit, circuit *G*.**

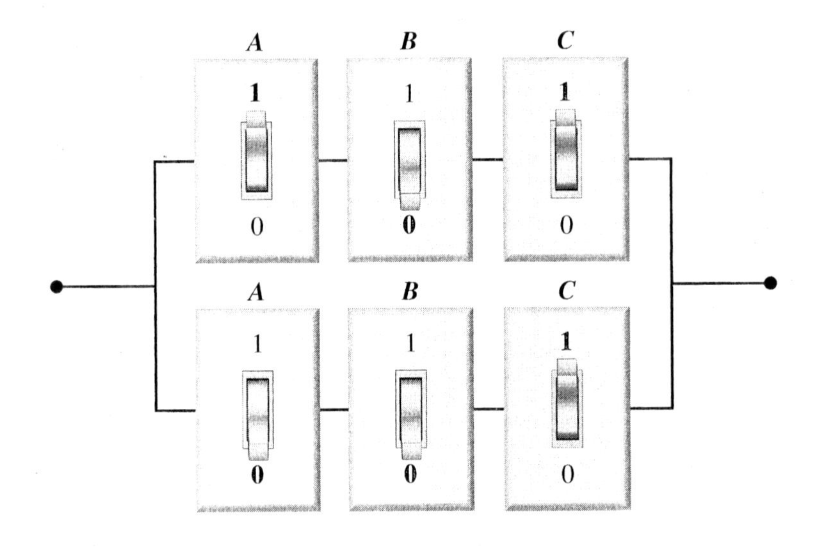

PRACTICE PROBLEM Answers
on page 268

EXAMPLE 4 **Find the truth table associated with the following circuit, circuit *F*.**

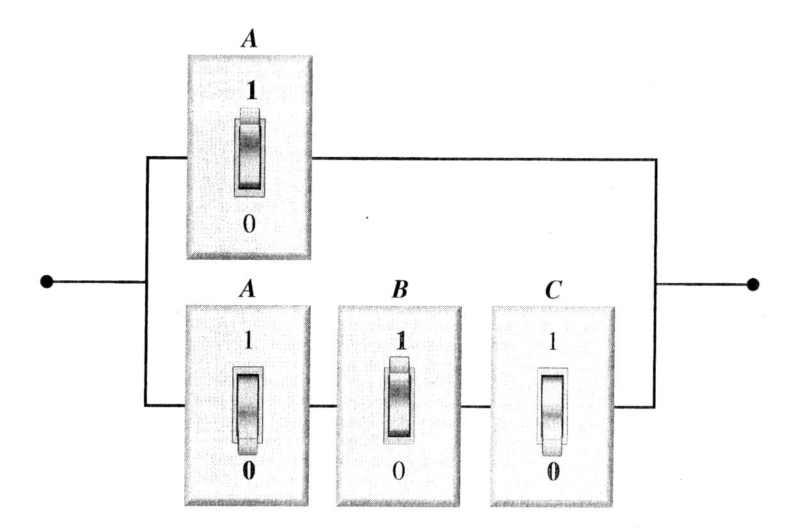

Solution

Again, there are two different criteria that allow the circuit to be complete. The first is that, as long as switch A is up, the circuit is complete. In the following truth table, this relates to the first four rows. In every one of those rows, A has a value of 1. The second setting that completes the circuit has A down, B up, and C down ($A'BC'$). We place a 1 in that row as well. All other rows get a 0.

A	*B*	*C*	CIRCUIT *F*
1	1	1	1
1	1	0	1
1	0	1	1
1	0	0	1
0	1	1	0
0	1	0	1
0	0	1	0
0	0	0	0

PRACTICE PROBLEM 4 Find the truth table associated with the following circuit, circuit G.

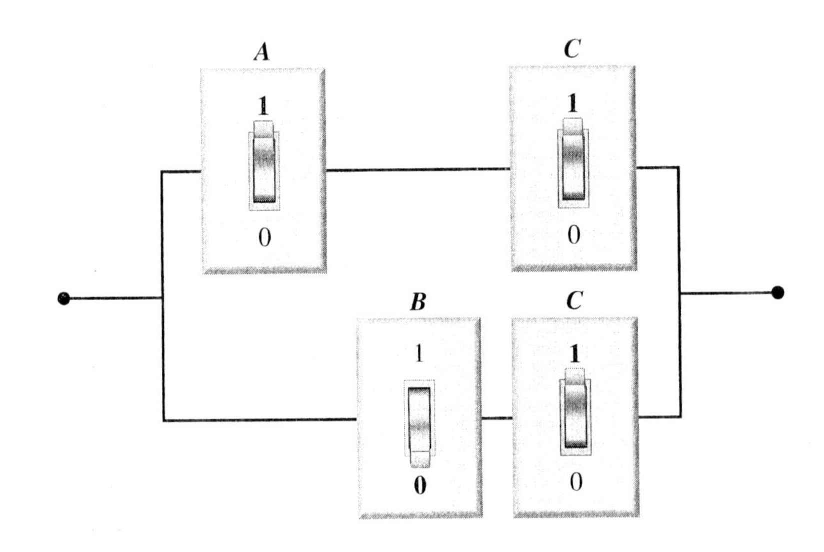

ANSWERS TO PRACTICE PROBLEMS

1. (a) $f(A, B) = AB \vee A'B \vee A'B'$

   (b) $f(A, B, C) = ABC \vee ABC' \vee A'BC \vee A'BC'$

2. (a)

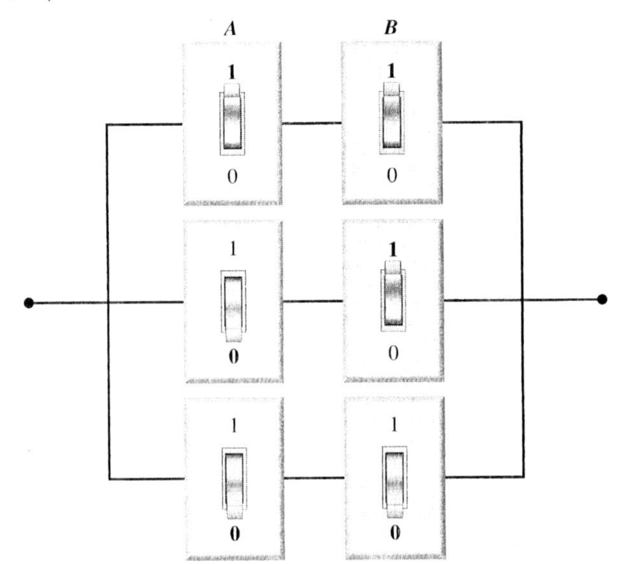

(b)

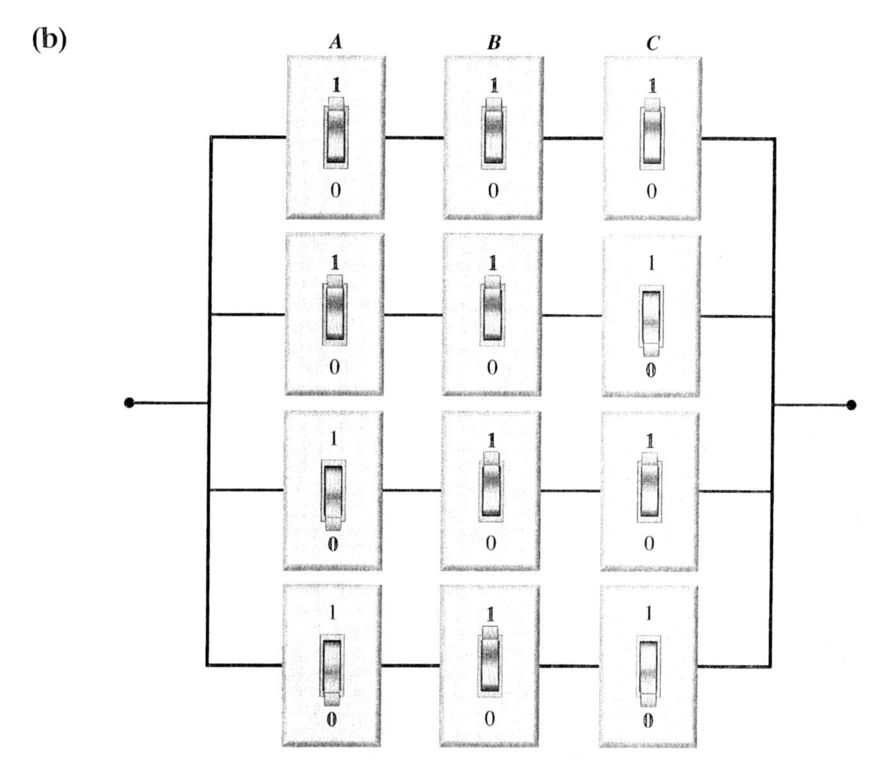

3.

A	B	C	$g(A, B, C)$
1	1	1	0
1	1	0	0
1	0	1	1
1	0	0	0
0	1	1	0
0	1	0	0
0	0	1	1
0	0	0	0

4.

A	B	C	$g(A, B, C)$
1	1	1	1
1	1	0	0
1	0	1	1
1	0	0	0
0	1	1	0
0	1	0	0
0	0	1	1
0	0	0	0

SECTION 5.3

Exercises

1. Write each of the following Boolean functions as a Boolean expression.

(a)

A	B	f(A, B)
1	1	0
1	0	0
0	1	1
0	0	0

(b)

A	B	f(A, B)
1	1	0
1	0	1
0	1	1
0	0	1

2. Write each of the following Boolean functions as a Boolean expression.

(a)

A	B	f(A, B)
1	1	1
1	0	0
0	1	0
0	0	1

(b)

A	B	f(A, B)
1	1	0
1	0	1
0	1	1
0	0	0

3. Sketch the switching circuit associated with each of the truth tables in Exercise 1.

4. Sketch the switching circuit associated with each of the truth tables in Exercise 2.

5. Write each of the following Boolean functions as a Boolean expression, and then sketch the equivalent switching circuit.

(a)

A	B	C	$f(A, B, C)$
1	1	1	1
1	1	0	1
1	0	1	0
1	0	0	0
0	1	1	1
0	1	0	1
0	0	1	0
0	0	0	0

(b)

A	B	C	$f(A, B, C)$
1	1	1	1
1	1	0	0
1	0	1	1
1	0	0	0
0	1	1	1
0	1	0	1
0	0	1	0
0	0	0	1

6. Write the Boolean expression for the following switching circuit, and then sketch the equivalent truth table.

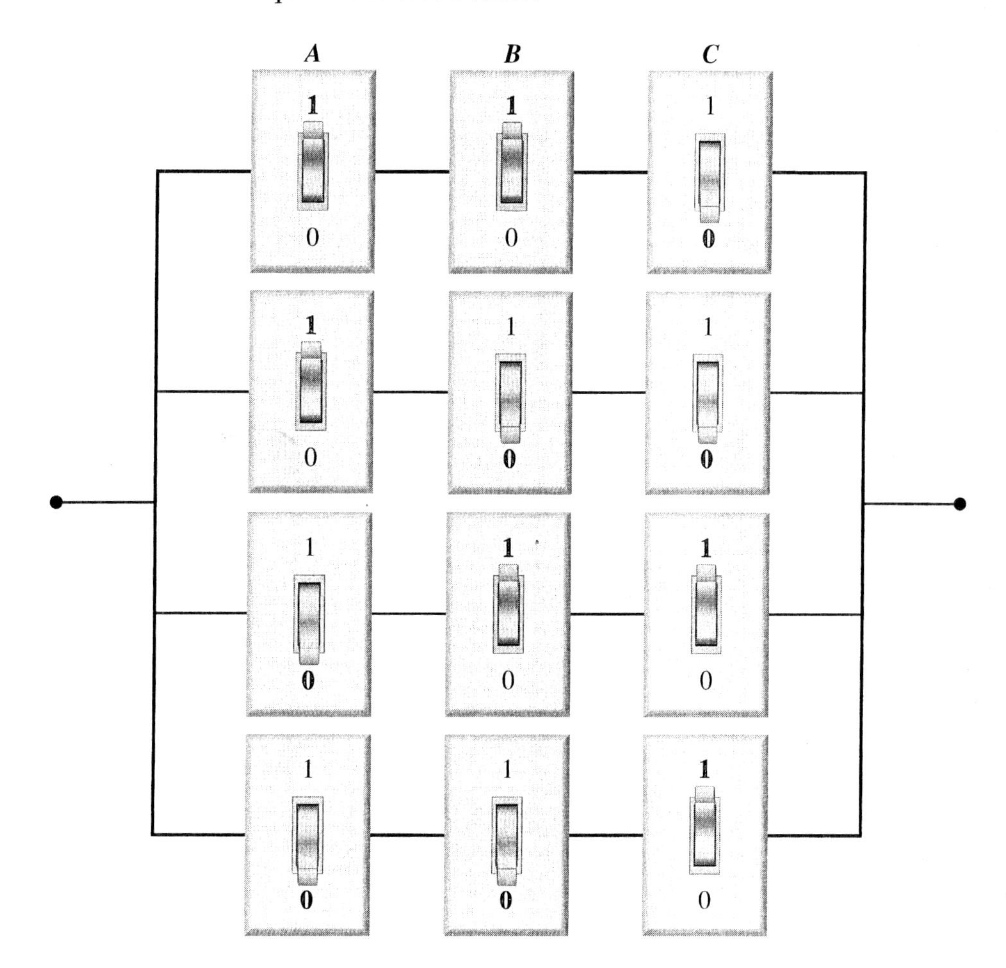

5.4 Logic Circuits Part II: Gated Circuits

We have seen logic circuits that allow us to see when an expression is **on** (true) or **off** (false). More useful in electronics are gated circuits.

> BITS of HISTORY
>
> **A.D. 1900** One of the most influential logicians of the twentieth century, Bertrand Russell's most important contribution is his paradox in set theory. A *paradox* refers to some idea that has the appearance of being logically true, but that is, in fact, so absurd that it cannot be true. This paradox is significant because it compelled mathematicians to reexamine previous results, thus prompting new developments in both set theory and logic.

Gated circuits use a different gate for each Boolean operator. There are three different gates.

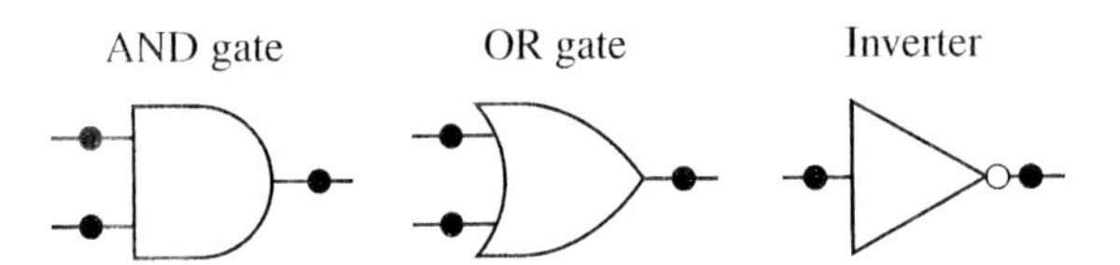

The AND gate and the OR gate each take two inputs and create a single output. Both inputs for the AND gate must be "on" for the output to be on. Only one input need be on for the OR gate output to be on. The inverter works like the NOT operator by turning off a circuit that is on and turning on a circuit that is off. In computer electronics, "on" is actually a high voltage and "off" is a low voltage.

EXAMPLE 1 **Given the Boolean expression $A \vee B$, find the related gated circuit.**

Solution

First, we begin with the two inputs, A and B. From each of those inputs, we draw paths leading into the OR gate. After the OR gate, the output is $A \vee B$.

A, B, $A \vee B$

PRACTICE PROBLEM 1 **Given the Boolean expression AB, find the related gated circuit.**

EXAMPLE 2 **Given the Boolean expression $A'B$, find the related gated circuit.**

Solution

Again, we first begin with the two inputs, A and B. However, notice that our Boolean expression contains A', which means we must use an inverter before we use the AND gate. We draw paths from A and B and then an inverter after A, changing this signal to A'. Finally, we lead both paths into an AND gate to produce the output $A'B$.

PRACTICE PROBLEM Answers on page 277

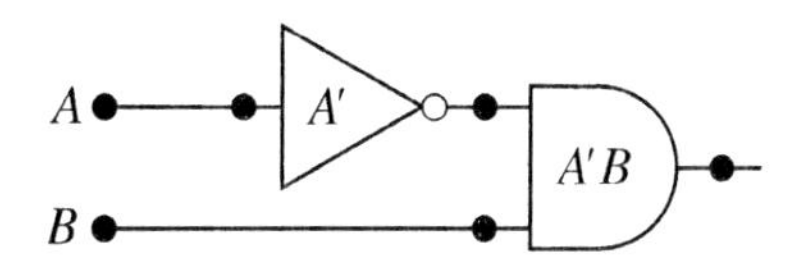

PRACTICE PROBLEM 2 **Given the Boolean expression $A \vee B'$, find the related gated circuit.**

EXAMPLE 3 **Given the Boolean expression $A'C \vee (BC)'$, find the related gated circuit.**

Solution

In this example, we begin with three inputs, A, B, and C. As in Example 2, our Boolean expression contains A', which means we must use an inverter before we use the AND gate. We draw paths from A and C and then an inverter after A, changing this signal to A'. Then we lead both paths into an AND gate to produce the output $A'C$.

To sketch the other part of the circuit, we draw paths from B and C and run them first through an AND gate and then through an inverter to get $(BC)'$. Note that because we use C twice, we need two paths coming from it and use a **hopper** to jump over the wires, so that our paths do not intersect. Finally, we run each of the remaining paths through an OR gate to get the expression $A'C \vee (BC)'$.

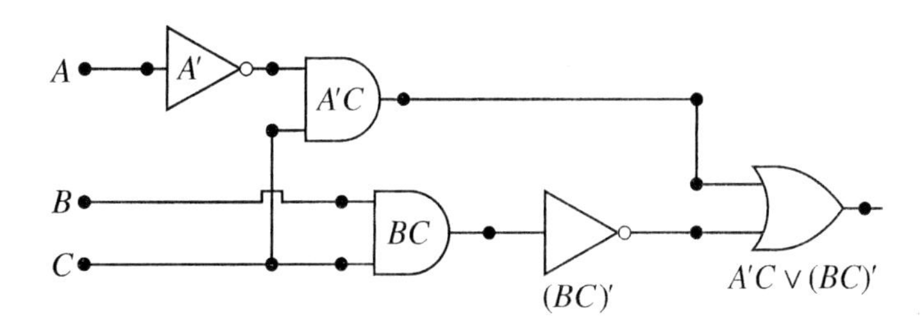

PRACTICE PROBLEM 1 **Given the Boolean expression $AB' \vee (AC)'$, find the related gated circuit.**

PRACTICE PROBLEM Answers
on page 277

EXAMPLE 4 **Given the Boolean expression $ABC' \vee B'C$, find the related gated circuit.**

Solution

Each gate accepts only two inputs, so the disjunctive ABC' requires two separate AND gates. Look at the circuit.

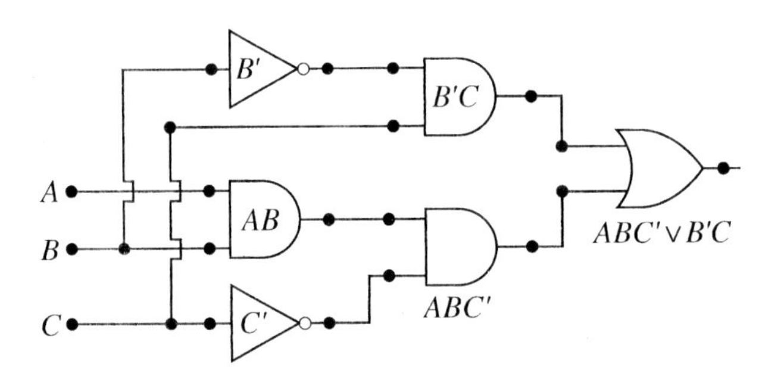

Note again that we treat branches differently from crossover points. Both B and C cross over the wire coming out of A. Both are shown to jump over the wire.

PRACTICE PROBLEM  **Given the Boolean expression $A'BC' \vee BC'$, find the related gated circuit.**

EXAMPLE 5 **Find the gated circuit that is equivalent to the Boolean expression**

$$ABC \vee AB'C \vee A'BC \vee A'BC'$$

Solution

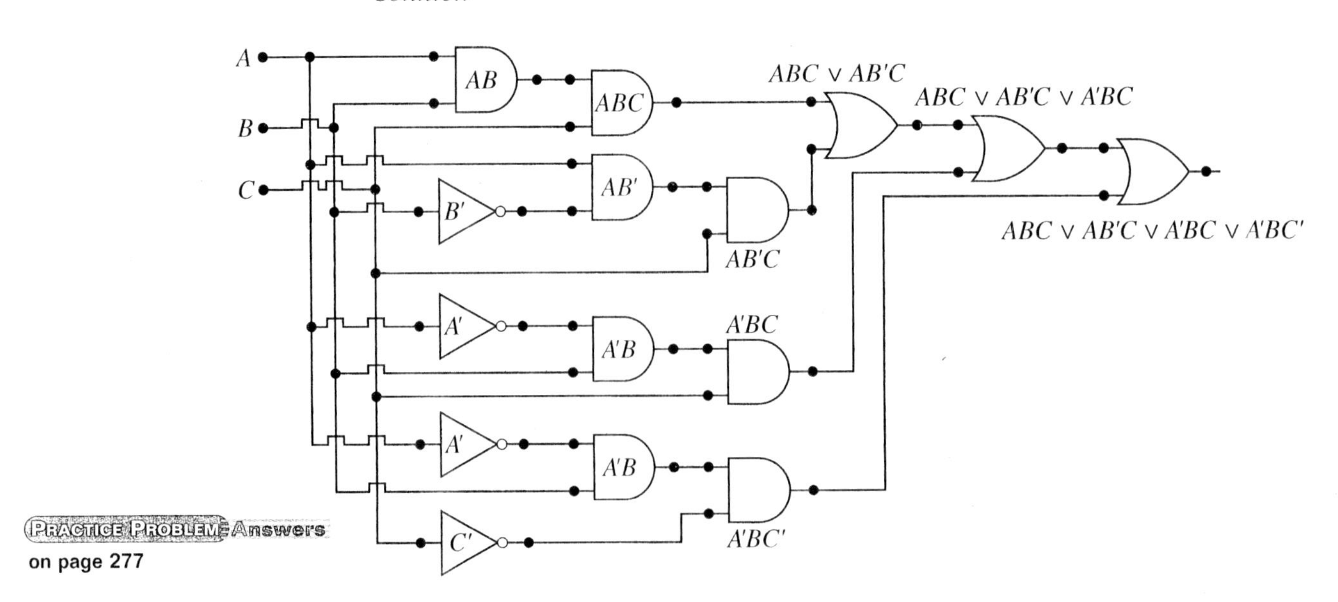

PRACTICE PROBLEM Answers on page 277

PRACTICE PROBLEM 5 **Find the gated circuit that is equivalent to the Boolean expression**

$$ABC \vee ABC' \vee AB'C' \vee A'BC.$$

In the next example, we use our Boolean algebra skills to simplify the expression from Example 5.

EXAMPLE 6 **Simplify the Boolean expression**

$$ABC \vee AB'C \vee A'BC \vee A'BC'.$$

Solution

$$\underbrace{ABC \vee AB'C}_{\textbf{Factor } AC} \vee \underbrace{A'BC \vee A'BC'}_{\textbf{Factor } A'B}$$

$= [AC \wedge (B \vee B')] \vee [A'B \wedge (C \vee C')]$ **distributive property**
$= [AC \wedge (1)] \vee [A'B \wedge (1)]$ **complement law**
$= AC \vee A'B$ **supplementary law**

By any reasonable measure, the expression

$$AC \vee A'B$$

is simpler than the expression

$$ABC \vee AB'C \vee A'BC \vee A'BC',$$

yet they are equivalent. How can that be? It is like asking how

$$7x^3y + 3x^2y - x^3y - 2x^2y + 3x^3y - 5x^3y$$

could be simplified to

$$4x^3y + x^2y.$$

By following the rules of algebra, we guarantee that the expressions are equivalent.

PRACTICE PROBLEM 6 **Simplify the Boolean expression**

$$ABC \vee ABC' \vee A'B'C' \vee A'B'C.$$

In the next example, we compare the circuit for the reduced expression in Example 6 to the circuit for the original expression in Example 5.

PRACTICE PROBLEM Answers on page 277

EXAMPLE 7 **Find the gated circuit that is equivalent to the Boolean expression**

$$AC \vee A'B$$

Solution

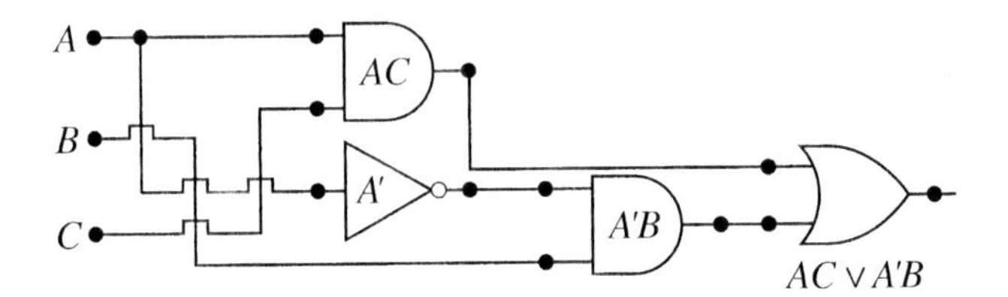

Compare this circuit to the equivalent circuit in Example 5. This is a much simpler circuit, requiring fewer gates. The problem is that the process for finding the simplified expression is very difficult. What we need is a better way to simplify Boolean expressions. We offer one in the next section!

PRACTICE PROBLEM 7 **Find the gated circuit that is equivalent to the simplified Boolean expression in Practice Problem 6.**

Answers to Practice Problems

1.

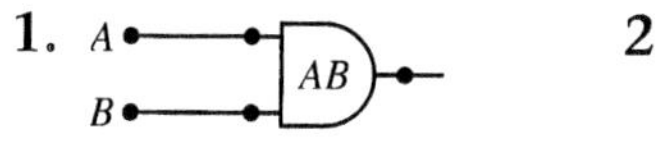

2.

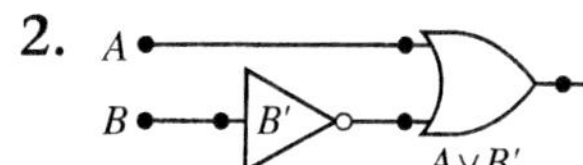

3.

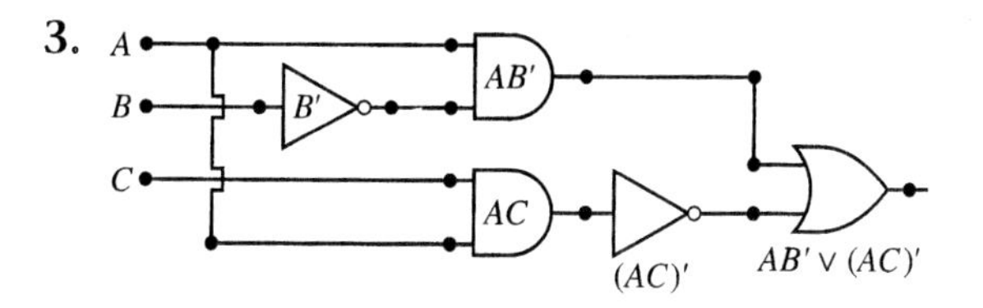

4.

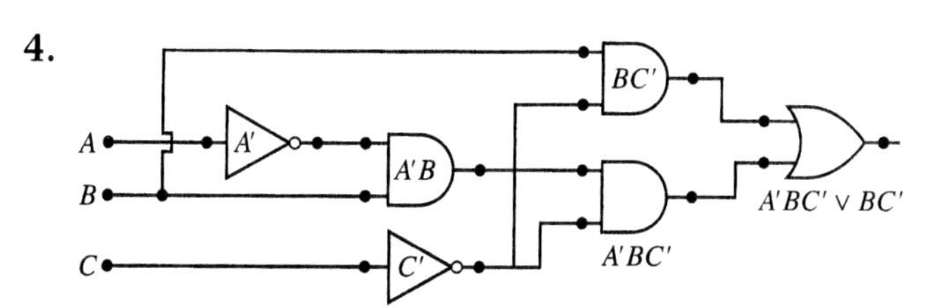

5.

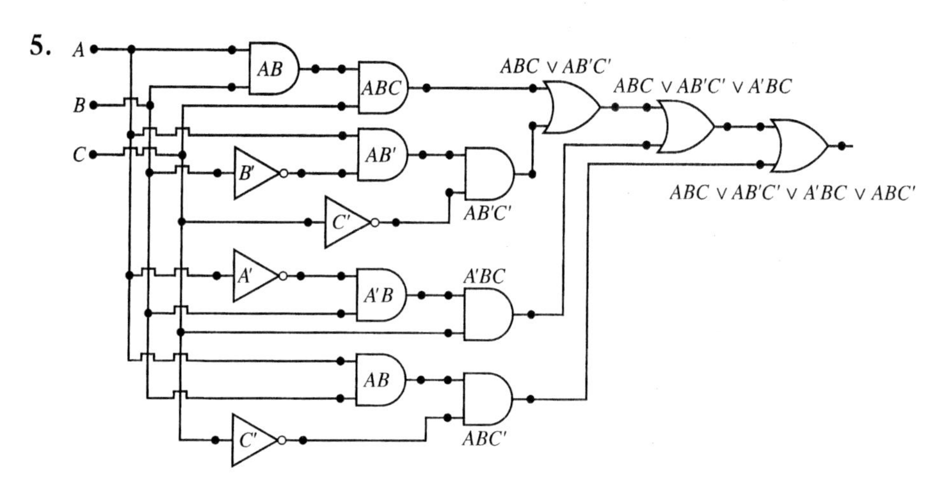

6. $AB \vee A'B'$

7.

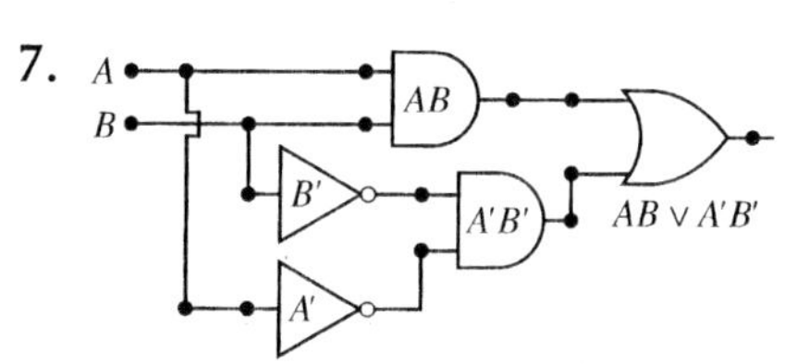

SECTION 5.4 Exercises

1. Sketch the gated circuit associated with the Boolean expression $A \vee B'$.

2. Sketch the gated circuit associated with the Boolean expression $A'B \vee C$.

3. Sketch the gated circuit associated with the Boolean expression $A'B \vee AB'$.

4. Sketch the gated circuit associated with the Boolean expression $B'C \vee ABC'$.

5. Give the Boolean expression associated with the following gated circuit.

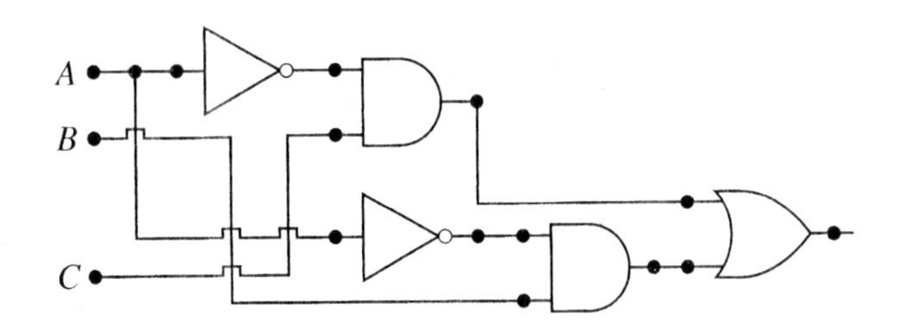

▲ **Represents additionally challenging problems.**

6. Give the Boolean expression associated with the following gated circuit.

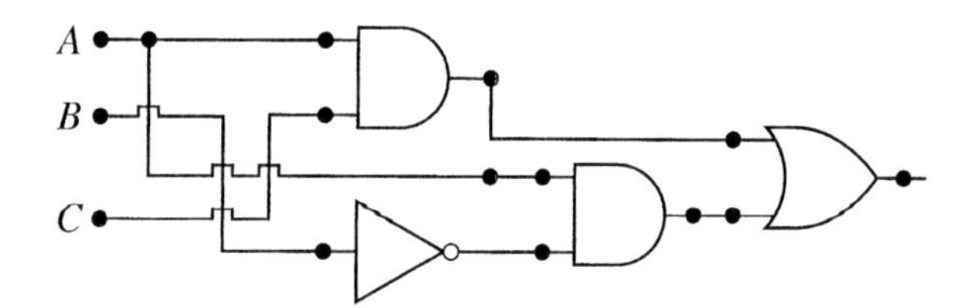

7. Give the Boolean expression associated with the following gated circuit.

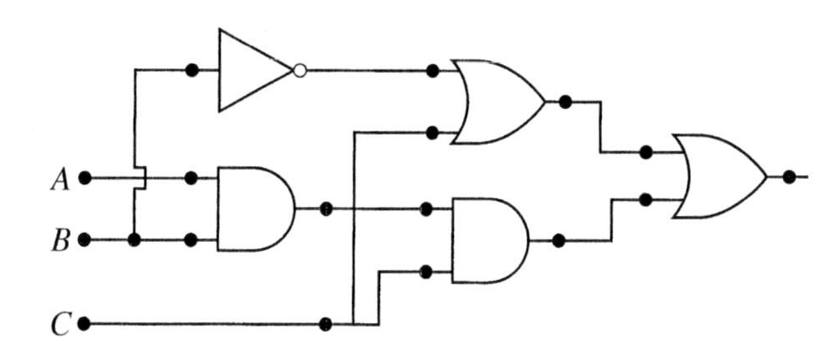

8. Use Boolean algebra to simplify the following expression:
$A'B' \vee AB \vee A'B$.

9. Sketch the gated circuit that is equivalent to the reduced expression in Exercise 8.

10. Sketch the gated circuit that is equivalent to the following Boolean expression: $ABC \vee ABC' \vee A'B'C$.

11. Use Boolean algebra to simplify the expression in Exercise 10.

12. Give the Boolean expression associated with the following gated circuit.

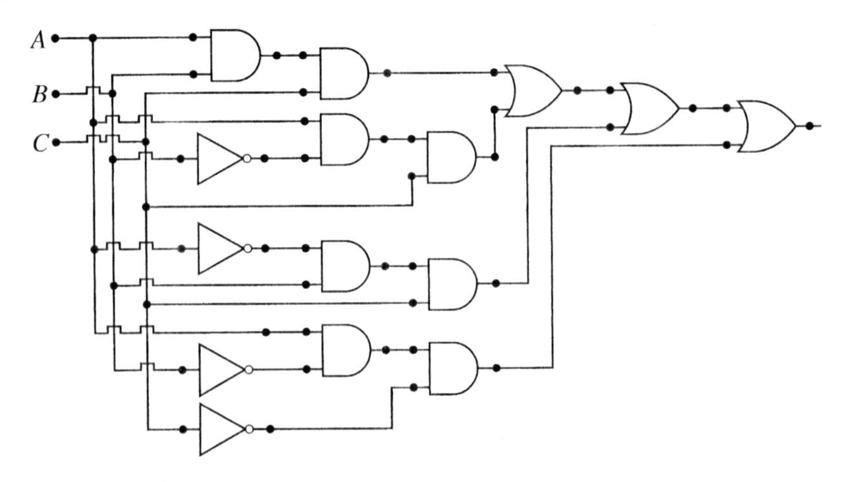

13. Give the Boolean expression associated with the following gated circuit.

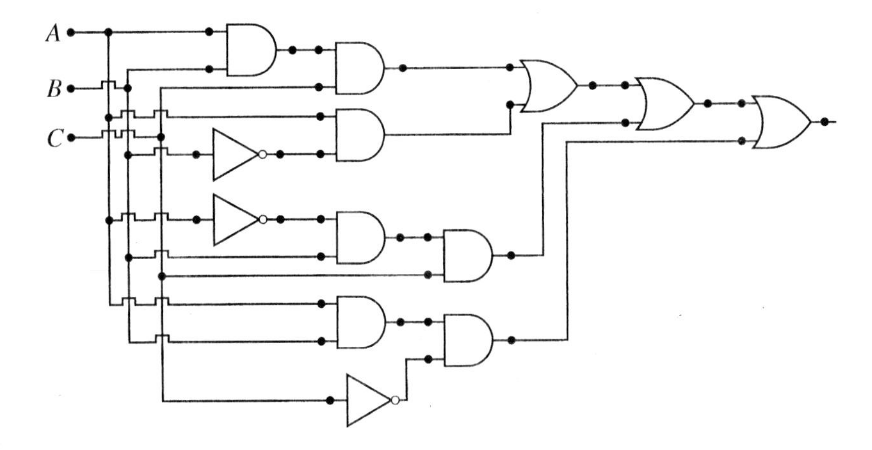

5.5 Karnaugh Maps

BITS of HISTORY

Maurice Karnaugh invented a tool in the 1950s that has become one of the most important instruments in digital logic. An alternative to truth tables, Karnaugh maps allow for reducing certain types of Boolean functions. This technique is quite valuable for circuit simplification.

In previous sections, we have seen that Boolean expressions can be simplified by using the rules of Boolean algebra. The simplified Boolean expressions can then be drawn as simpler circuits. This process can save time and, in a production environment, much money. Unfortunately, the properties of Boolean algebra can be challenging to master.

In this section, we examine a visual tool that allows for the reduction of any Boolean expression that is written in disjunctive normal form. Recall that dnf requires that each variable appear in every disjunctive term. Let us look first at the reduction of a Boolean expression in two variables.

EXAMPLE 1 **Reduce the Boolean expression**

$$AB \vee AB' \vee A'B$$

Solution

First, we sketch the two-variable **Karnaugh map**. A Karnaugh map is a visual representation of a Boolean expression that is written in dnf. It consists of rectangles that are divided into cells, each of which represents one of the possible terms in the expression.

The diagram below illustrates how a two-variable Karnaugh map may be set up. The two states of one variable (in this case A or A') form the columns, and the two states of the other variable (in this case B or B') form the rows.

	A	**A'**
B		
B'		

Next, we put a "1" (or a check) in each cell that is associated with one of the disjunctives that appear in the given expression.

	A	**A'**
B	1	1
B'	1	

Then we circle each pair of adjacent cells. Cells are considered to be adjacent if they connect vertically or horizontally, not diagonally.

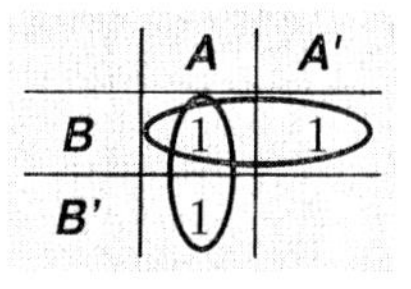

One variable may be eliminated from each pair. (Note that cell AB is used twice.)

1. Each time we circle a group, we write down the variable values that the cells have in common. The common values then become a term in the simplified expression. In this example, the vertical pair can be reduced to A and the horizontal pair can be reduced to B.

2. These terms are then connected with disjunctives. Thus, the expression $AB \vee AB' \vee A'B$ reduces to

$$A \vee B$$

Comparing both expressions in a truth table, we see that the final columns are identical.

A	B	$AB \vee AB' \vee A'B$	$A \vee B$
1	1	1	1
1	0	1	1
0	1	1	1
0	0	0	0

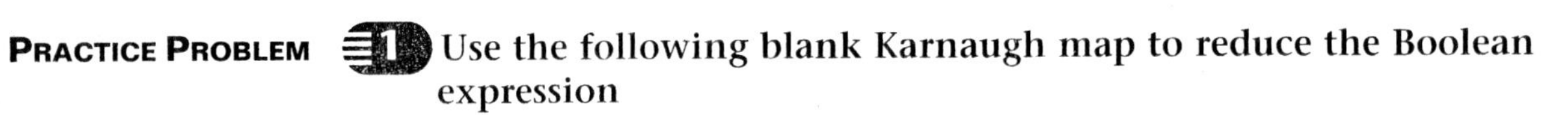

PRACTICE PROBLEM 1 **Use the following blank Karnaugh map to reduce the Boolean expression**

$$AB \vee A'B \vee A'B'$$

	A	A'
B		
B'		

If a cell cannot be grouped with any other cells, the term represented by this cell is already reduced and should be written in the final answer.

EXAMPLE 2 **Use a Karnaugh map to reduce the Boolean expression**

$$AB' \vee A'B$$

PRACTICE PROBLEM Answers on page 288

We start by placing a one in each of the two cells of the Karnaugh map that relate to a term in the Boolean expression.

	A	*A'*
B		1
B'	1	

There is no pair of cells that can be grouped vertically or horizontally. The Boolean expression cannot be reduced.

PRACTICE PROBLEM 2 **Use the following blank Karnaugh map to reduce the Boolean expression**

$$AB \vee A'B'$$

	A	*A'*
B		
B'		

In the next example, we examine an expression with three variables.

EXAMPLE 3 **Reduce the Boolean expression**

$$ABC \vee AB'C \vee AB'C' \vee A'BC \vee A'BC' \vee A'B'C.$$

This time, we begin with a three-variable Karnaugh map.

	AB	*AB'*	*A'B'*	*A'B*
C				
C'				

Again, we put a "1" in the cells related to terms that appear in the original expression and circle adjacent sets. The sets can be in groups of any power of two, but they must be connected horizontally, vertically, in a square, or in corners.

	AB	*AB'*	*A'B'*	*A'B*
C	1	1	1	1
C'		1		1

PRACTICE PROBLEM Answers on page 288

Now, we have a group of four, so we can eliminate two variables. The common variable of the entire horizontal group of four is C. The two vertical groups reduce to AB' and $A'B$.

	AB	AB'	A'B'	A'B
C	1	1	1	1
C'		1		1

The entire expression can be rewritten as

$AB' \vee A'B \vee C.$

How do we know this reduced expression is equivalent to the original? A Venn diagram illustrates.

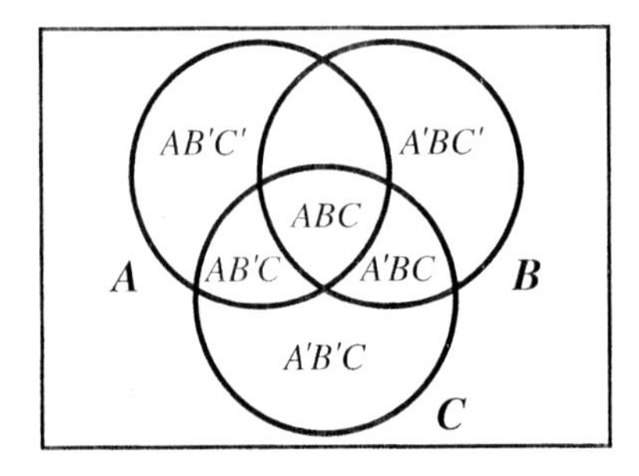

$ABC \vee AB'C \vee AB'C' \vee A'BC \vee A'BC' \vee A'B'C$

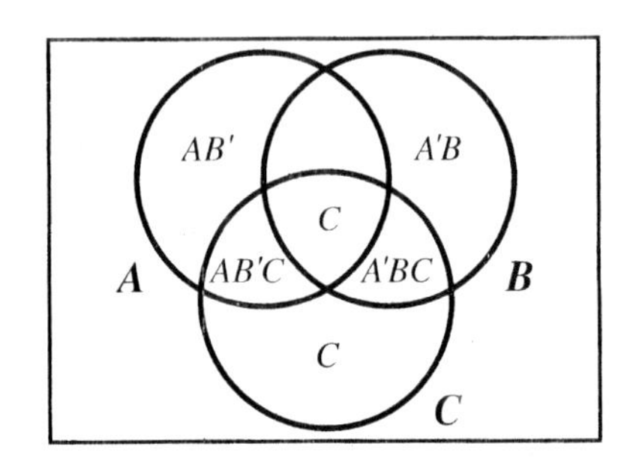

$AB' \vee A'B \vee C$

PRACTICE PROBLEM 3 **Use the following blank Karnaugh map to reduce the Boolean expression**

$ABC \vee AB'C \vee AB'C' \vee A'B'C \vee A'B'C'$

	AB	AB'	A'B'	A'B
C				
C'				

We next examine a four-variable expression.

PRACTICE PROBLEM Answers on page 288

EXAMPLE 4 **Reduce the expression**

$ABCD \vee ABC'D \vee ABC'D' \vee AB'CD' \vee A'BCD \vee A'BC'D \vee A'B'CD \vee A'B'CD' \vee A'B'C'D.$

Solution

First, we sketch the four-variable Karnaugh map.

	AB	AB'	$A'B'$	$A'B$
CD				
CD'				
$C'D'$				
$C'D$				

Now we put a 1 in the appropriate cells.

	AB	AB'	$A'B'$	$A'B$
CD	1		1	1
CD'		1	1	
$C'D'$	1			
$C'D$	1		1	1

Note the many groups of two that can be circled; however, we should look first for groups of four. The way the table is constructed, it may seem that there are no groups of four, but there are. To illustrate, we extend our Karnaugh map, adding on to the right side and below, keeping the order of the variables the same. We also put a "1" in all of the appropriate cells, which quadruples the number in our original Karnaugh map.

	AB	AB'	$A'B'$	$A'B$	AB	AB'	$A'B'$	$A'B$
CD	1		1	1	1		1	1
CD'		1	1			1	1	
$C'D'$	1				1			
$C'D$	1		1	1	1		1	1
CD	1		1	1	1		1	1
CD'		1	1			1	1	
$C'D'$	1				1			
$C'D$	1		1	1	1		1	1

The Karnaugh map that is formed in the shaded region is equivalent to our original, containing sixteen different cells, nine of which are represented in the given Boolean expression. By regrouping, we can now see the two groups of four and two groups of two.

	A B	A B'	A'B'	A'B	A B	A B'	A'B'	A'B
CD	1		1	1	1		1	1
CD'		1	1			1	1	
C'D'	1				1			
C'D	1		1	1	1		1	1
CD	1		1	1	1		1	1
CD'		1	1			1	1	
C'D'	1				1			
C'D	1		1	1	1		1	1

Circling those groups, we see that we have

$$A'D \vee BD \vee ABC' \vee B'CD'.$$

These same groups exist in our original Karnaugh map, but they are harder to see unless one thinks "outside the box."

The four corners (eliminating two variables) yield BD. The connected pairs on the top and bottom rows yield $A'D$. The pair on the left is ABC', and the middle pair is $B'CD'$.

	A B	A B'	A'B'	A'B
CD	1		1	1
CD'		1	1	
C'D'	1			
C'D	1		1	1

The reduced expression is

$$BD \vee A'D \vee ABC' \vee B'CD'$$

PRACTICE PROBLEM 4 **Use the following Karnaugh map to reduce the expression**

$ABCD \vee ABCD' \vee ABC'D' \vee AB'CD' \vee AB'C'D \vee AB'C'D' \vee A'BCD' \vee A'BC'D' \vee A'B'C'D.$

	AB	AB'	A'B'	A'B
CD				
CD'				
C'D'				
C'D				

What if we have five variables? Our final example will illustrate.

EXAMPLE 5 **Simplify the expression represented in the following map.**

E:

	AB	AB'	A'B'	A'B
CD			1	1
CD'	1	1	1	1
C'D'				
C'D				

E':

	AB	A B'	A'B'	A'B
CD				
CD'	1	1	1	1
C'D'				
C'D			1	

Solution

Note first the group of eight. The second row of the top Karnaugh map and the second row of the bottom Karnaugh map together contain a group of eight checks. That allows us to remove three variables ($2^3 = 8$). The eight boxes have CD' in common. There is a group of four in the top Karnaugh map. This group can be reduced to $A'CE$. Finally, the cell denoted by $A'B'C'DE'$ cannot be grouped with any other cells, so we list it in our final reduced expression.

E:

	AB	AB'	A'B'	A'B
CD			1	1
CD'	1	1	1	1
C'D'				
C'D				

E':

	AB	AB'	A'B'	A'B
CD				
CD'	1	1	1	1
C'D'				
C'D			1	

The eleven disjunctives can be reduced to the expression

$$CD' \vee A'CE \vee A'B'C'DE'.$$

PRACTICE PROBLEM Answers on page 288

PRACTICE PROBLEM 5 Simplify the expression represented in the following map.

	AB	AB'	$A'B'$	$A'B$
CD	1	1	1	1
CD'	1	1	1	1
$C'D'$		1	1	
$C'D$				

E

	AB	AB'	$A'B'$	$A'B$
CD	1	1		
CD'	1	1	1	1
$C'D'$				
$C'D$				

E'

ANSWERS TO PRACTICE PROBLEMS

1. $A' \vee B$
2. $AB \vee A'B'$
3. $B' \vee AC$
4. $ABC \vee AD' \vee BD' \vee B'C'D$
5. $CE \vee CD' \vee B'D'E \vee ACE$

SECTION 5.5

Exercises

1. Sketch the Karnaugh map for the Boolean expression $A'B' \vee AB' \vee A'B$, and then use it to find an equivalent reduced expression.

2. Sketch the Karnaugh map for the Boolean expression $AB' \vee AB \vee A'B'$, and then use it to find an equivalent reduced expression.

3. Sketch the Karnaugh map for the Boolean expression $AB'C \vee ABC \vee A'B'C$, and then use it to find an equivalent reduced expression.

4. Sketch the Karnaugh map for the Boolean expression $A'B'C' \vee AB'C' \vee A'B'C \vee AB'C$, and then use it to find an equivalent reduced expression.

5. Write the reduced expression for the following Karnaugh map.

	AB	AB'	A'B'	A'B
C	1			1
C'		1	1	

6. Write the reduced expression for the following Karnaugh map.

	AB	AB'	A'B'	A'B
C	1			1
C'	1		1	

7. Sketch the Karnaugh map for the Boolean expression $AB'CD \vee ABCD \vee A'B'CD' \vee A'B'C'D' \vee AB'CD'$, and then use it to find an equivalent reduced expression.

8. Write the reduced expression for the following Karnaugh map.

	AB	AB'	A'B'	A'B
CD		1		
CD'				
C'D'	1	1	1	1
C'D	1			1

9. Given the following truth table, complete (a) through (d).

A	B	f(A, B)
1	1	1
1	0	1
0	1	0
0	0	1

(a) Write the Boolean expression in dnf.
(b) Sketch the switching circuit that is equivalent to the Boolean expression.
(c) Construct a Karnaugh map for this expression, and find an equivalent reduced expression.
(d) Sketch the gated circuit that is equivalent to the reduced expression.

10. Given the following truth table, complete (a), (b), and (c).

A	B	f(A, B)
1	1	1
1	0	1
0	1	1
0	0	0

(a) Write the Boolean expression in dnf.

(b) Construct a Karnaugh map for this expression, and find an equivalent reduced expression.

(c) Sketch the gated circuit that is equivalent to the reduced expression.

11. Simplify the expression represented by the following Karnaugh map.

E

	AB	*AB'*	*A'B'*	*A'B*
CD	1			
CD'	1			
C'D'	1	1		
C'D	1	1		

E'

	AB	*AB'*	*A'B'*	*A'B*
CD				
CD'		1	1	
C'D'	1	1	1	1
C'D	1	1	1	1

12. The following is a Boolean Rosetta stone in which the Boolean expression is given. Complete the Venn diagram and the truth table so that they are equivalent to the Boolean expression.

Venn diagram

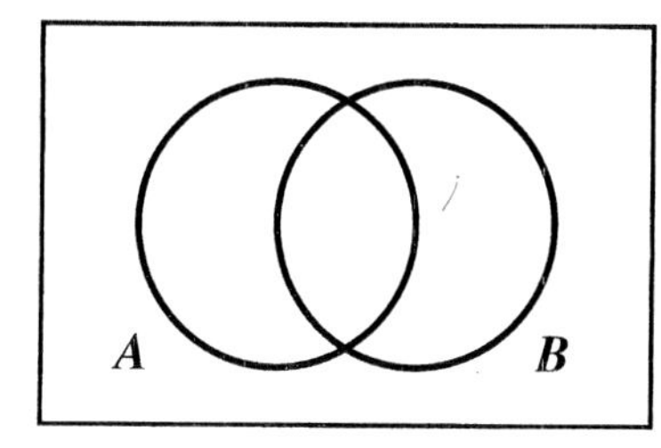

Truth table

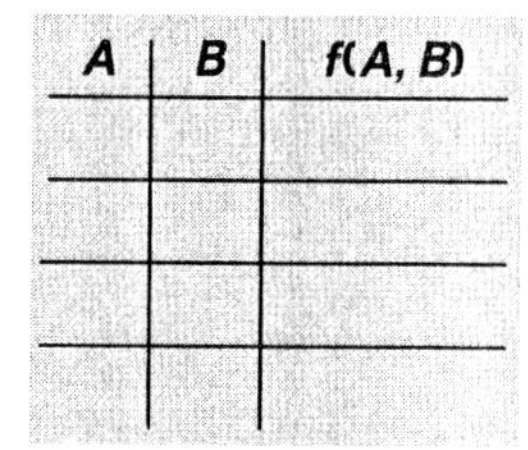

A	B	f(A, B)

Boolean expression

$A'B' \vee AB \vee A'B$

13. Circling a group of two in a Karnaugh map allows you to eliminate how many variables? How about circling a group of eight?

SUMMARY

5.1 Boolean Logic

- Boolean algebra is concerned with any system that has only two states, e.g., 0 and 1.
- $'$ represents negation a' means **not** a or the complement of a.
- $\vee$ represents *or* (also called join)
- $\wedge$ represents *and* (also called meet) xy is the same as $x \wedge y$.
- A Boolean set is usually referred to as ***B***, and two distinct elements of B are denoted as **0** and **1**. **0** is the **zero element** and **1** is the **unit element**.
- If a Boolean expression has several operators, the parentheses take precedence. Then, in **Order of Operation**, we perform negation, intersection, and then union.

Commutative Properties

$x \vee y = y \vee x$

$x \wedge y = y \wedge x (xy = yx)$

Complement Property

$x \wedge x' = 0$

$x \vee x' = 1$

Identity Properties

$x \vee 0 = x$

$x \wedge 1 = x$

Reduction

$x \vee 0 = x$

$x \wedge 1 = x$

Distributive Properties

$x \wedge (y \vee z) = (x \wedge y) \vee (x \wedge z)$

$x \vee (y \wedge z) = (x \vee y) \wedge (x \vee z)$

These properties can be used to simplify Boolean expressions.

5.2 Logic Circuits Part I: Switching Circuits

- A **logic circuit** is a set of symbols that relate to a Boolean expression.

- The Boolean expression $A \vee B$ can be represented by the following circuit. This is called a **parallel circuit**.

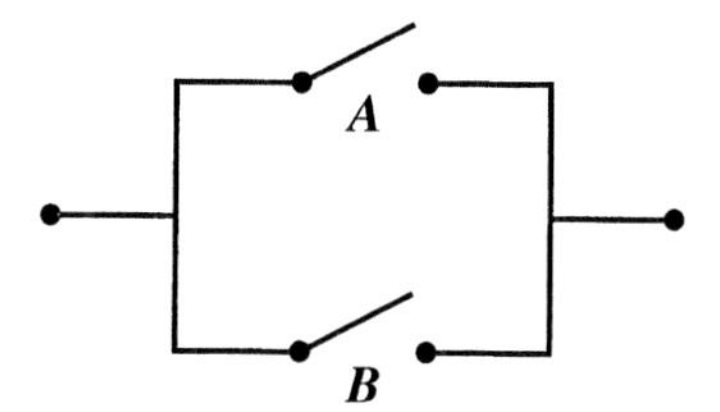

- The Boolean expression $A \wedge B$ can be represented by the following circuit. This is called a **series circuit**.

- A Boolean expression can also be represented by a set of switches, which we call a **switching circuit**. Each variable is represented by a two-position switch. The following graphic represents the expression $A \wedge B$.

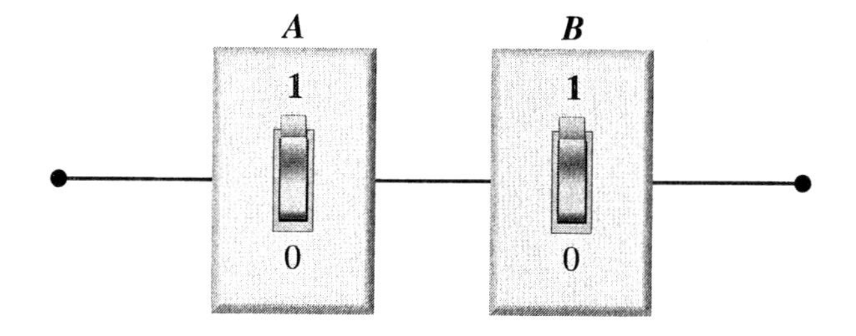

5.3 Truth Tables and Disjunctive Normal Form

- A Boolean expression that consists of a series of terms in which every variable appears in every term is said to be in **disjunctive normal form (dnf)**.
- Every truth table represents a Boolean expression and every Boolean expression can be rewritten as a truth table.
- Because the values in the final column of a truth table are a function of the values of the variables, we sometimes use function notation to abbreviate that column.

5.4 Logic Circuits Part II: Gated Circuits

Gated circuits use a different gate for each Boolean operator. There are three different gates.

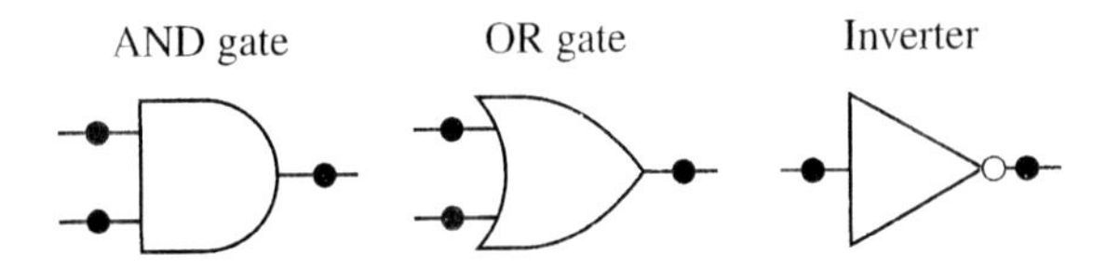

5.5 Karnaugh Maps

A Karnaugh map is a visual representation of a Boolean expression that is written in disjunctive normal form. It consists of rectangles that are divided into cells, each of which represents one of the possible terms in the expression. The diagram below illustrates how a two-variable Karnaugh map may be set up.

	A	**A'**
B		
B'		

The following illustrates the relationship between a Venn diagram and a Karnaugh map.

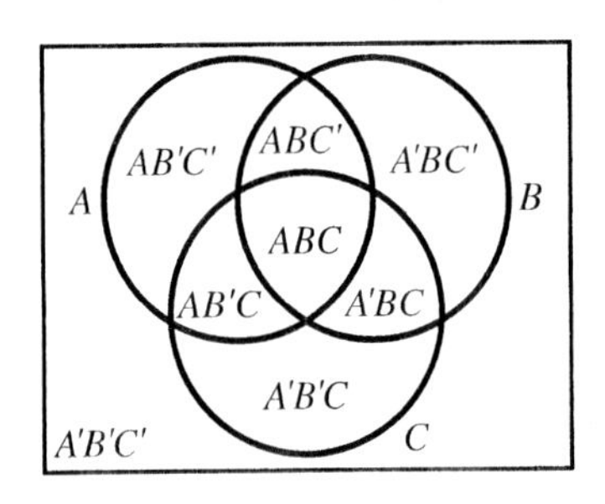

	AB	**AB'**	**A'B'**	**A'B**
C	ABC	AB'C	A'B'C	A'BC
C'	ABC'	AB'C	AB'C'	A'BC'

CHAPTER 5 GLOSSARY QUIZ

The following terms were introduced in Chapter 5 of the text. Match each with one of the definitions that follows.

parallel circuit ______

series circuit ______

gated circuits ______

Karnaugh map ______

switching circuit ______

the commutative property of Boolean algebra ______

the distributive property of Boolean algebra ______

logic circuit ______

disjunctive normal form (dnf) ______

(a) $x \vee y = y \vee x$

(b) $x \wedge (y \vee z) = (x \wedge y) \vee (x \wedge z)$

(c) a set of symbols that relate to a Boolean expression

(d) a circuit related to the Boolean expression $A \vee B$

(e) a circuit related to the Boolean expression $A \wedge B$

(f) a Boolean expression represented by a set of switches

(g) a Boolean expression that consists of a series of terms in which every variable appears in every term

(h) a circuit that uses AND gates, OR gates, and the inverter

(i) a visual representation of a Boolean expression that is written in dnf

CHAPTER 5 REVIEW EXERCISES

[5.1] **Name the property demonstrated in each real variable expression.**

1. $2(x + y) = 2x + 2y$

2. $x + y = y + x$

3. $5 \cdot (y + z) = (y + z) \cdot 5$

4. $x \cdot (y \cdot z) = (x \cdot y) \cdot z$

[5.1] **Name the property demonstrated in each Boolean variable expression.**

5. $x \vee x' = 1$

6. $(x \vee y) \vee z = x \vee (y \vee z)$

7. $x \wedge (y \vee z) = (x \wedge y) \vee (x \wedge z)$

8. $x \vee y = y \vee x$

[5.1] **Use real variable algebra to simplify the following expressions.**

9. $5(x + 3y)$

10. $(x + y)(x + y)$

11. $4x + 2(x + 6y) - 3y$

[5.1] Simplify each of the following Boolean expressions.

12. $x \wedge x'$

13. $x \vee 0$

14. $(x \vee y)'$

15. $y' \vee y$

16. $1 \vee x$

[5.1] Use Boolean algebra to simplify the following expressions, and then evaluate the expressions for $x = 1$ and $y = 0$.

17. $x \wedge (x' \vee y)$

18. $xy \vee xy'$

[5.2] Sketch the switching circuit associated with each of the following Boolean expressions.

19. $A \vee B'$

20. $AB' \vee A'B$

21. $BC' \vee A'B'C$

[5.2] Give the Boolean expression associated with each of the following switching circuits.

22.

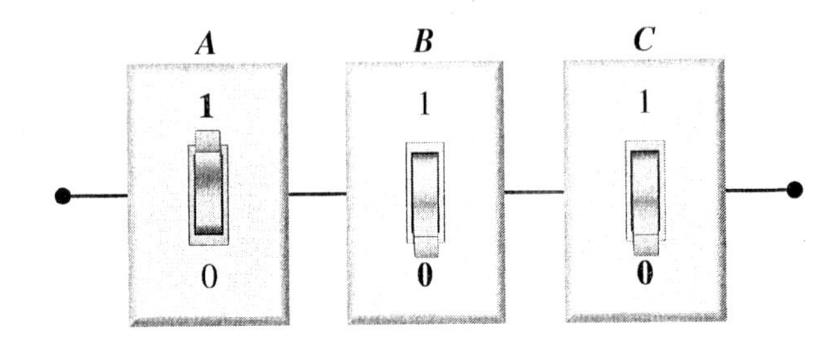

23.

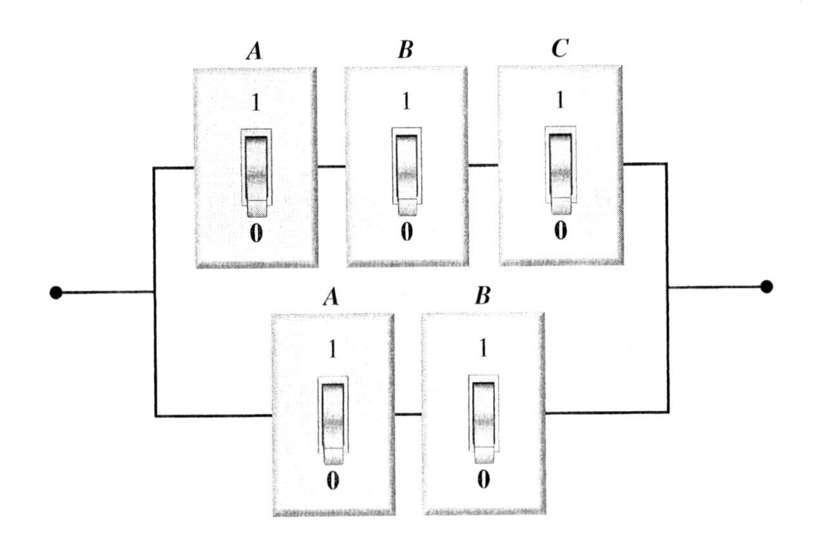

24. Find a Boolean expression that is equivalent to the one you found in Exercise 23 and reduce it using Boolean algebra.

[5.3] Write each of the Boolean functions as a reduced Boolean expression.

25.

A	B	$f(A, B)$
1	1	1
1	0	1
0	1	0
0	0	1

26.

A	B	$f(A, B)$
1	1	0
1	0	1
0	1	0
0	0	1

27. Sketch the switching circuit associated with the reduced Boolean expression you found Exercise 26.

[5.3] Write each of the following Boolean functions as a reduced Boolean expression, and then sketch the switching circuit that is equivalent to the reduced expression.

28.

A	B	C	f(A, B, C)
1	1	1	0
1	1	0	0
1	0	1	1
1	0	0	1
0	1	1	0
0	1	0	0
0	0	1	1
0	0	0	0

29.

A	B	C	f(A, B, C)
1	1	1	0
1	1	0	1
1	0	1	0
1	0	0	0
0	1	1	1
0	1	0	1
0	0	1	0
0	0	0	1

30. Sketch the equivalent truth table for the following Boolean expression,

$$AB' \vee C'$$

[5.4] Sketch the gated circuits associated with each of the following Boolean expressions.

31. $A' \vee B$

32. $AB' \vee C'$

33. $ABC' \vee A'C$

[5.4] Give the Boolean expression associated with each of the following gated circuits.

34.

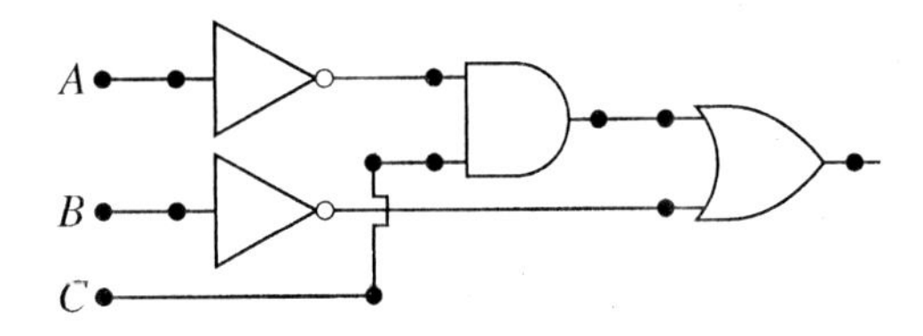

35.

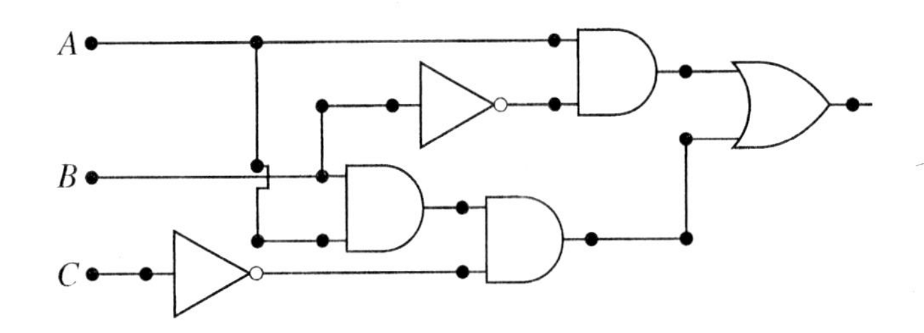

[5.4] Use Boolean algebra to simplify the following expression, and then sketch the gated circuit that is equivalent to the reduced expression.

36. $A'BC' \vee A'B'C' \vee AB'C$

[5.5] **37.** Sketch the Karnaugh map for the Boolean expression $AB \vee A'B' \vee A'B$, and then use it to find an equivalent reduced expression.

38. Sketch the Karnaugh map for the Boolean expression $ABC \vee ABC' \vee A'B'C \vee A'B'C'$, and then use it to find an equivalent reduced expression.

[5.5] Write the reduced expression for each of the following Karnaugh maps.

39.

	AB	AB'	A'B'	A'B
C	1			1
C'		1		1

40.

	AB	AB'	A'B'	A'B
CD		1		
CD'		1	1	1
C'D'			1	1
C'D	1			

41.

E

	AB	AB'	A'B'	A'B
CD				1
CD'				
C'D'		1	1	
C'D				

E'

	AB	AB'	A'B'	A'B
CD				
CD'				
C'D'	1	1	1	1
C'D	1			

42. Use the following truth table to complete the exercises.

A	B	f(A, B)
1	1	1
1	0	0
0	1	1
0	0	1

(a) Write the Boolean expression in dnf.

(b) Sketch the switching circuit that is equivalent to the Boolean expression.

(c) Sketch the Venn diagram that is equivalent to the Boolean expression.

(d) Construct a Karnaugh map for this expression and find an equivalent reduced expression.

(e) Sketch the gated circuit that is equivalent to the reduced expression.

43. Use the following truth table to complete the exercises.

A	B	C	$f(A, B, C)$
1	1	1	1
1	1	0	0
1	0	1	1
1	0	0	1
0	1	1	0
0	1	0	0
0	0	1	1
0	0	0	1

(a) Write the boolean expression in dnf.

(b) Sketch the Venn diagram that is equivalent to the Boolean expression.

(c) Construct a Karnaugh map for this expression and find an equivalent reduced expression.

(d) Sketch the gated circuit that is equivalent to the reduced expression.

5 Self-Test

1. **Name the property demonstrated in each real variable expression.**

   (a) $x \cdot y = y \cdot x$

   (b) $x + (y + z) = (x + y) + z$

   (c) $4(x + 2y) = 4x + 8y$

   (d) $5 + (x + y) = (x + y) + 5$

2. **Name the property demonstrated in each Boolean variable expression.**

   (a) $(x \wedge y) \wedge z = x \wedge (y \wedge z)$

   (b) $x \wedge x' = 0$

   (c) $x \vee (y \wedge z) = (x \vee y) \wedge (x \vee z)$

   (d) $1 \vee x = x \vee 1$

3. **Use real variable algebra to simplify the following expressions.**

   (a) $7(4x + y)$

   (b) $(x + 4)(x - 4)$

   (c) $8x + 3(y + 2x) - 6x$

4. **Simplify each of the following Boolean expressions.**

(a) $x \wedge 1$

(b) $(x' \vee y')'$

(c) $(y')' \vee y'$

(d) $x \wedge x'$

5. **Use Boolean algebra to simplify the following expressions, and then evaluate the expressions for $x = 0$ and $y = 1$.**

(a) $x \wedge (x' \vee y)$

(b) $xy' \vee x'y'$

6. **Sketch the switching circuit associated with each of the following Boolean expressions.**

(a) $AB' \vee A'B$

(b) $A'BC \vee AC'$

7. **Write each of the Boolean functions as a Boolean expression in dnf, and then sketch the switching circuit for the expression.**

(a)

A	B	f(A, B)
1	1	0
1	0	0
0	1	1
0	0	1

(b)

A	B	C	f(A, B, C)
1	1	1	0
1	1	0	0
1	0	1	0
1	0	0	1
0	1	1	0
0	1	0	1
0	0	1	1
0	0	0	0

8. **Write the Boolean expression for the following switching circuit, then sketch the equivalent truth table.**

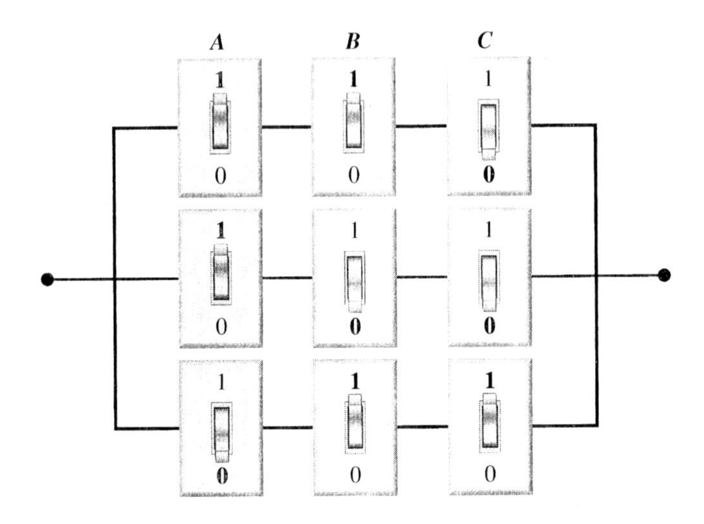

9. **Sketch the gated circuits associated with each of the following Boolean expressions.**

   (a) $AB \vee B'$ (b) $AB' \vee A'C$

10. **Give the Boolean expression associated with the following gated circuit.**

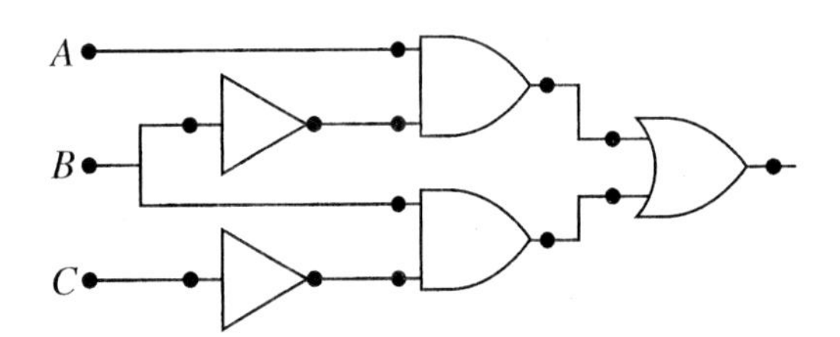

11. **Sketch the Karnaugh map for the Boolean expression $AB'C \vee A'B'C \vee A'BC' \vee A'BC$, and then use it to find an equivalent reduced expression.**

12. Write the reduced expression for the following Karnaugh map.

	AB	AB'	$A'B'$	$A'B$
CD	1			
CD'			1	
$C'D'$	1	1	1	1
$C'D$	1			

13. Use the following truth table to complete the exercises.

(a) Write the boolean expression in dnf.

A	B	C	$f(A, B, C)$
1	1	1	0
1	1	0	1
1	0	1	0
1	0	0	1
0	1	1	1
0	1	0	1
0	0	1	0
0	0	0	0

(b) Sketch the switching circuit that is equivalent to the Boolean expression.

(c) Sketch the Venn diagram that is equivalent to the Boolean expression.

(d) Construct a Karnaugh map for this expression and find an equivalent reduced expression.

(e) Sketch the gated circuit that is equivalent to the reduced expression.

Graphs

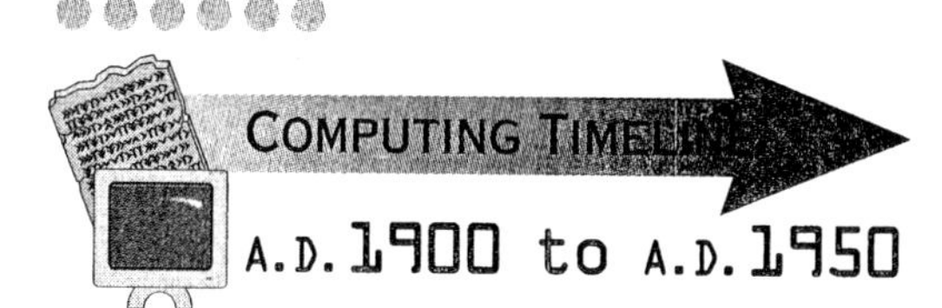

This is the beginning of the digital era. Starting with Clifford Berry's small, binary, digital calculator in 1939, almost every subsequent computer development was based on digital and electronic principles. The decoding of German messages in World War II focused many material and intellectual resources in the areas of coding and computing. By the end of the war, the principles of programming and computer architecture were an integral part of computer research and development.

6.1 Color Schemes

BITS of HISTORY

A.D. 1900 Konrad Zuse was a German engineer who reinvented Babbage's analytical engine. His original plan was to develop a universal calculator that could solve any equation. His Z1 machine employed a binary system rather than a decimal one, distinguishing it from its predecessors. Zuse was commissioned to build computers for the German government during World War II. Zuse, who seemed more concerned with computation, designed the Z3, which was capable of calculating airplane wing flutter and much more.

At the paint store, you see hundreds of different colors on swatches, yet it may surprise you that most of those colors can be made by mixing the three primary colors, blue, yellow, and red, in different ratios. A slight adjustment to this color scheme allows inkjet printers to produce thousands or even millions of distinct hues.

Many inkjet printers use a color scheme known as **CMYK**, which is composed of four colors: cyan, magenta, yellow, and black. The three CMYK primary colors, cyan, yellow, and magenta, can be mixed to get many colors, including black; however, a separate black cartridge is often supplied to conserve ink.

This type of color scheme is known as **subtractive color**, since the pigments absorb certain wavelengths of light to produce color. The following Venn diagram illustrates the CMYK color scheme. Mixing two of the three CMYK primary colors produces the other basic colors, such as red, green, and blue. If all three colors are mixed together, the result is black. This is denoted as the intersection of all three primary colors. White is the region outside the circles, since it is the absence of pigment.

CMYK Color Scheme

C
Y
Cyan
Green
Yellow
K
(Black)
Blue
Red
White
Magenta
M

EXAMPLE 1 **For each of the following, identify the color represented in the CMYK Venn diagram.**

(a) CMY' (b) $CM'Y'$ (c) $C'M'Y'$

Solution

(a) CMY' is the region inside cyan and yellow, but outside magenta. This is the blue region on the diagram.

(b) $CM'Y'$ is the region inside cyan, but outside yellow and magenta. This is the cyan region on the diagram.

(c) $C'M'Y'$ is the region that is outside cyan, yellow, and magenta. This region is white.

Practice Problems 1 **For each of the following, identify the color represented in the CMYK Venn diagram.**

(a) $C'MY$ (b) $C'M'Y$ (c) CMY

There is another important color scheme that relates to the computer. The computer monitor uses a color model that is based upon light rather than pigment. The **RGB** color scheme is composed of three primary colors, red, green, and blue, which can be mixed together to produce the other colors.

The RGB scheme is an **additive color** scheme, since augmenting different proportions of light produces the colors. The following Venn diagram illustrates the RGB color scheme. Mixing two of the three RGB primary colors produces the other basic colors, such as cyan, yellow, and magenta. If all three colors are mixed together, the result is white. This is denoted as the intersection of all three primary colors. Black is outside the circles, since it is the absence of light.

RGB Color Scheme

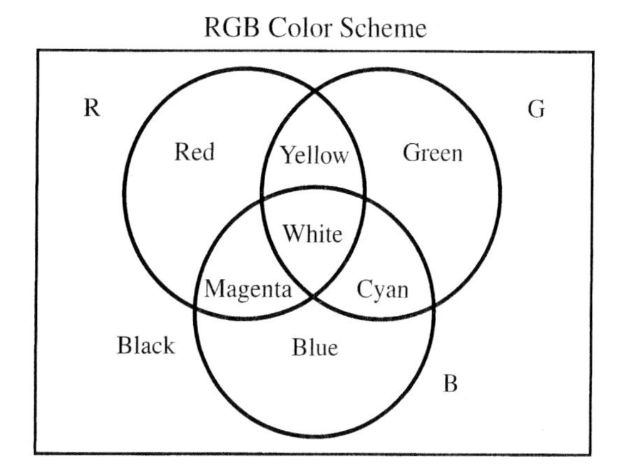

Although both color systems are important to the computer, we will focus our attention on the RGB color scheme, since it has important applications to HTML.

Example 2 **For each of the following, identify the color represented in the RGB Venn diagram.**

(a) RGB' (b) $RG'B$ (c) $R'G'B'$

Solution

(a) RGB' is the region inside red and green, but outside blue. This is the yellow region on the diagram.

Practice Problem Answers on page 313

(b) $RG'B$ is the region inside red and blue, but outside green. This is the magenta region on the diagram.

(c) $R'G'B'$ is the region outside red, green, and blue. This is the black region on the diagram.

PRACTICE PROBLEMS 2 **For each of the following, identify the color represented in the RGB Venn diagram.**

(a) $R'GB$ (b) RGB (c) $R'G'B$

In some computer programs, colors are specified as a percentage of the colors red, green, and blue. The following table gives the eight basic colors using either 0% or 100% of the primary colors.

COLOR	RED%	GREEN%	BLUE%
BLACK	0	0	0
BLUE	0	0	100
GREEN	0	100	0
CYAN	0	100	100
RED	100	0	0
MAGENTA	100	0	100
YELLOW	100	100	0
WHITE	100	100	100

EXAMPLE 3 **Using an RGB triplet, specify the percentages of red, green, and blue in the following colors.**

(a) blue (b) yellow

Solution

(a) blue = 0%, 0%, 100% (b) yellow = 100%, 100%, 0%

PRACTICE PROBLEMS 3 **Using an RGB triplet, specify the percentages of red, green, and blue in the following colors.**

(a) red (b) cyan

We can form other colors by mixing equal ratios of RGB primary colors. To produce orange, a mixture of both red and yellow, we average the two colors.

EXAMPLE 4 **Using an RGB triplet, state the percentages of red, green, and blue for the color orange.**

Solution

Because orange is a combination of red and yellow, we start with the RGB percentages for both red and yellow.

red = 100%, 0%, 0%   yellow = 100%, 100%, 0%

Next we average each color separately.

the red component:	**(100% + 100%)/2 = 100% Both colors contain full amounts of red.**
the green component:	**(0% + 100%)/2 = 50% Only yellow contains green, so the amount of green in orange is halved.**
the blue component:	**(0% + 0%)/2 = 0% Neither color contains any blue.**

Therefore, orange = 100%, 50%, 0%.

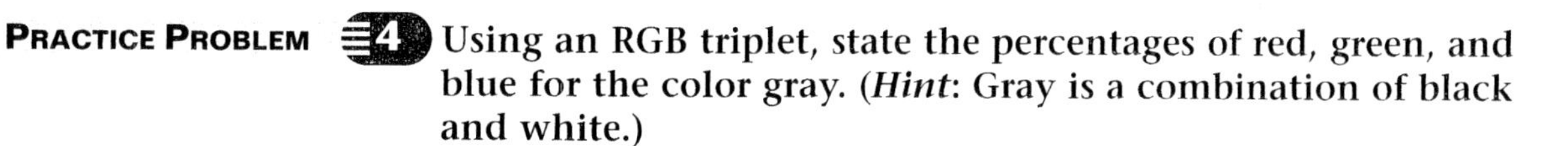

PRACTICE PROBLEM 4 **Using an RGB triplet, state the percentages of red, green, and blue for the color gray. (*Hint*: Gray is a combination of black and white.)**

ANSWERS TO PRACTICE PROBLEMS

1. (a) $C'MY$ is the region inside yellow and magenta, but outside cyan. This is the red region on the diagram.
   (b) $C'M'Y$ is the region inside yellow, but outside cyan and magenta. This is the yellow region on the diagram.
   (c) CMY is the region inside cyan, yellow, and magenta. This region is black.
2. (a) $R'GB$ is the region inside green and blue, but outside red. This is the cyan region on the diagram.
   (b) RGB is the region inside red, green, and blue. This is the white region on the diagram.
   (c) $R'G'B$ is the region inside blue, but outside red and green. This is the blue region on the diagram.
3. (a) red = 100%, 0%, 0%   (b) cyan = 0%, 100%, 100%
4. gray = 50%, 50%, 50%

Exercises

1. Complete the Venn diagram for the CMYK color scheme.

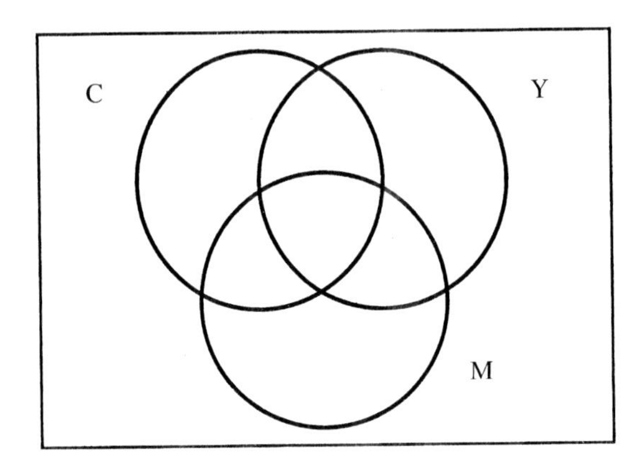

2. For each of the following, identify the color represented in the CMYK Venn diagram.

   (a) $C'MY'$ (b) $CM'Y$ (c) $C'M'Y'$

3. Complete the following table for the CMYK color scheme.

COLOR	CYAN%	MAGENTA%	YELLOW%
BLACK			
	0	0	100
	0	100	0
CYAN			
BLUE			
	0	100	100
	100	0	100
WHITE			

4. Complete the Venn diagram for the RGB color scheme.

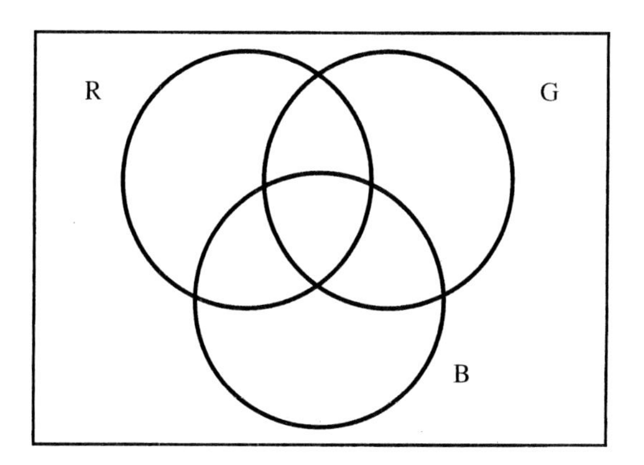

5. Complete the following table for the RGB color scheme.

COLOR	RED	GREEN	BLUE
	100	100	100
YELLOW			
	100	0	100
RED			
	0	100	100
GREEN			
	0	0	100
BLACK			

6. For each of the following, identify the color represented in the RGB Venn diagram.

   (a) $R'GB'$ **(b)** $RG'B'$ **(c)** RGB

7. Suppose one wanted to make a shade of green that is darker than the standard green for RGB. What would be the percentages of red, green, and blue for such a color?

8. Suppose one wanted to make a shade of gray that is lighter than the standard gray for RGB. What would be the percentages of red, green, and blue for such a color?

6.2 Hexadecimal RGB Codes

BITS of HISTORY

A.D. 1900 Navy Rear Admiral Grace Hopper was instrumental in the development of the modern computer. Admiral Hopper was a strong proponent of COBOL, the first computer language that allowed a person to communicate with a computer using words instead of numbers. She is also credited with coining the term "bug," as it relates to computing, when she discovered a moth that had gotten stuck inside a relay of the Mark II computer.

In the previous section, we learned that some programs use percentages of red, green, and blue to produce various colors. Colors in HTML are often expressed as RGB triplets such as "CC00FF" or "00FF00." These triplets are made up of 3 two-character hexadecimal numbers (one for red, one for green, and one for blue), which can take any value between 00 and FF. If you recall from Section 3.2, this produces 256 different numbers (0 to 255), decimally speaking. These RGB triplets are analogous to the percentages we used earlier: 00 would be 0% and FF would be 100%.

EXAMPLE 1 **State the color represented by each hexadecimal RGB triplet.**

(a) FF0000 **(b)** 00FFFF

Solution

(a) The first pair of digits represents red, and FF is the maximal amount. The following pairs of zeros mean our color contains no green and no blue, so FF0000 is red.

(b) The first pair of zeros means there is no red. The middle pair, FF, means we have the maximal amount of green, and the last pair, FF, means we have the maximal amount of blue. Since we have equal parts of green and blue, 00FFFF is cyan.

PRACTICE PROBLEMS 1 **State the color represented by each hexadecimal RGB triplet.**

(a) 00FF00 **(b)** FF00FF

We can form other colors by mixing equal ratios of RGB primary colors. To produce orange, a mixture of both red and yellow, we average the two colors.

EXAMPLE 2 **State the hexadecimal RGB triplet for orange.**

Solution

We start with the RGB triplets for both red and yellow:

red = FF0000 and yellow = FFFF00.

Next, we average each component separately.

PRACTICE PROBLEMS Answers on page 317

the red component: $(FF + FF)/2 = FF$ — **Both colors contain full amounts of red.**

the green component: $(00 + FF)/2 = 80$ or $7F$ — **Only yellow contains green, so the amount of green in orange is halved. Because that result is not a whole number, either 80 or 7F is acceptable.**

the blue component: $(00 + 00)/2 = 00$ — **Neither color contains any blue.**

Therefore, orange = FF8000 or FF7F00.

PRACTICE PROBLEM 2 **State the hexadecimal RGB triplet for gray. (*Hint*: Gray is equal parts black and white.)**

To obtain even more colors, a Web scheme was developed that split each primary color into six different amounts. For example, the palette consists of the digits 00, 33, 66, 99, CC, and FF for red, green, and blue, producing $6 \cdot 6 \cdot 6 = 216$ different colors. This palette is often used to optimize images for display on the Web and is useful for images containing a limited number of colors captured from Windows programs.

However, still many more colors can be formed. Since there are actually six hexadecimal digits for each code and each digit has sixteen different possibilities, there are $16 \cdot 16 \cdot 16 \cdot 16 \cdot 16 \cdot 16 = 16{,}777{,}216$ distinct colors available, even though most humans cannot distinguish them all.

EXAMPLE 3 **State the hexadecimal RGB triplet for medium blue (50% as intense as pure blue).**

Solution

Since we want a shade of blue, we do not want to add any red or green, but we need to decrease the amount of blue, so that the color is darker. For a medium blue, we need about 50% of the original concentration, so the RGB code would be 000080 or 00007F.

PRACTICE PROBLEM 3 **State the hexadecimal RGB triplet for medium yellow (50% as intense as pure yellow).**

ANSWERS TO PRACTICE PROBLEMS

1. **(a)** green **(b)** magenta
2. gray = 808080 or 7F7F7F
3. medium yellow = 808000 or 7F7F00

SECTION 6.2

Exercises

1. Complete the following table for the hexadecimal RGB color scheme.

COLOR	RED	GREEN	BLUE
	00	00	00
BLUE			
GREEN			
	00	FF	FF
	FF	00	00
	FF	00	FF
YELLOW			
WHITE			

2. What is the hexadecimal RGB code for magenta?

3. What is the hexadecimal RGB code for cyan?

4. What color is represented by the hexadecimal RGB code FFFFFF?

5. What color is represented by the hexadecimal RGB code FFFF00?

6. Suppose that purple is an equal ratio of magenta (FF00FF) and blue (0000FF). What is the hexadecimal RGB code for purple?

7. Suppose that you want a shade of purple that is 50% as intense as pure purple. How would you change the hexadecimal RGB code in Exercise 6 to make it?

8. Suppose that you want a reddish-purple that has 50% more red than does pure purple. How would you change the hexadecimal RGB code in Exercise 6 to make it?

9. Suppose that red-orange is an equal ratio of red (FF0000) and orange (FF8000). What is the hexadecimal RGB code for red-orange?

10. Suppose that you want a shade of red-orange that is 50% as intense as red-orange. How would you change the hexadecimal RGB code in Exercise 9 to make it?

11. Describe the color given by the following hexadecimal RGB code in terms of percentages of red, green, and blue: 00FFFF.

12. Describe the color given by the following hexadecimal RGB code in terms of percentages of red, green, and blue: 80FF00.

13. Describe the color given by the following hexadecimal RGB code in terms of percentages of red, green, and blue: 0F0F0F.

14. Describe the color given by the following hexadecimal RGB code in terms of percentages of red, green, and blue: F0F0F0.

15. In many color computer games, a 16-bit block is divided among three colors: 5 bits for red, 6 bits for green, and 5 bits for blue. How many different colors can be obtained using this system?

6.3 Cartesian and Monitor Coordinates

BITS of HISTORY

A.D.1900 British mathematician Alan Turing was instrumental in breaking the codes sent by the Nazi's Enigma machines during World War II. His invention of the Bombe in 1940 enabled the British to decode the messages transmitted by the German bombers, allowing evacuations that saved numerous lives. Turing was also interested in artificial intelligence and in 1950 developed a test that could be used to determine if a computer is intelligent. The Turing Test is still used today as computers become more and more automated.

More than two thousand years ago, two branches of mathematics, algebra and geometry, began to develop entirely separately. No one discovered that these two fields could be combined until the seventeenth century, when René Descartes used a tool to combine them: the coordinate plane. In Descartes's honor, this has come to be known as the **Cartesian plane.**

The Cartesian plane consists of two axes (the x-axis and the y-axis) that are perpendicular.

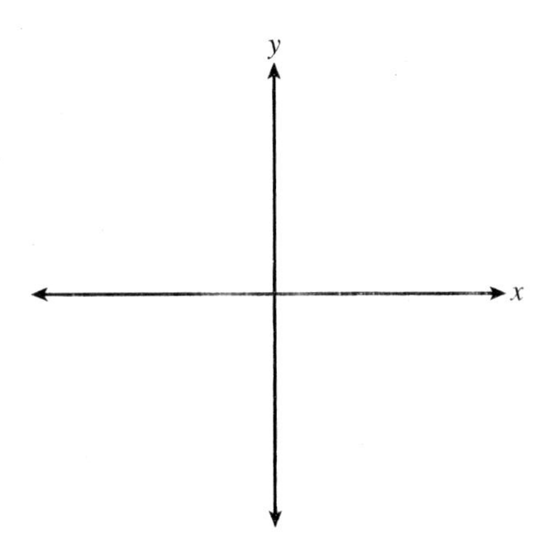

After marking each axis, we can identify any point in the plane.

EXAMPLE 1 **Find the coordinates of each indicated point.**

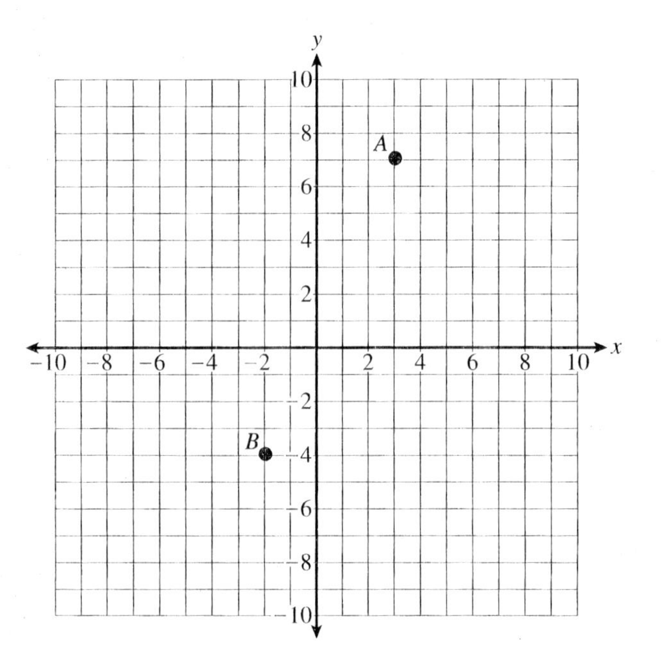

Solution

Beginning at the origin, which can be labeled (0, 0), we move to the right three units and up seven units to find point A. We label that point (3, 7). To arrive at point B from the origin, we move two units to the left and four units down. We label that point (−2, −4).

PRACTICE PROBLEM  **Find the coordinates for each indicated point.**

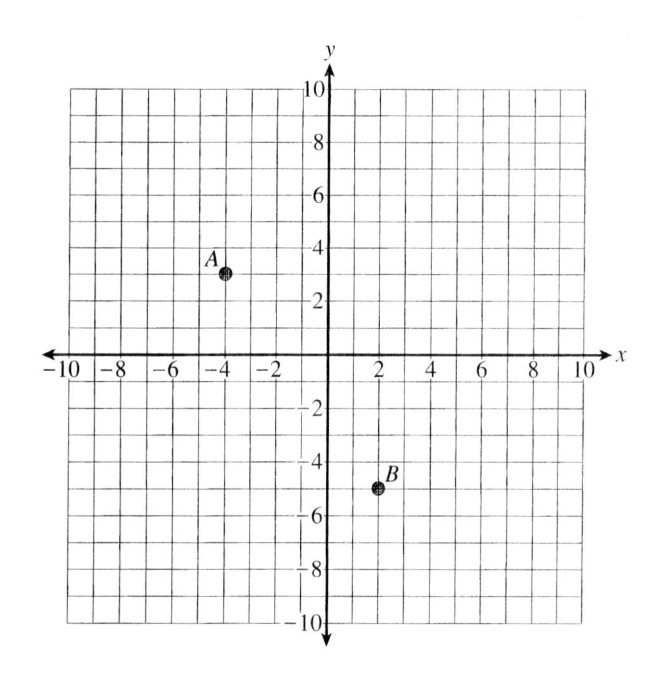

Similarly, we can plot a point, given its coordinates. In the next example, we restrict the coordinate system to the first quadrant, the upper right section in which all values are positive.

EXAMPLE 2 **Plot the two points A (4, 4) and B (6, 2).**

Solution

As before, we begin at the origin (now the lower left corner) and move first in the horizontal direction and then in the vertical direction.

PRACTICE PROBLEM Answers
on page 325

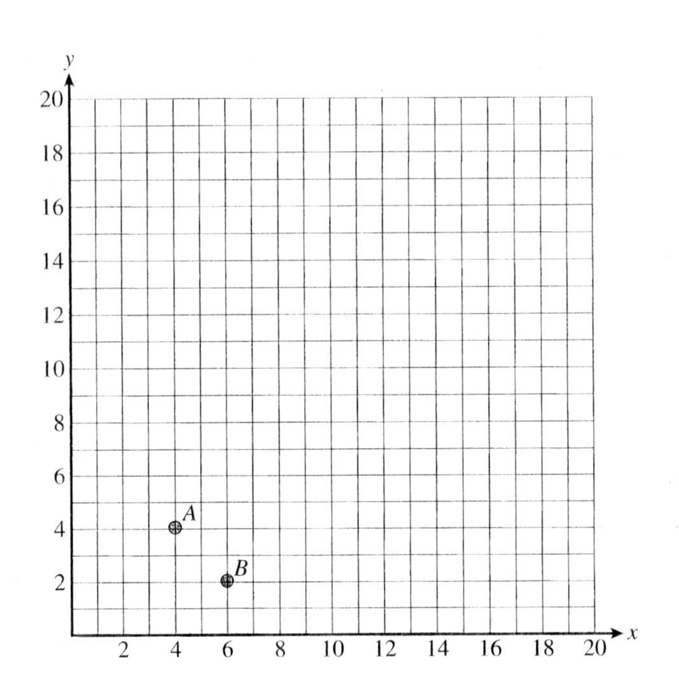

PRACTICE PROBLEM 2 Plot the two points A (1, 7) and B (5, 1).

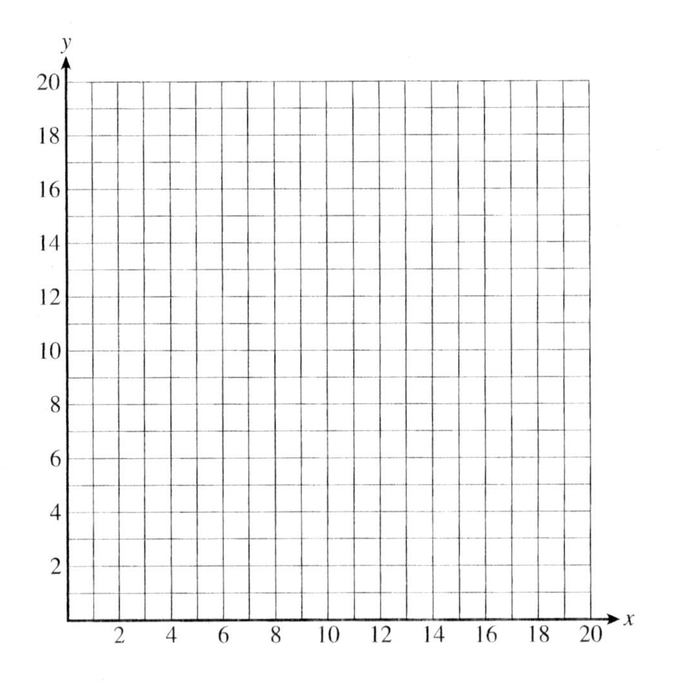

PRACTICE PROBLEM Answers
on page 325

In the first two examples of this section, we have used a coordinate system for which each hatch mark represented one unit. It is frequently necessary to have each hatch mark represent a larger increment, as in the next example.

EXAMPLE 3 **Identify (approximately) the indicated point.**

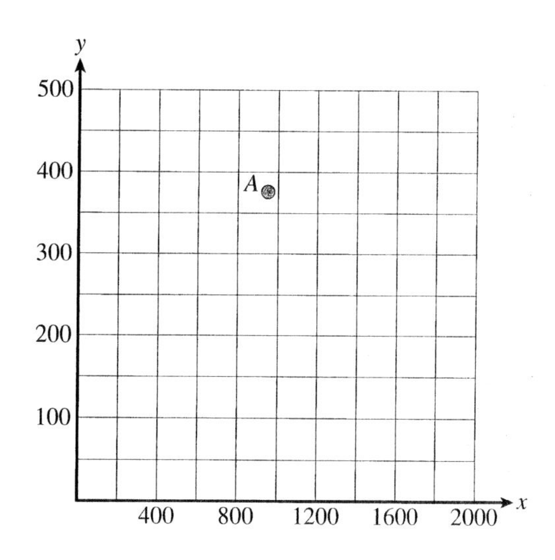

Solution

We can only approximate the coordinates in this case. The x-coordinate is closer to 1000 than it is to 800, and the y-coordinate appears to be halfway between 350 and 400. The coordinates for point A are approximately (950, 375).

PRACTICE PROBLEM 3 **Identify (approximately) the indicated point.**

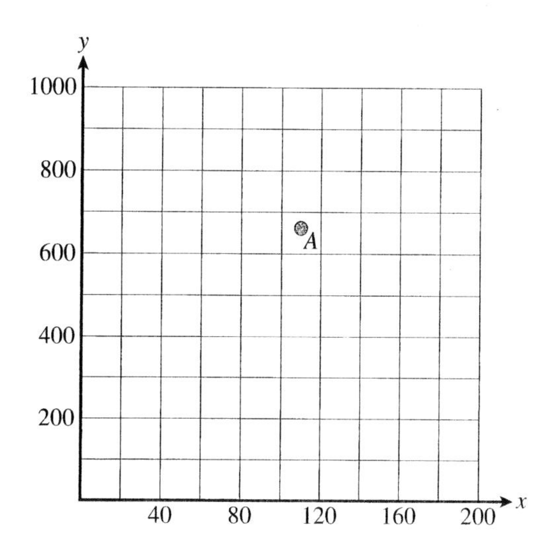

PRACTICE PROBLEM Answers
on page 325

The location of a pixel on a computer monitor can be determined using the same principles that Descartes used. The biggest difference is that, on the computer screen, we start in the upper-left corner.

EXAMPLE 4 **Assume that the following box represents a computer screen. With standard resolution, there are 1024 pixels in the horizontal direction and 768 in the vertical direction. Use that information to approximate the coordinates of the indicated pixel.**

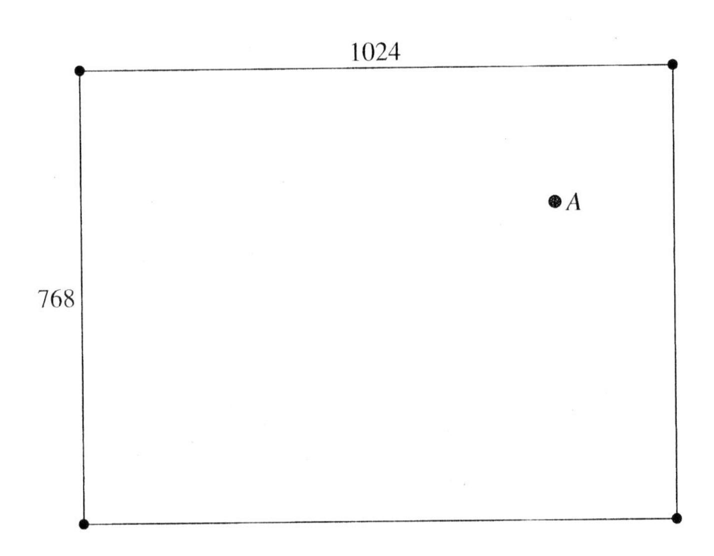

Solution

A number of math skills come into play here. First, we find the horizontal distance of the point as a percentage of the total horizontal distance of the box. In this case, the box is 80 mm long, and point A is 64 mm from the left edge. We find that percentage to be $64 \div 80 = 0.80 = 80\%$. Using the same technique with the vertical distance, we find that

$$18 \div 60 = 0.30 = 30\%.$$

The coordinates become

$$(80\% \text{ of } 1024, 30\% \text{ of } 768) = (0.8 \times 1024, 0.3 \times 768) = (899.2, 230.4).$$

The approximate coordinates for point A are (899, 230). (Of course, based on your measurement, you may have slightly different coordinates.)

PRACTICE PROBLEM 4 Assume that the following box represents a computer screen. With standard resolution, there are 1024 pixels in the horizontal direction and 768 in the vertical direction. Use that information to approximate the coordinates of the indicated pixel.

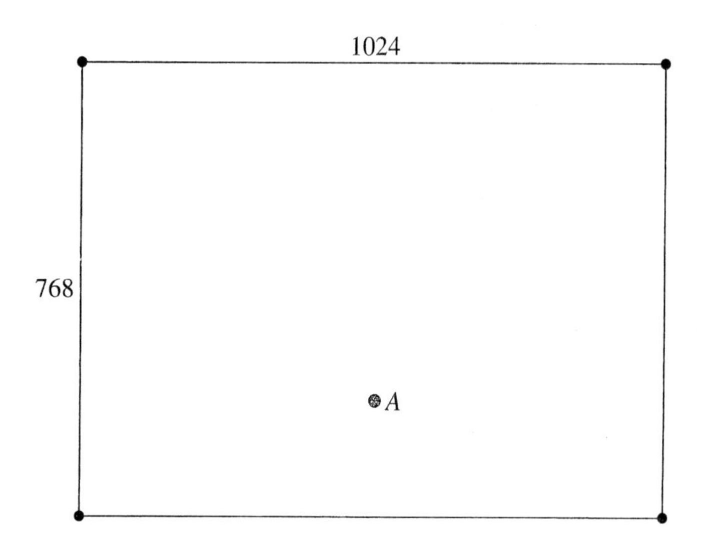

ANSWERS TO PRACTICE PROBLEMS

1. $A(-4, 3)$ and $B(2, -5)$

2. 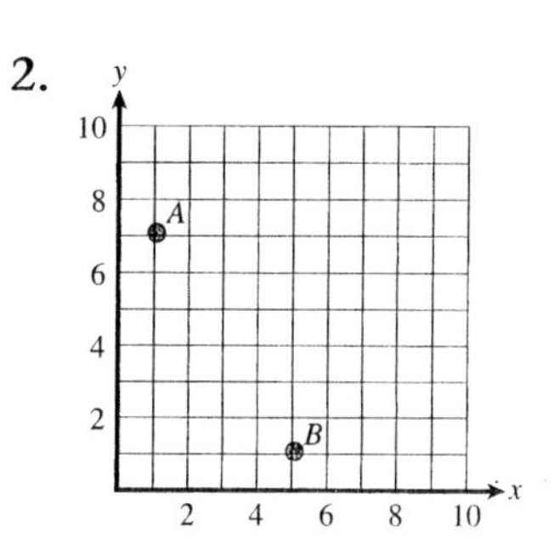

3. $A(110, 675)$

4. $A(512, 576)$

SECTION

6.3 Exercises

1. Find the coordinates of each of the indicated points.

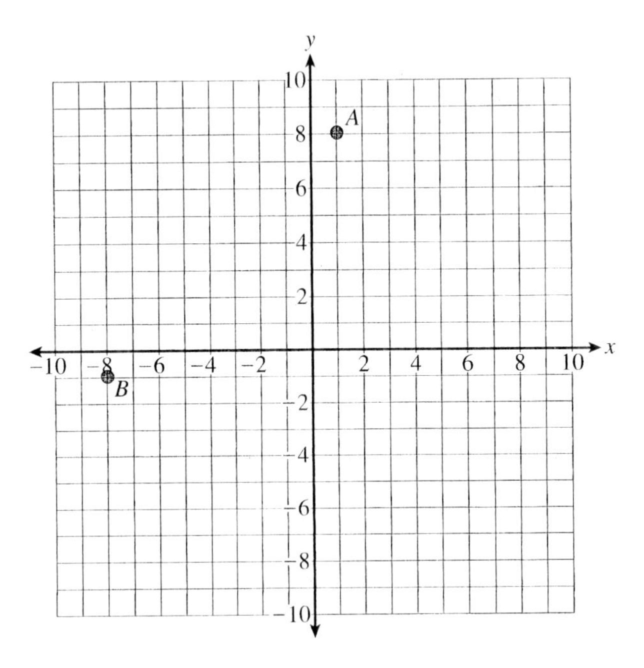

2. Find the coordinates of each of the indicated points.

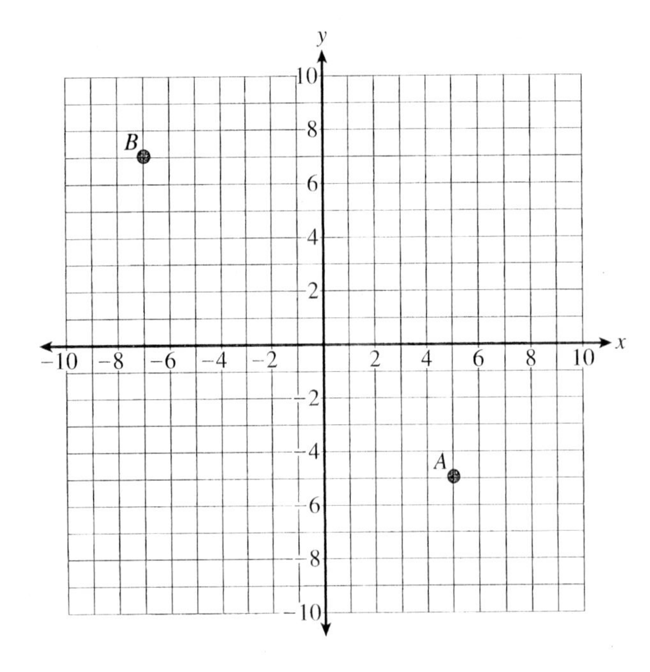

3. On the following grid, plot the two points A (8, 6) and B (6, 8).

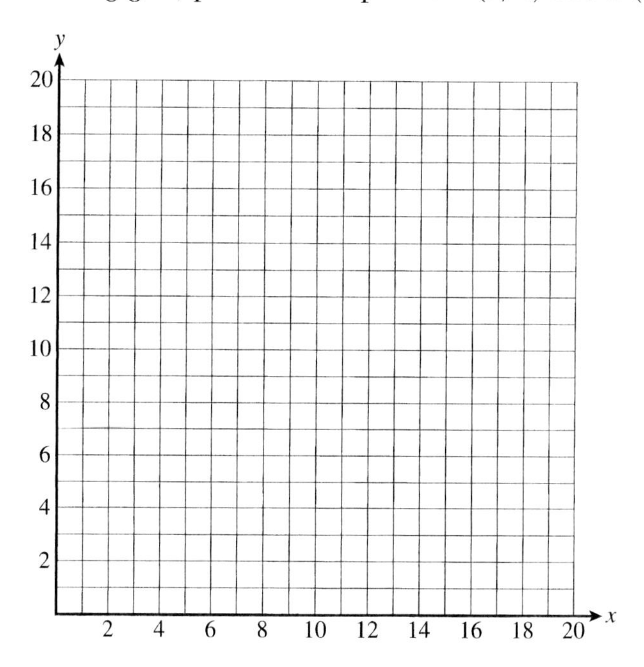

4. On the following grid, plot the two points A (0, 4) and B (4, 0).

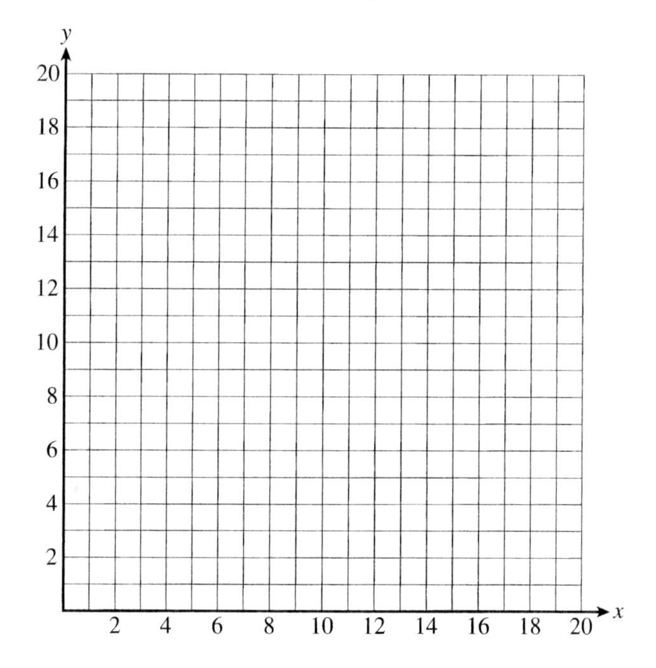

For Exercises 5–8, identify the indicated point. Approximate the point's coordinates if they are not exact.

5. A 6. B 7. C 8. D

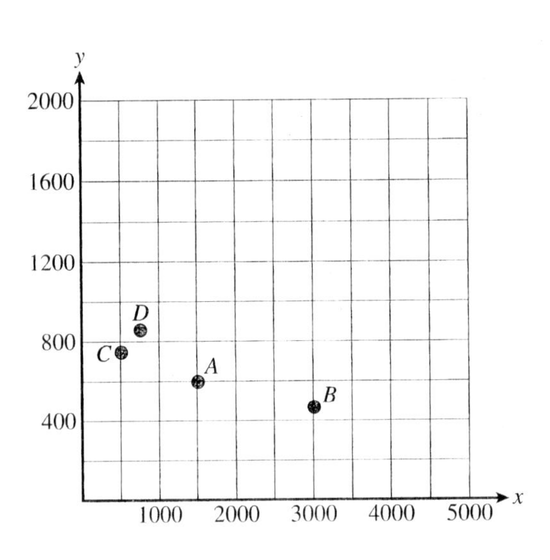

Assume that the following box represents a computer screen. Use it to complete Exercises 9–12.

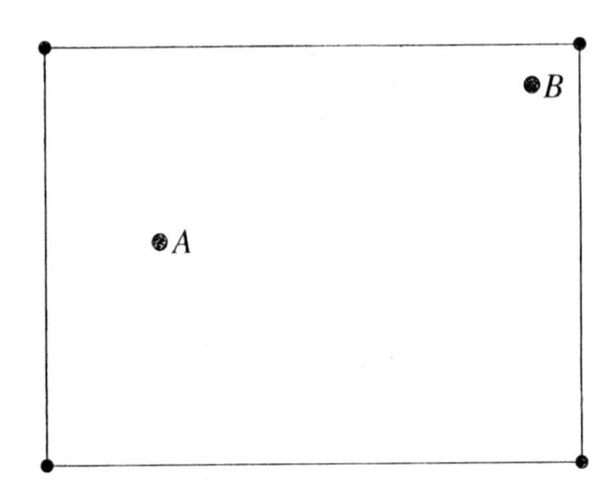

9. With standard resolution, there are 1024 pixels in the horizontal direction and 768 in the vertical direction. Use the information to approximate the coordinates of pixel A.

10. Again, assume that there are 1024 pixels in the horizontal direction and 768 in the vertical direction. Use the information to approximate the coordinates of pixel B.

11. Assume that the resolution has 1280 pixels in the horizontal direction and 1024 in the vertical direction. What would be the coordinates of pixel A?

12. Assume that the resolution has 1280 pixels in the horizontal direction and 1024 in the vertical direction. What would be the coordinates of pixel B?

6.4 Elements of Computer Animation

BITS of HISTORY

A.D. 1900 Hungarian mathematician John Von Neumann immigrated to the United States prior to World War II and joined an elite group of researchers at Princeton University. There Von Neumann developed the study of game theory and automata theory, which further advanced computer science. Von Neumann was also a strong supporter of the implementation of the binary system for computers and advocated for the bit as the fundamental memory storage unit.

An **algorithm** is a series of steps (or instructions) that performs a task. If the task is to use the ATM to retrieve cash from your savings account, what follows is one possible algorithm.

```
Insert bank card
Enter PIN
Select WITHDRAWAL
Select SAVINGS
Enter amount
Take cash
Take receipt
Take bank card
```

The level of detail can be expanded, but the more it is expanded the less universal the algorithm becomes. At some level of detail, the algorithm becomes relevant to only one particular bank's machines.

In this section, we use algorithms to represent computer instructions. These algorithms are very general, as we are not intending to give instructions for any particular computer or computer language.

Our first example is designed to help you see the role of a variable in computer algorithms.

EXAMPLE 1 **Determine the output for the following algorithm.**

```
Let x = 1
  Print x
Let x = 2
  Print x
Let x = 2 × 3
  Print x
```

Solution

The output will be the numbers

1

2

6

The calculation $2 \times 3 = 6$ is done before anything is printed.

PRACTICE PROBLEM 1 **Determine the output for the following algorithm.**

```
Let x = 5
  Print x
Let x = 9
  Print x
Let x = 3 x 7
  Print x
```

Determine the output for the following algorithm.

```
Let x = 1
  Print x
Let x = 2
  Print x
Let x = 3
  Print x
Let x = 4
  Print x
Let x = 5
  Print x
Let x = 6
  Print x
Let x = 7
  Print x
Let x = 8
  Print x
Let x = 9
  Print x
Let x = 10
  Print x
```

Obviously, the output will be the counting numbers from 1 to 10. The algorithm basically consists of the same step (`Print x`) repeated ten times with only the variable changing. In our second example, we discover a tool for combining repeated steps in an algorithm.

EXAMPLE 2 **What output would result from the following loop?**

```
For x = 1 to 10
  Print x
Next x
```

This is called a *loop* because it starts at the top (x begins with 1), goes to the bottom (x becomes 2), and then loops back to the top. This process continues until we encounter the last possible value of x (in this case, 10).

PRACTICE PROBLEM Answers on page 337

Solution

The output would be

```
1
2
3
4
5
6
7
8
9
10
```

PRACTICE PROBLEM 2 **What output would result from the following loop?**

```
For x = 6 to 8
  Print x
Next x
```

The definition of "animate" that is relevant to computer animation is "to give motion to." The kind of motion that we examine in this section is linear motion. Linear motion is movement along a straight line. In a computer program, linear motion is frequently achieved with the help of a programming loop. Even a loop as simple as the one in Example 2 can be used to create the illusion of motion, which is the essence of animation.

Before we continue, let's define a couple of terms that we will use to illustrate animation.

DEFINITIONS: **Plot Pixel (*x*, *y*)** turns on the pixel that is *x* pixels to the right of the upper-left corner and *y* pixels down from the top of the screen. **ClearScreen** turns off every pixel, leaving a dark screen.

EXAMPLE 3 **Assume that the following code is applied to a computer monitor with a resolution of 1024 by 768. Describe the output.**

```
For x = 1 to 1024
  Plot Pixel (x, 384)
Next x
```

First, a pixel would be plotted at (1, 384). That is the first column in the middle of the screen. The next time through the loop, a pixel would plot at (2, 384), then at (3, 384), and so on. We would see a horizontal line being drawn across the middle of the screen.

If we want to change this from a line being drawn to the illusion of a point moving across the monitor, we would need to add another line of code that would erase the previous pixel before the next one was plotted. One way to do this would be with code similar to the following.

```
For x = 1 to 1024
  Plot Pixel (x, 384)
  ClearScreen
Next x
```

Solution

That code would result in a point running horizontally across the middle of the screen.

PRACTICE PROBLEM 3 **Assume that the following code is applied to a computer monitor with a resolution of 1024 by 768. Describe the output.**

```
For x = 1 to 768
  Plot Pixel (512, x)
  ClearScreen
Next x
```

It is also possible to use variables for both coordinates of the pixel that is to be plotted. This occurs in the next example.

PRACTICE PROBLEM Answers on page 337

EXAMPLE 4 **Assume that the following code is applied to a computer monitor with a resolution of 1024 by 768. Describe the output.**

```
For x = 1 to 768
   Plot Pixel (x, x)
   ClearScreen
Next x
```

The plotted pixel begins at (1, 1) and moves through (2, 2), (3, 3), and so on until it ends at (768, 768). The output would be a pixel appearing to move along a dotted line from the upper-left corner to a point on the bottom about three-quarters of the way across.

PRACTICE PROBLEM 4 **Assume that the following code is applied to a computer monitor with a resolution of 1024 by 768. Describe the output.**

```
For x = 256 to 768
   Plot Pixel (x, x)
   ClearScreen
Next x
```

Because computers perform calculations so quickly, it is very easy to include arithmetic in computer commands.

EXAMPLE 5 **What output would result from the following loop?**

```
For x = 1 to 4
   Print 2x - 5
Next x
```

Solution

The first time through, $x = 1$, so $2x - 5 = 2(1) - 5 = -3$.

The second time through, $x = 2$, so $2x - 5 = -1$.

The third time, $x = 3$ and $2x - 5 = 1$.

Finally, $x = 4$ so $2x - 5 = 3$.

The output would be:

```
-3
-1
1
3.
```

PRACTICE PROBLEM Answers on page 337

Practice Problem 5 What output would result from the following loop?

```
For x = 2 to 5
  Print 3x - 4
Next x
```

You may recall from algebra that graphing a linear equation (an equation with no exponents) such as $y = 2x - 5$ results in a straight line. This idea is used to create computer motion in a straight line.

Example 6 What output would result from the following loop?

```
For x = 16 to 399
  y = 2x - 31
  Plot Pixel (x, y)
  ClearScreen
Next x
```

Solution

The first few pixels to be plotted would be (16, 1), (17, 3), (18, 5), (19, 7). The last couple would be (398, 765) and (399, 767). The resulting output would be a pixel appearing to move in a straight line from (16, 1), near the upper-left corner, to (399, 767), on the bottom row less than halfway across the screen. The box below indicates the path of movement.

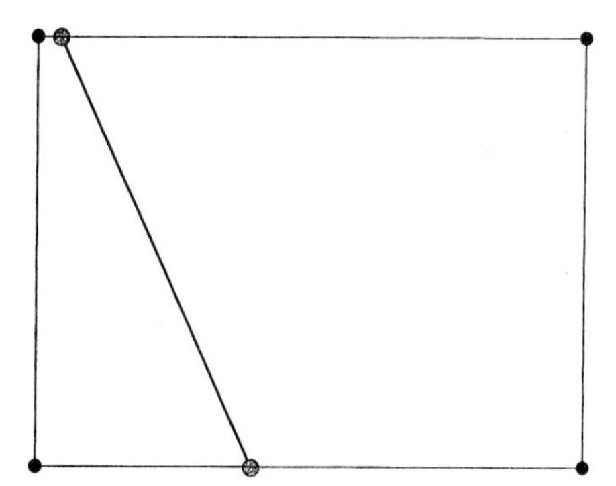

Practice Problem 6 What output would result from the following loop?

```
For x = 8 to 199
  y = 4x - 31
  Plot Pixel (x, y)
  ClearScreen
Next x
```

Practice Problem Answers on page 337

We have seen that a larger value for the coefficient of x (which we call the slope) results in a steeper line. What happens if we make the slope smaller? The final example illustrates.

EXAMPLE 7 What output would result from the following loop?

```
For x = 8 to 1024 step 8
  y = (1/8)x
  Plot Pixel (x, y)
  ClearScreen
Next x
```

Solution

In the first line, "`step 8`" indicates that x will be counted by 8s. It will include only 8, 16, 24, 32, 40, . . . , 1024.

The plotted points would be (8, 1), (16, 2), (24, 3), . . . , (1024, 128). The resulting motion would start near the upper left corner and move to the right edge, about one-sixth of the way down. That line of motion is indicated in the box below.

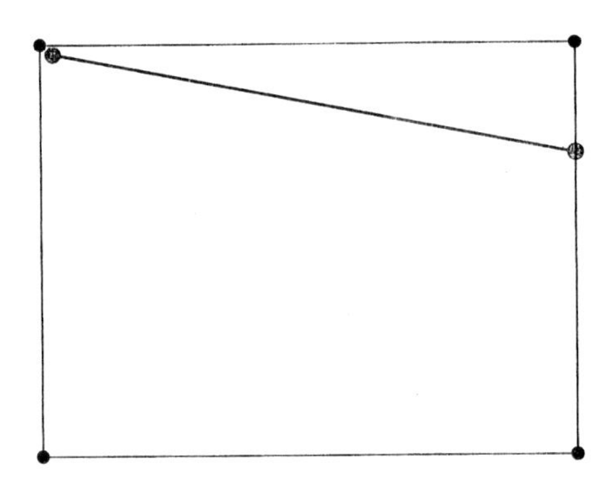

PRACTICE PROBLEM 7 What output would result from the following loop?

```
For x = 8 to 1024 step 2
  y = (1/2)x
  Plot Pixel (x, y)
  ClearScreen
Next x
```

PRACTICE PROBLEM Answers on page 337

ANSWERS TO PRACTICE PROBLEMS

1. output:
```
5
9
21
```

2. output:
```
6
7
8
```

3. a point running vertically down the middle of the screen

4. a point running diagonally from the middle of the screen to a point on the right side of the bottom row

5. output:
```
2
5
8
11
```

6. a point running from (8, 1) (upper left) to (199, 767) (bottom, closer to the right side)

7. a point running from (8, 4) (upper left) to (1024, 512) (left side, closer to the bottom)

Exercises

1. What output would result from the following loop?

```
For x = 1 to 7
  Print x
Next x
```

2. What output would result from the following loop?

```
For x = 2 to 10
  Print x
Next x
```

3. What output would result from the following loop?

```
For x = 10 to 120 step 10
  Print x
Next x
```

4. What output would result from the following loop?

```
For x = 20 to 36 step 2
  Print x
Next x
```

5. What output would result from the following loop?

```
For x = 5 to 35 step 5
  Print x - 2
Next x
```

6. What output would result from the following loop?

```
For x = 4 to 64 step 4
  Print x + 3
Next x
```

▲ Represents additionally challenging problems.

7. What output would result from the following loop?

```
For x = 100 to 200 step 25
  Print 1/5x
Next x
```

8. What output would result from the following loop?

```
For x = 0 to 1024 step 64
  Print x
Next x
```

9. What output would result from the following loop?

```
For x = 128 to 768 step 128
  Print x - 1
Next x
```

10. What output would result from the following loop?

```
For x = 24 to 36
  Print 2x + 7
Next x
```

11. Assume that the following code is applied to a computer monitor with a resolution of 1024 by 768. Describe the output.

```
For x = 1 to 768
  Plot Pixel (256, x)
Next x
```

12. Assume that the following code is applied to a computer monitor with a resolution of 1024 by 768. Describe the output.

```
For x = 1 to 1024
  Plot Pixel (x, 384)
Next x
```

13. Assume that the following code is applied to a computer monitor with a resolution of 1024 by 768. Describe the output.

```
For x = 1 to 1024
  Plot Pixel (1025 - x, 384)
  ClearScreen
Next x
```

14. Assume that the following code is applied to a computer monitor with a resolution of 1024 by 768. Describe the output.

```
For x = 1 to 768
  Plot Pixel (512, 769 - x)
  ClearScreen
Next x
```

15. Assume that the following code is applied to a computer monitor with a resolution of 1280 by 1024. Describe the output.

```
For x = 16 to 256 step 16
  y = 4x
  Plot Pixel (x, y)
Next x
```

16. Assume that the following code is applied to a computer monitor with a resolution of 1280 by 1024. Describe the output.

```
For x = 40 to 1280 step 40
  y = (2/5)x
  Plot Pixel (x, y)
Next x
```

17. Assume that the following code is applied to a computer monitor with a resolution of 1280 by 1024. Describe the output.

```
For x = 16 to 512 step 16
  y = 2x - 16
  Plot Pixel (x, y)
Next x
```

Exercises 18–22 refer to Sections 6.1 and 6.2.

Assume that we are working with a 24-bit true color scheme, in which each pixel is assigned an RGB color with values for red, green, and blue that each range from 0 to 255. The command for a solid blue pixel would be `RGBColor = "0, 0, 255"` (no red, no green, and full concentration of blue).

18. Assume that the following code is applied to a computer monitor with a resolution of 1024 by 768. Describe the output.

```
RGBColor = "0, 255, 0"
For x = 1 to 1024
   Plot Pixel (x, 384)
Next x
```

19. Assume that the following code is applied to a computer monitor with a resolution of 1024 by 768. Describe the output.

```
For x = 4 to 1024 step 4
   RGBColor = "0, .25x - 1, 0"
   Plot Pixel (x, 384)
   ClearScreen
Next x
```

20. Assume that the following code is applied to a computer monitor with a resolution of 1024 by 768. Describe the output. How would the output differ if a `ClearScreen` command were added after the pixel was plotted?

```
For x = 1 to 255
   RGBColor = "x, 0, 256 - x"
   Plot Pixel (x, 384)
Next x
```

21. Write a pseudocode that would display a vertical line centered horizontally on a computer screen with a resolution of 1024 by 768. The line's color would change gradually from black (at the top) to white. Modify your code to change the color from black (at the top) to yellow.

22. Write a pseudocode that would fill in the entire screen (1024 by 768) with a different color for each pixel on the screen.
(Hint: assume that fractional color codes truncate.)

Summary

6.1 Color Schemes

Many inkjet printers use a color scheme known as **CMYK**, which is composed of four colors: cyan, magenta, yellow, and black. This type of color scheme is known as **subtractive color**, since the pigments absorb certain wavelengths of light to produce color.

The following Venn diagram illustrates the CMYK color scheme.

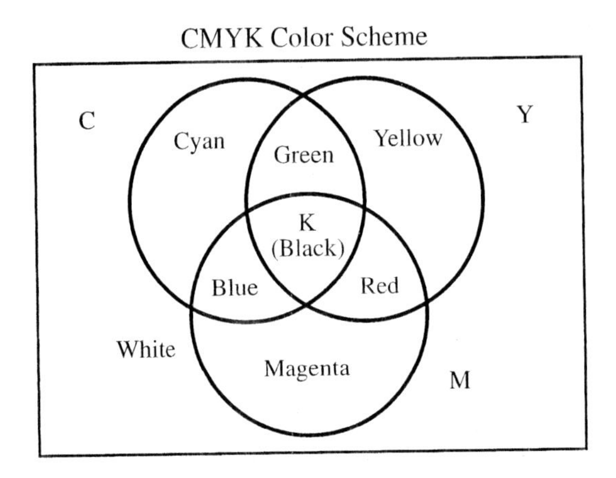

The RGB (red, green, blue) scheme is an **additive color** scheme since augmenting different proportions of light produces the colors. The following Venn diagram illustrates the RGB color scheme.

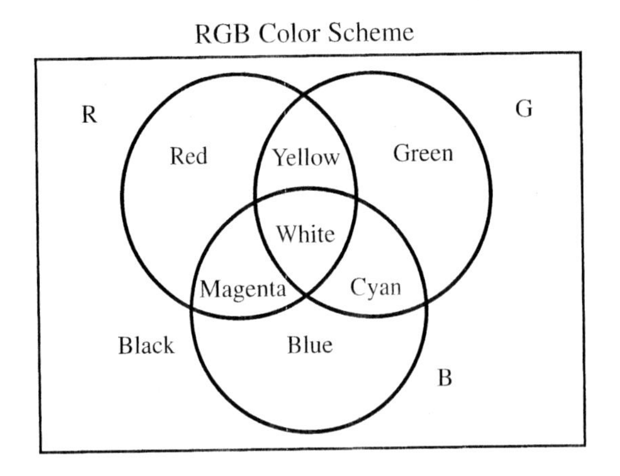

The following table gives the eight basic colors using either 0% or 100% of the primary colors.

COLOR	RED%	GREEN%	BLUE%
BLACK	0	0	0
BLUE	0	0	100
GREEN	0	100	0
CYAN	0	100	100
RED	100	0	0
MAGENTA	100	0	100
YELLOW	100	100	0
WHITE	100	100	100

6.2 Hexadecimal RGB Codes

Colors in HTML are often expressed as RGB triplets such as "CC00FF" or "00FF00." These triplets are made up of 3 two-character hexadecimal numbers (one for red, one for green, and one for blue), which can take the value of 00 to FF.

Mixing equal ratios of colors can form other colors. Since orange is a mixture of both red and yellow, we average the two colors. Orange = "FF8000" or "FF7F00."

6.3 Cartesian and Monitor Coordinates

The Cartesian plane consists of two axes (the x-axis and the y-axis) that are perpendicular. By marking each axis, we can identify any point in the plane.

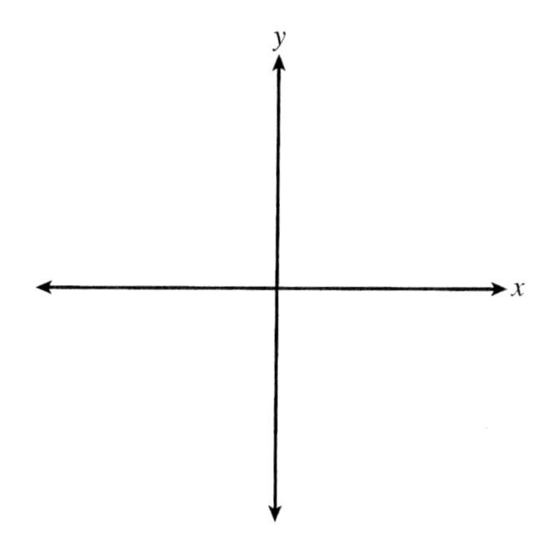

The location of a pixel on a computer monitor can be determined using the same principles used by Descartes. The biggest difference is that, on the computer screen, we start in the upper left-hand corner.

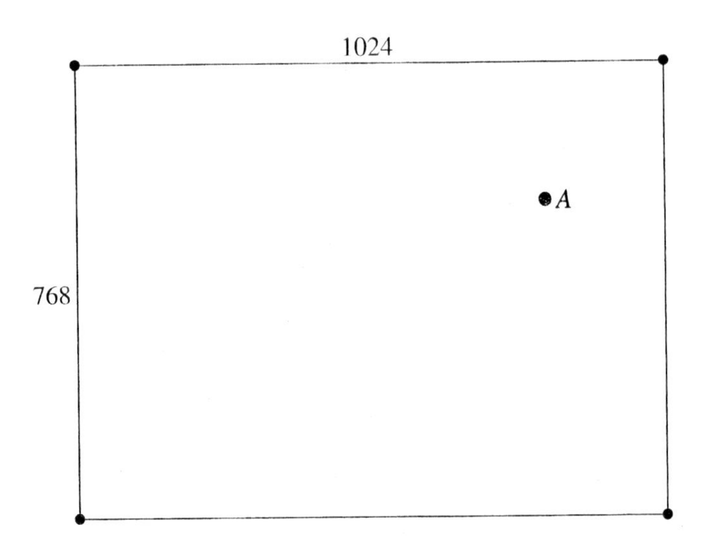

The approximate coordinates for the point A are (900, 230).

6.4 Elements of Computer Animation

An **algorithm** is a series of steps (or instructions) that performs a task.

```
For x = 1 to 10
  Print x
Next x
```

This is called a loop because it starts at the top (x begins with 1), goes to the bottom (x becomes 2), then loops back to the top. This process continues until we encounter the last possible value of x (in this case, x goes to 10).

The output would be the counting numbers from one to ten.

We will define a couple of terms that were used to illustrate animation.

> **Plot Pixel (x, y)** will turn on the pixel that is x pixels to the right of the upper left-hand corner and y pixels down from the top of the screen.

ClearScreen will turn off every pixel, leaving a dark screen.

```
For x = 1 to 1024
  Plot Pixel (x, 384)
  ClearScreen
Next x
```

That code would result in a point running horizontally across the middle of the screen.

A linear equation can be used to create a sloping line on the monitor. The following algorithm uses a linear equation to compute *x*.

```
For x = 1 to 100
  y = 4x - 32
  Plot Pixel (x, y)
  ClearScreen
Next x
```

A larger absolute value for the coefficient of *x* (the slope) results in a steeper line. A smaller value results in a flatter line.

Chapter 6 Glossary Quiz

The following terms were introduced in Chapter 6 of the text. Match each with one of the definitions that follows.

Cartesian plane ______ CMYK ______ RGB ______

(a) a color scheme used by many inkjet printers—it is called *subtractive color,* since the pigments absorb certain wavelengths of light to produce color

(b) an additive color scheme used by most computer monitors

(c) two scaled axes (the x-axis and the y-axis) that are perpendicular

REVIEW EXERCISES

[6.1] **1.** Complete the Venn diagram for the RGB color scheme.

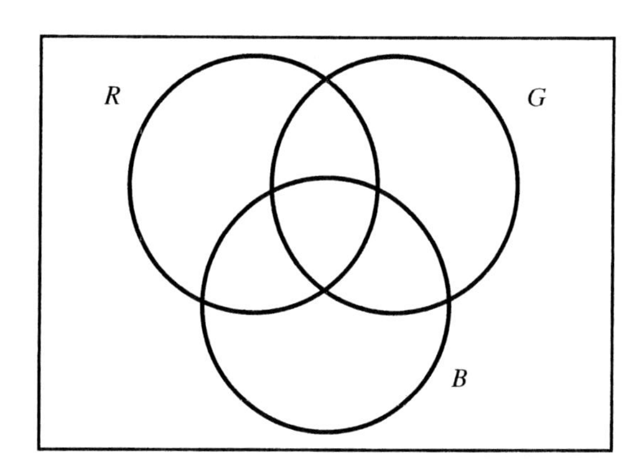

2. For each of the following schemes, identify the color represented in the RGB Venn diagram.

(a) $R'G'B$ **(b)** RGB' **(c)** $R'G'B'$ **(d)** RGB

3. Complete the following table for the RGB color scheme.

COLOR	RED	GREEN	BLUE
WHITE			
	100	100	0
	100	0	0
BLUE			
MAGENTA			
	0	0	0
CYAN			
	0	100	0

[6.2] 4. What color is represented by the hexadecimal RGB code "00FF00"?

[6.2] 5. What is the hexadecimal RGB code for yellow?

6. Suppose medium blue is an equal ratio of blue = "0000FF" and black = "000000." What is the hexadecimal RGB code for medium blue?

7. Describe the color given by the hexadecimal RGB code "00FF00" in terms of percentages of red, green, and blue.

8. Describe the color given by the hexadecimal RGB code "8000FF" in terms of percentages of red, green, and blue.

[6.3] **9.** Find the coordinates for each of the indicated points.

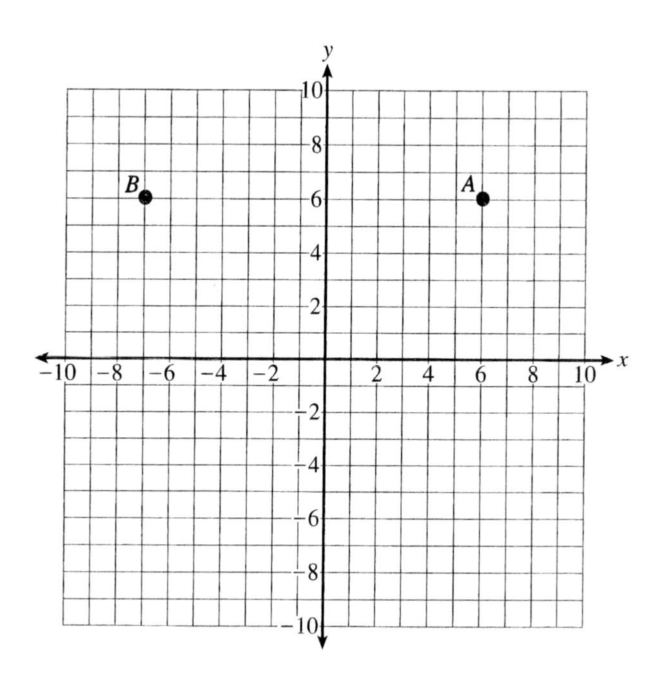

[6.3] **10.** On the grid below, plot the two points A (10, 4) and B (2, 18).

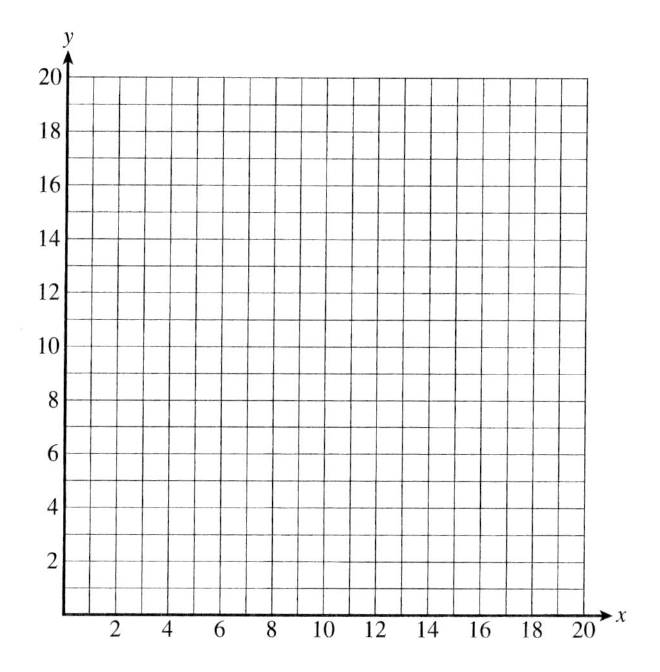

For Exercises 11–14, identify the indicated point. Approximate if it is not exact.

11. A **12.** B **13.** C **14.** D

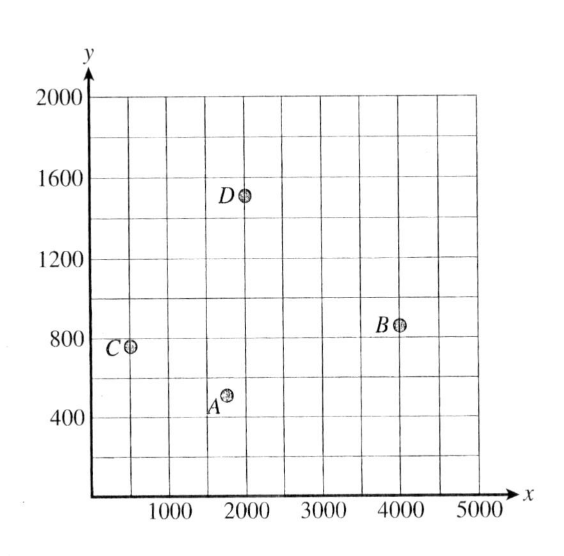

[6.3] **To complete Exercises 15–17, assume that the following box represents a computer screen.**

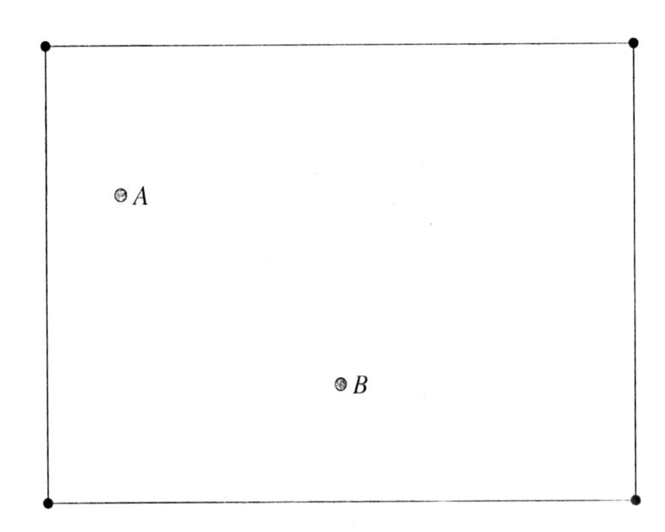

15. With standard resolution, there are 1024 pixels in the horizontal direction and 768 in the vertical direction. Use that information to approximate the coordinates of the pixel labeled *A*.

16. Again, assume there are 1024 pixels in the horizontal direction and 768 in the vertical direction. Use that information to approximate the coordinates of the pixel labeled *B*.

17. Assume that the resolution has 1280 pixels in the horizontal direction and 1024 in the vertical direction. What would the coordinates of the pixel labeled *A* become?

[6.4] **18.** What output would result from the following loop?

```
For x = 0 to 5
  Print x
Next x
```

[6.4] **19.** What output would result from the following loop?

```
For x = 0 to 50 step 5
  Print x
Next x
```

20. What output would result from the following loop?

```
For x = 1 to 5
  Print x + 3
Next x
```

21. What output would result from the following loop?

```
For x = 2 to 24 step 2
  Print 4x - 3
Next x
```

22. Assume that the following code is applied to a computer monitor with resolution 1024 by 768. Describe the output.

```
For x = 1 to 1024
  Plot Pixel (x, 384)
Next x
```

23. Assume that the following code is applied to a computer monitor with resolution 1024 by 768. Describe the output.

```
For x = 1 to 768
  Plot Pixel (512, 769 - x)
  ClearScreen
Next x
```

[6.4] **24.** Assume that the following code is applied to a computer monitor with resolution 1280 by 1024. Describe the output.

```
For x = 512 to 1024 step 16
  y = 2x - 1024
  Plot Pixel (x, y)
Next x
```

[6.4] **Assume that we are working with a 24-bit true color scheme in which each pixel is assigned an RGB color with values for red, green, and blue, that each range from 0 to 255. The command for a solid blue pixel would be** `RGBColor = "0, 0, 255"` **(no red, no green, and full concentration of blue).**

25. Assume that the following code is applied to a computer monitor with resolution 1024 by 768. Describe the output.

```
RGBColor = "0, 255, 0"
For x = 1 to 768
   Plot Pixel (512, x)
Next x
```

26. Assume that the following code is applied to a computer monitor with resolution 1024 by 768. Describe the output. How would the output differ if a ClearScreen command were added after the pixel was plotted?

```
For x = 1 to 255
   RGBColor = "x, 256 - x, 0"
   Plot Pixel (x, 384)
Next x
```

SELF-TEST

1. Complete the Venn diagram for the CMYK color scheme.

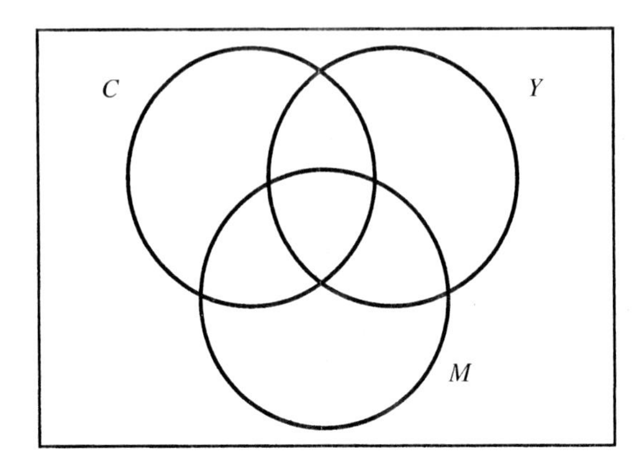

2. For each of the following, identify the color represented in the CMY Venn diagram.

   (a) $C'M'Y$ (b) CMY' (c) $C'M'Y'$ (d) CMY

3. Complete the following table for the RGB color scheme.

COLOR	RED	GREEN	BLUE
	100	100	100
BLUE			
	0	0	0
CYAN			
YELLOW			
	100	0	100
GREEN			
	100	0	0

4. What color is represented by the hexadecimal RGB code "FFFF00"?

5. What is the hexadecimal RGB code for cyan?

6. Suppose orange is an equal ratio of red = "FF0000" and yellow = "FFFF00". What is the hexadecimal RGB code for orange?

7. Describe the color given by the hexadecimal RGB code "FF00FF" in terms of percentages of red, green, and blue.

8. Describe the color given by the hexadecimal RGB code "0080FF" in terms of percentages of red, green, and blue.

9. Find the coordinates for each of the indicated points.

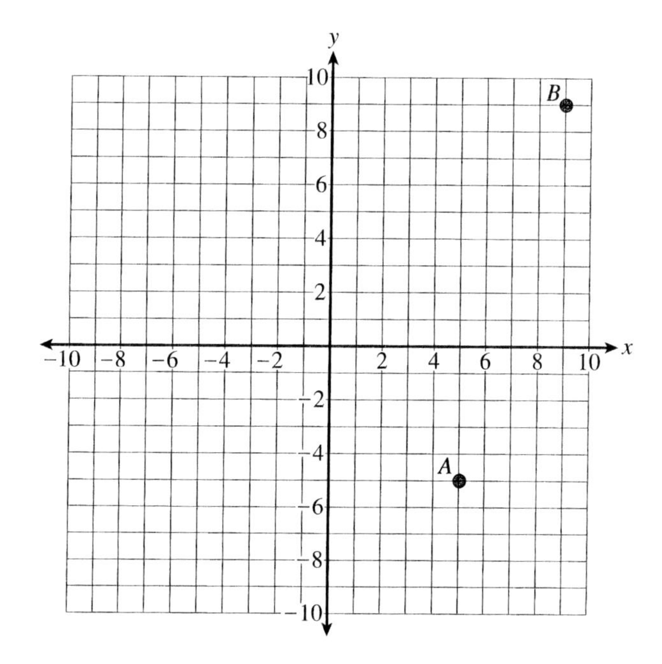

10. On the grid below, plot the two points A (10, 4) and B (2, 18).

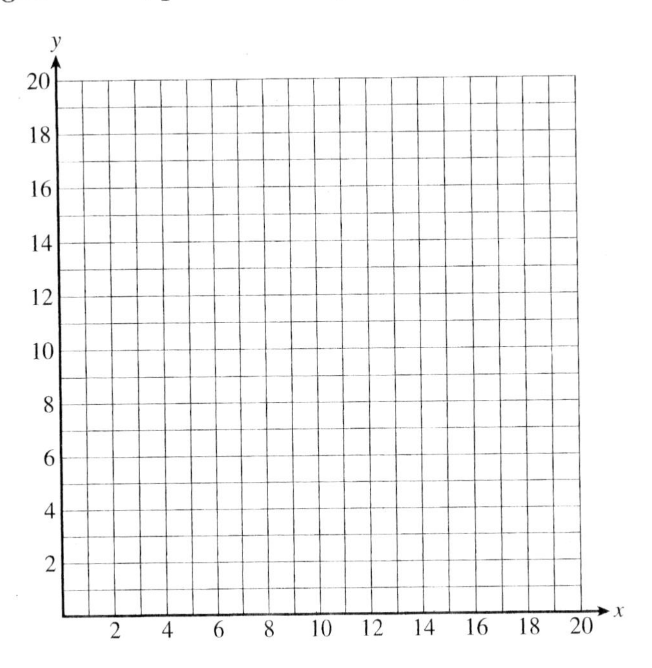

For Exercises 11–14, identify the indicated point. Approximate it if the point is not exact.

11. A **12.** B **13.** C **14.** D

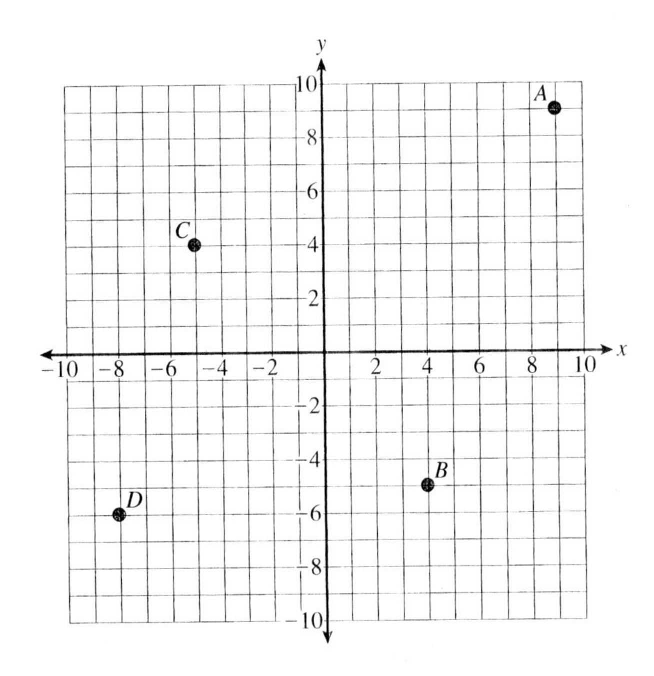

Assume that the box below represents a computer screen. Use it to complete Exercises 15–17.

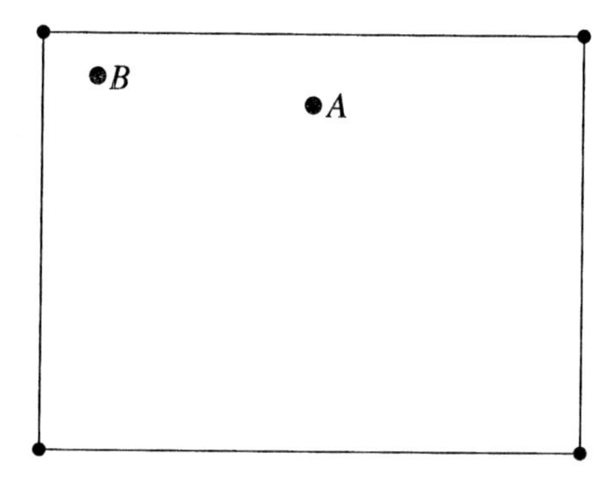

15. With standard resolution, there are 1024 pixels in the horizontal direction and 768 in the vertical direction. Use that information to approximate the coordinates of the pixel labeled A.

16. Again, assume there are 1024 pixels in the horizontal direction and 768 in the vertical direction. Use that information to approximate the coordinates of the pixel labeled B.

17. Assume that the resolution has 1280 pixels in the horizontal direction and 1024 in the vertical direction. What would be the coordinates of the pixel labeled *A*?

18. What output would result from the following loop?

```
For x = 1 to 7
  Print x
Next x
```

19. What output would result from the following loop?

```
For x = 0 to 250 step 25
  Print x
Next x
```

20. What output would result from the following loop?

```
For x = 8 to 14
  Print x - 5
Next x
```

21. What output would result from the following loop?

```
For x = 6 to 24 step 3
  Print 5x + 1
Next x
```

22. Assume that the following code is applied to a computer monitor with resolution 1024 by 768. Describe the output.

```
For x = 1 to 768
  Plot Pixel (512, x)
Next x
```

23. Assume that the following code is applied to a computer monitor with resolution 1024 by 768. Describe the output.

```
For x = 1 to 1024
  Plot Pixel (1025 - x, 384)
  ClearScreen
Next x
```

24. Assume that the following code is applied to a computer monitor with resolution 1280 by 1024. Describe the output.

```
For x = 512 to 1024 step 16
  y = 2x - 1024
  Plot Pixel (x, y)
Next x
```

Assume that we are working with a 24-bit true color scheme, in which each pixel is assigned an RGB color with values for red, green, and blue, that each range from 0 to 255. The command for a solid blue pixel would be `RGBColor = "0, 0, 255"` (no red, no green, and full concentration of blue).

25. Assume that the following code is applied to a computer monitor with resolution 1024 by 768. Describe the output.

```
RGBColor = "255, 0, 255"
For x = 1 to 768
  Plot Pixel (512, x)
Next x
```

26. Assume that the following code is applied to a computer monitor with resolution 1024 by 768. Describe the output. How would the output differ if a ClearScreen command were added after the pixel was plotted?

```
For x = 1 to 255
  RGBColor = "0, 255 - x, x "
  Plot Pixel (x, 384)
Next x
```

CHAPTER 6 FINAL EXAM

Name: ______________________________

1. Evaluate each of the following. Write each answer in decimal form. If the operation is not defined, so state.

a. $2^3 \cdot 2^{-3}$ b. $2^3 + 2^{-3}$ c. $\frac{12}{0}$ d. $\sqrt{-9}$

e. $-\sqrt{9}$ f. $\sqrt{8}$ g. $\frac{2.3 \times 10^{-15}}{4.81 \times 10^{-20}}$ h. $\frac{4^8}{2^{13}}$

2. The following numbers are in 8-bit, two's complement notation. Give the decimal equivalent for each.

a. 00001111_{2^*} b. 10001111_{2^*} c. 11001100_{2^*}

3. In the computer industry, 1000 bytes is approximated with one kilobyte, which is 2^{10} bytes. What is the absolute and relative error associated with this approximation?

4. Complete the following table so that the values in each row are all equivalent.

BINARY	DECIMAL	OCTAL	HEXADECIMAL
111101100111_2			
	599		
		101.01_8	
			$3.F2_{16}$

5. Complete the following arithmetic problems:

a.
$$\begin{array}{r} E5B_{16} \\ +\ B0A_{16} \\ \hline \end{array}$$

b.
$$\begin{array}{r} EBA_{16} \\ -\ 5C_{16} \\ \hline \end{array}$$

c.
$$\begin{array}{r} C3F_{16} \\ +\ 101_{16} \\ \hline \end{array}$$

d.
$$\begin{array}{r} 9A_{16} \\ -\ 13_{16} \\ \hline \end{array}$$

e.
$$\begin{array}{r} 732_{8} \\ +\ 317_{8} \\ \hline \end{array}$$

f.
$$\begin{array}{r} 21 \\ -\ 23 \\ \hline \end{array}$$

6. Let $U = \{0, 1, 2, 3, 4, 5, 6, 7, 8, 9\}$
$A = \{2, 5, 6, 7, 8\}$
$B = \{1,6,7,9\}$
Find the following.

a. $A \cup B$

b. A'

c. $(A \cup B)'$

d. $|B|$

e. $|A'|$

7. The following bytes are coded for even parity. Cross out all bytes that have visible errors.

a. 01010101 **b.** 10110110 c. 11111100

d. 11111111 **e.** 11111110 **f.** 00001000

8. Construct the Karnaugh map to reduce the Boolean expression in the following truth table, write down this reduced expression, then find a gated circuit for the reduced expression.

A	B	C	D	f(A, B, C, D)
T	T	T	T	T
T	T	T	F	F
T	T	F	T	F
T	T	F	F	T
T	F	T	T	T
T	F	T	F	F
T	F	F	T	F
T	F	F	F	T
F	T	T	T	T
F	T	T	F	F
F	T	F	T	T
F	T	F	F	T
F	F	T	T	T
F	F	T	F	F
F	F	F	T	F
F	F	F	F	T

9. Use a Karnaugh map to simplify the following expression:

$ABC'D \vee A'BC'D \vee AB'C'D$

10. Suppose you are designing a Web page and you are using the standard *hexadeximal* RGB display that allows for more than 16 million colors.

a. Describe the color given by the *hexadecimal* RGB: 80FF00

b. Assuming that the red is not used, how many different colors can you obtain by only changing the amount of green and/or blue (i.e., just by changing the last four characters in the RGB)?

11. Complete the following truth table.

P	Q	$\sim P$	$\sim Q$	$\sim Q \wedge \sim P$	$Q \vee P$	$(\sim Q \wedge \sim P) \wedge (Q \vee P)$

12. What is one-half of 2^{320}?

13. Shade the following region on the Venn diagram:

$ABC' \vee A'BC \vee AB'C$

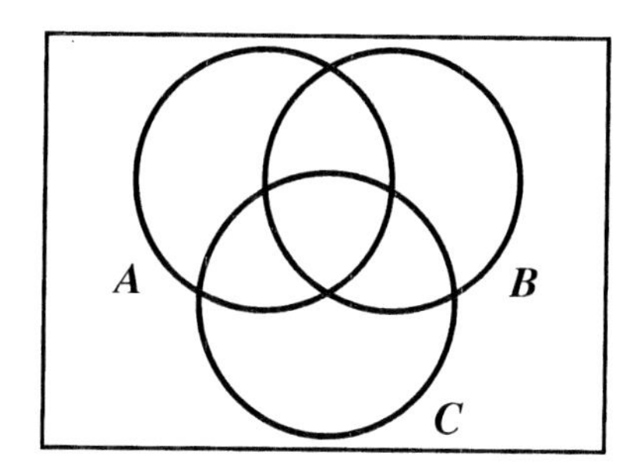

14. Write the Boolean expression (in dnf) that is represented by the following gated circuit.

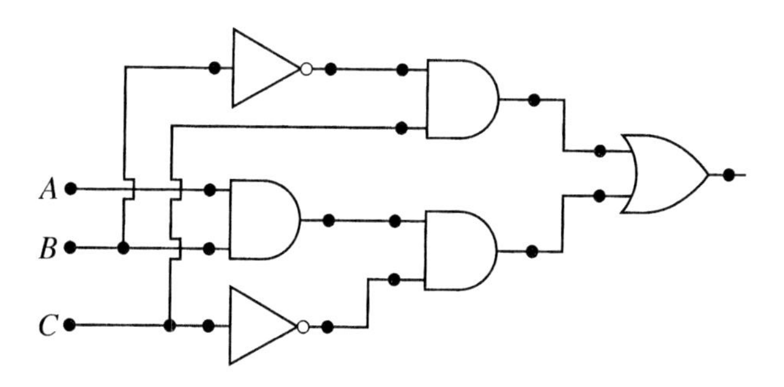

15. Find the following:

(a) The ASCII bit pattern for K

(b) The *character* represented by the ASCII bit pattern 0 1 1 0 1 1 0 1

Describe the output for each of the following. Assume that all commands are valid. Assume screen dimensions of 1024 by 768 pixels.

16.
```
For x = 1 to 5
    Print x
Next x
```

17.
```
For x = 1 to 5
   y = 2x - 5
   Print x,y
Next x
```

18.
```
For x = 2 to 1024 step 2
   y = 768 - 0.5x
   Plot Pixel (x, y)
Next x
```

19.
```
For x = 4 to 1024 step 4
   y = 1/4x + 100
   Plot Pixel (x, y)
   ClearScreen
Next x
```

For Problems 20 and 21, assume the command Dcolor sets a decimal color triple.

20.
```
For x = 0 to 255
   Dcolor = "x, x, x"
   Plot Pixel (x, 300)
Next x
```

21.
```
For x = 0 to 255
   Dcolor = "x, 255, 255 - x"
   Plot Pixel (x + 1, 255 - x)
Next x
```

Selected Answers

Chapter 1

Section 1.1 Odd-Numbered Exercise Answers

1. 9 tens 8 ones **3.** 12 tens 1 one **5.** 16 tens 1 one **7.** -10 **9.** 4 **11.** -34 **13.** 27 **15.** -60 **17.** 14 **19.** 1 **21.** -33

Section 1.2 Odd-Numbered Exercise Answers

1. $a \cdot a \cdot a \cdot a \cdot a \cdot a \cdot a \cdot a \cdot a = a^9$ **3.** $2 \cdot 2 \cdot 2 \cdot 2 = 2^4$ **5.** $2 \cdot 2 \cdot 2 \cdot b \cdot b \cdot b \cdot b \cdot b \cdot b = 2^3b^6$

7. $x^4 \cdot x^3 = x^7$ **9.** $2^5 \cdot 2^7 = 2^{12} = 4096$ **11.** $2^6 \cdot 2^{-1} = 2^5 = 32$ **13.** $\frac{x^8}{x^2} = x^6$

15. $\frac{2^{10}}{2^3} = 2^7 = 128$ **17.** $\frac{x^{16}}{x^{-5}} = x^{21}$ **19.** $x^0 = 1$ **21.** $6x^0 = 6$

23. $(-6x)^0 = 1$ **25.** $x^{-4} = \frac{1}{x^4}$ **27.** $2^{-2} = \frac{1}{4}$ **29.** $2^6 \cdot 2^{-10} = \frac{1}{2^4} = \frac{1}{16}$ **31.** $\frac{x^4}{x^5} = \frac{1}{x}$

33. $\frac{x^{-6}}{x^5} = \frac{1}{x^{11}}$ **35.** $(x^3)^5 = x^{15}$ **37.** $(2^2)^2 = 2^4 = 16$ **39.** $(4x)^3 = 64x^3$

41. $(2x^2y)^0 \cdot (2x)^4 = 2^4x^4 = 16x^4$ **43.** $2^9, 2^{99}$

Section 1.3 Odd-Numbered Exercise Answers

1. $2 \cdot 2^{-1} = 1$ **3.** $-5^2 = -25$ **5.** $\sqrt{16} = 4$ **7.** $\sqrt[3]{16^2} \approx 6.3496$

9. $\frac{6.022 \times 1.6}{1.41} \approx 6.8335$ **11.** $2^{2^2} = 16$ **13.** $\pi \approx 3.141593$ **15.** $\frac{1 + \sqrt{5}}{2} \approx 1.618034$

17. $200 - (100 \div 4 \times (5 - 1))^3 \times 2 - 9 = -1{,}999{,}809$

Section 1.4 Odd-Numbered Exercise Answers

1. $120{,}000{,}000 = 1.2 \times 10^8$ **3.** $0.0000000000024 = 2.4 \times 10^{-12}$

5. $0.0000037 \text{ m} = 3.7 \times 10^{-6}\text{m}$ **7.** $5.27 \times 10^9 = 5{,}270{,}000{,}000$

9. $2.217 \times 10^{-6} = 0.000002217$ **11.** $\$5.87 \times 10^{10} = \$58{,}700{,}000{,}000$

13. $\frac{3.14 \times 10^4}{2.72 \times 10^2} \approx 1.154411765 \times 10^2$ **15.** $\frac{4.36 \times 10^3 - 1.2 \times 10^{11}}{3.01 \times 10^7} \approx -3.986710819 \times 10^3$

17. $\frac{6.022 \times 10^{23}}{4.174 \times 10^{-17}} \approx 1.442740776 \times 10^{40}$ **19.** 1.246872×10^9 **21.** 1.0×10^{-11}

23. 1.457664×10^6 **25.** \$66,114,000 **27.** 39,629,000,000,000

Section 1.5 Odd-Numbered Exercise Answers

1. 2 **3.** 2 **5.** 2 **7.** $5.145 \text{ ft} \times 7.92 \text{ ft} \approx 40.7 \text{ ft}^2$ **9.** AE ≈ 0.00143, RE $\approx .33\%$ **11.** $a \approx 0.0091$, $r \approx 10\%$ **13.** mean is 55 min. **15.** $a \approx 6$ min., $r \approx 10\%$ **17. (a)** 25 **(b)** 25.5 **19.** Answers will vary.

Section 1.6 Odd-Numbered Exercise Answers

1. $\frac{19 \text{ miles}}{\text{gallon}} \times 18.5 \text{ gallons} = 351.5$ miles or 350 miles (two significant digits)

3. $\frac{24 \text{ hours}}{\text{day}} \times \frac{365 \text{ days}}{\text{year}} = \frac{8760 \text{ hours}}{\text{year}}$,

$\frac{60 \text{ min}}{\text{hour}} \times \frac{60 \text{ sec}}{\text{min}} \times \frac{8760 \text{ hours}}{\text{year}} = \frac{31{,}536{,}000 \text{ sec}}{\text{year}}$

5. $\frac{1.660 \text{ AUD}}{\text{USD}} \times \frac{1.119 \text{ CAD}}{\text{AUD}} \times \frac{71.108 \text{ JPY}}{\text{CAD}} = \frac{132.086 \text{ JPY}}{\text{USD}}$

7. $\frac{.62 \text{ miles}}{\text{km}} \times \frac{36 \text{ inches}}{\text{yard}} \times \frac{1760 \text{ yards}}{\text{mile}} = \frac{39283.2 \text{ inches}}{\text{km}}$

9. $\frac{1 \text{ page}}{260 \text{ words}} \times 10{,}000 \text{ words} \approx 39$ pages 11. $260 \times 115\% = 299$, so

$\frac{1 \text{ page}}{299 \text{ words}} \times 10{,}000 \text{ words} \approx 34$ pages 12. $12 \text{ pages} \times \frac{245 \text{ Arial words}}{\text{page}} = 2940$ words

$12 \text{ pages} \times \frac{260 \text{ Times New Roman words}}{\text{page}} = 3120$ words The difference is 180 words.

Chapter 1 Glossary Quiz

accuracy G absolute error H relative error I the mean J units K exponent A *base* B the *exponent*, or *power* C scientific D significant digits F mantissa E

Chapter 1 Review Exercises

1. 0 hundreds; 4 tens; 3 ones 2. 0 hundreds; 9 tens; 1 ones 3. 1 hundreds; 9 tens; 1 ones 4. 7 hundreds; 8 tens; 8 ones 5. 13 hundreds; 2 tens; 0 ones 6. 40 hundreds; 0 tens; 7 ones 7. -42 8. -16 9. -72 10. -11 11. -26 12. a^8 13. 16^5 14. $2^3a^5 = 8a^5$ 15. a^{17} 16. b^5 17. 8192 18. 256 19. x^8 20. 64 21. 1 22. 2 23. 1 24. $\frac{1}{x^{12}}$ 25. .04 26. -32 27. $\frac{1}{x^5}$ 28. $\frac{1}{x^2}$ 29. x^8 30. $\frac{1}{8} = .125$ 31. $\frac{1}{256} = .00390625$ 32. 16 33. 1 34. $16x^5$ 35. 256 36. $\frac{53}{42} = 1\frac{11}{42}$ 37. -4 38. 4 39. ≈ 1.4142 40. ≈ 1.2599 41. ≈ 2.1629 42. 12.67 43. ≈ 0.6180 44. ≈ 1.5708 45. 5.52×10^{-6} 46. 1.2305×10^{-13} 47. 2.0327×10^7 km^2 48. 9.6002×10^4 miles^2 49. 8,780,000,000,000 50. .00000562 51. .00000001125 52. 4,447,100 people 53. 1.220736×10^0 54. $\approx 2.304 \times 10^{20}$ 55. mean: 4.6 min; absolute error: .65 min; relative error: 12.38% 56. minimum of 5 57. 3 58. 6 59. absolute error: 2 mL; relative error: 1.35% 60. absolute error: .36 cm; relative error: 0.89% 61. 673.2 miles 62. 525,600 minutes 63. $\frac{7.467 \text{ kroner}}{\$1 \text{ U.S.}}$ 64. $\frac{12{,}318{,}750 \text{ minutes of sleep}}{1 \text{ lifetime}}$ 65. 8.2677 inches by 11.6929 inches 66. 47 pages 67. 42 pages

Chapter 1 Self-Test

1. (a) 10 hundreds; 7 tens; 3 ones **(b)** a^6b^8 **2. (a)** $4y^{21}$ **(b)** x^5 **(c)** $2^{11} = 2048$ **3. (a)** $\frac{1}{x^{17}}$ **(b)** $\frac{1}{32} = .03125$ **(c)** $\frac{1}{y^{11}}$ **(d)** $\frac{1}{x^{10}}$ **(e)** 5 **(f)** $1000x^5$

4. (a) $\frac{49}{60}$ **(b)** ≈ 1.7321 **(c)** ≈ 1.0472 **5. (a)** 2.015×10^{12} **(b)** 1.36×10^{-6}

6. (a) .0000000512 **(b)** 2,715,000,000 **7. (a)** 4.5348×10^7 **(b)** 8.3145×10^{-28}

8. mean: 53.6; absolute error: 5.6 secs; relative error: $11\frac{2}{3}\%$ **9. (a)** 4 **(b)** minimum of 7 **10. (a)** 571.2 miles **(b)** 25,500 sunsets

Chapter 2

Section 2.1 Odd-Numbered Exercise Answers

1. 5 **3.** 13 **5.** 109 **7.** 170 **9.** 3855 **11.** 2049 **13.** 0111_2 **15.** 1100_2 **17.** 1111_2 **19.** 111001_2 **21.** 1101011_2 **23.** 10000000000_2 **25.** 867-5309 **27.** 1-800-876-5353 **29.** 9-10-4 **31.** $0010_2 = 2$ **33.** $1011_2 = 11$ **35.** $10101100_2 = 172$

Section 2.2 Odd-Numbered Exercise Answers

1. 0110_2 **3.** 1011_2 **5.** 10100_2 **7.** 100010_2 **9.** 100111100_2 **11.** 100011011_2

13. +

1	0	1	1	0	1	1	1

15. +

1	0	1	0	0	0	0	0

17. 11010_2 **19.** 101010_2 **21.** 1001101_2

Section 2.3 Odd-Numbered Exercise Answers

1. 0100_2 **3.** 1001_2 **5.** 0011_2 **7.** 01111011_2 **9.** 5 times **11.** 101 **13.** 100001 **15.** 111000_2 R 010

Section 2.4 Odd-Numbered Exercise Answers

1. 0010_{2*} **3.** 1010_{2*} **5.** 5 **7.** −5 **9.** 00000110_{2*} **11.** 11000010_{2*} **13.** 10010100_{2*} **15.** 29 **17.** −70 **19.** −1 **21.** 0111_{2*} **23.** $(1)0000_{2*}$ **25.** $(1)0100_{2*}$ error **27.** $(1)00111000_{2*}$ error **29.** 11110001_{2*} **31.** 256 different numbers; Minimum: −128; Maximum: 127

Section 2.5 Odd-Numbered Exercise Answers

1. 0.5 **3.** 0.375 **5.** 0.421875 **7.** 11.625 **9.** 15.390625 **11.** 0.625 **13.** 0.9921875 **15.** 0.6640625 **17.** 0.00001_2 **19.** 0.1111_2 **21.** 0.01010101_2 **23.** 1100.101_2 **25.** 0.001_2 **27.** $0.\overline{10}_2$ **29.** Fractions repeat in base 10: Exercise 27; Fractions repeat in base 2: 27, 28; You can tell by rounding **31. (a)** AE: 0.0125; RE: 6.25% **(b)** AE: 0.00078125; RE: 0.4% **(c)** AE: 0.0046875; RE: 2.34%; b. is closest to the actual value. **33.** $\frac{1}{8}, \frac{3}{16}, \frac{43}{256}$ **35.** 1.375 **37.** 5.25

Section 2.6 Odd-Numbered Exercise Answers

1. 1024 **3.** 5,242,880 **5.** 65,536 **7.** 8192 **9.** 167,772,160 **11.** 5,242,880 **13.** 1000 **15.** 0.01 **17.** 0.000001 **19.** 20 documents **21.** $2.56\ \frac{\text{Kb}}{\text{sec}}$ **23.** 576,000 bits **25.** 3,145,728 bytes **27.** 144 MB **29.** 703.125 MB **31.** AE: 99,511,628,000; RE: 9%

Chapter 2 Glossary Quiz

pixel L monochrome M binary code A decimal representation B gigabyte K Mb O digit D kilobyte I megabyte J bit E byte F pico- S overflow error H nano- R milli- P MB N binary representation C micro- Q two's complement notation G

Chapter 2 Review Exercises

1. 4 **2.** 15 **3.** 177 **4.** 228 **5.** 2879 **6.** 1001_2 **7.** 1110_2 **8.** 11101_2 **9.** 10101100_2 **10.** 111111101_2 **11.** 503−657−6958 **12.** 781−944−3700 **13.** 229 **14.** 44 **15.** 102 **16.** 111_2 **17.** 10001_2 **18.** 11100_2 **19.** 11101_2 **20.** 101101000_2 **21.** 1010_2 **22.** 100111_2 **23.** 110010_2 **24.** 1000_2 **25.** 111_2 **26.** 10001100_2 **27.** 10_2 **28.** 5 times **29.** 101_2 **30.** 1011_2 **31.** 101_2 **32.** 101000_2 R 11 **33.** 10010001_2 R 1 **34.** 00000011_{2*} **35.** 11111101_{2*} **36.** 00011101_{2*} **37.** 10110101_{2*} **38.** 01100110_{2*} **39.** 10001101_{2*} **40.** 2 **41.** −3 **42.** −111 **43.** 95 **44.** −75 **45.** −35 **46.** $(1)1000_{2*}$ **47.** $(1)0010_{2*}$ **48.** 1101_{2*} error **49.** 0.3125 **50.** 0.8125 **51.** 0.8359375 **52.** 14.6875 **53.** .875 **54.** .28125 **55.** .6640625 **56.** .9296875 **57.** 0.001_2 **58.** $.01011_2$ **59.** 1101.010011_2 **60.** 1001.0100011_2 **61.** 0.0001_2 **62.** 0.111_2 **63.** $0.\overline{1100}_2$ **64.** $\frac{5}{8}, \frac{10}{16}, \frac{154}{256}$ **65.** 2^{30} **66.** 12×2^{20} **67.** 640×2^{10} **68.** 2^{23} **69.** 120×2^{33} **70.** 2^{43} **71.** 7.4 or 7 **72.** 2,359,296

Chapter 2 Self-Test

1. (a) 13 **(b)** 191 **(c)** 161 **2. (a)** 10_2 **(b)** 11011_2 **(c)** 1110001_2 **3. (a)** 10011_2 **(b)** 100010011_2 **4. (a)** 1110_2 **(b)** 100001_2 **(c)** 10000010_2 **5. (a)** 1000_2 **(b)** 111_2 **6. (a)** 1101_2 **(b)** 11100 1 R 10 **(c)** 110110 R 1_2 **7. (a)** 01100010_{2*} **(b)** 111100110_{2*} **(c)** 10001100_{2*} **8. (a)** 14 **(b)** −3 **(c)** −239 **9. (a)** $(1)0010_{2*}$ **(b)** 10011_{2*} error **(c)** $(1)1011_{2*}$ **10. (a)** .78125 **(b)** 6.5625 **(c)** .1328125 **(d)** .8515625 **11. (a)** 0.0001_2 **(b)** 101.01011_2 **(c)** 1011.01011_2 **12. (a)** 0.00001_2 **(b)** 0.000011_2 **(c)** 0.0111001_2 **13. (a)** 2^{20} **(b)** 5120 **14. (a)** 40×2^{33} **(b)** 200×2^{13} **15.** 6

Chapter 1–2 Cumulative Review Exercises

1. (a) 25.04 **(b)** 1 **(c)** 1×10^{-600} **(d)** 8.5094 **(e)** undefined **(f)** $5i$ or undefined **2.** a, d, c, b **3.** a, c, d, b **4. (a)** 10110110_2 **(b)** 101010_2 **5.** AE: 3.28; RE: 2.9% **6. (a)** 2,869,776,000 **(b)** $\frac{\$1 \text{ U.S.}}{132 \text{ yen}}$ **7.** 52,428,800 **8. (a)** 1100_2 **(b)** 101101_2 **9.** 2^{999} **10. (a)** .671875 **(b)** 29.390625 **11. (a)** 0.00011_2 **(b)** 1111100.0111_2 **(c)** 10000101010_2 **12.** (−128,127) **13. (a)** 2.48136×10^{13} **(b)** 2×10^{-8}

14. (a) 1509949.44 **(b)** 2^{34} **15. (a)** 00001001_{2*} **(b)** overflow **16. (a)** 0.000011_2
(b) $0.11\overline{01}_2$

Chapter 3

Section 3.1 Odd-Numbered Exercises

1. 7 **3.** 9 **5.** 32 **7.** 49 **9.** 174 **11.** 566 **13.** 5_8 **15.** 12_8 **17.** 40_8 **19.** 101_8 **21.** 200_8 **23.** 3605_8 **25.** 52_8 **27.** 110_8 **29.** 252_8 **31.** 7452_8 **33.** 7265_8 **35.** 011001_2 **37.** 001010110_2 **39.** 011000111_2 **41.** 001000101111_2 **43.** an octal digit ending with 1, 3, 5, 7 **45.** 0, 1, 2, 3, 4, 5, 6, 7, 8

Section 3.2 Odd-Numbered Exercises

1. 5 **3.** 19 **5.** 161 **7.** 112 **9.** 3346 **11.** 43962 **13.** 1_{16} **15.** C_{16} **17.** $2F_{16}$ **19.** 48_{16} **21.** 79_{16} **23.** $6F7_{16}$ **25.** 48_{16} **27.** $2A_{16}$ **29.** EF_{16} **31.** $EB5_{16}$ **33.** $F2A_{16}$ **35.** 00110111_2 **37.** 11111111_2 **39.** 110000101001_2 **41.** 0011011010011100_2
43. 0.5_8, $0.A_{16}$ **45.** 0.25_8, 0.54_{16} **47.** 0.6551_8, $0.D69_{16}$

Section 3.3 Odd-Numbered Exercises

1. 9_{16} **3.** 11_{16} **5.** 101_{16} **7.** $DB2_{16}$ **9.** $1A9E_{16}$ **11.** $A780_{16}$ **13.** 6_{16} **15.** B_{16} **17.** $C3_{16}$ **19.** 542_{16} **21.** 450_{16} **23.** $2BF_{16}$ **25.** 18_{16} **27.** BC_{16} **29.** $6ACD_{16}$ **31.** 6_8 **33.** 20_8 **35.** 117_8 **37.** 5 **39.** 20 **41.** 218 **43.** H_{20} **45.** $9C_{20}$ **47.** $28B_{20}$ **49.** 19_{20} **51.** HC_{20} **53.** $1F4_{20}$ **55.** 20; total of fingers and toes

Section 3.4 Odd-Numbered Exercise Answers

1. 88 **3.** 71 **5.** 57 **7.** 01011001_2 **9.** 01000010_2 **11.** 00110000_2 **13.** $4F_{16}$ **15.** $4B_{16}$ **17.** 71_{16} **19.** R 2 D 2 **21.** N e o **23.** S p o c k **25.** 4096 **27.** ERROR **29.** OK **31.** OK **33. (a)** OK **(b)** OK **(c)** ERROR **35.** The "1" in column B row 3. **37.** Make all 3-blocks the same number as the majority in the block; you would be certain as long as each 3-block had at most 1 error.

Chapter 3 Glossary Quiz

hexadecimal number D ASCII E octal number A algorithm C parity bit F
tribble B

Chapter 3 Review Exercises

1. 6 **2.** 15 **3.** 53 **4.** 88 **5.** 161 **6.** 835 **7.** 3_8 **8.** 20_8 **9.** 61_2 **10.** 146_8 **11.** 545_8 **12.** 2724_8 **13.** 16_8 **14.** 75_8 **15.** 304_8 **16.** 6237_8 **17.** 001101_2 **18.** 111100_2 **19.** 010011110_2 **20.** 001000100011_2 **21.** 6 **22.** 12 **23.** 178 **24.** 2781 **25.** 3054 **26.** 61817 **27.** 3_{16} **28.** F_{16} **29.** $2B_{16}$ **30.** 61_{16} **31.** 260_{16} **32.** 714_{16} **33.** 9_{16} **34.** 33_{16} **35.** 99_{16} **36.** 81_{16} **37.** $2FB_{16}$ **38.** $AD7_{16}$ **39.** 01000001_2 **40.** 11111000_2 **41.** 001000111110_2 **42.** 110010101011_2 **43.** 0101001010111001_2 **44.** 1111000001111101_2 **45.** 0.6_8, $0.C_{16}$ **46.** 0.3_8, 0.6_{16} **47.** 0.506_8, $0.A3_{16}$

48. 8_{16} 49. $1E_{16}$ 50. 121_{16} 51. $77F_{16}$ 52. 19090_{16} 53. $EE6C_{16}$ 54. 5_{16} 55. 37_{16} 56. $EC6_{16}$ 57. ABE_{16} 58. $C99F_{16}$ 59. 41_{16} 60. 1230_{16} 61. $C27C_{16}$ 62. 54 63. 105 64. 80 65. 01100101_2 66. 00110111_2 67. 01010001_2 68. Y 69. 4 70. m 71. 49_{16} 72. 39_{16} 73. 72_{16} 74. S t i n g 75. B l i n k 1 8 2 76. E n i g m a 77. P . D i d d y 78. ERROR 79. OK 80. ERROR

Chapter 3 Self-Test

1. (a) 11 (b) 38 (c) 175 2. (a) 35_8 (b) 116_8 (c) 245_8 3. (a) 13_8 (b) 35_8 (c) 276_8 4. (a) 61 (b) 180 (c) 2586 5. (a) ERROR (b) ERROR (c) OK 6. (a) $3E_{16}$ (b) 69_{16} (c) ED_{16} 7. (a) $B3_{16}$ (b) 28_{16} (c) $0.C_{16}$ (d) 83.3_{16} 8. (a) 10100100_2 (b) 101100001110_2 (c) 111111000010_2 9. (a) 84 (b) 52 10. (a) 95_{16} (b) $10B_{16}$ (c) 15_{16} (d) 44_{16} (e) $32A_{16}$ (f) $5FE0_{16}$ 11. (a) 00111000_2 (b) 01100100_2 12. (a) Phoenix (b) Seattle (c) Detroit

Chapter 4

Section 4.1 Odd-Numbered Exercise Answers

1. (a) {Alaska, Arkansas, Alabama, Arizona} (b) {Monday, Tuesday, Wednesday, Thursday, Friday, Saturday, Sunday} (c) {10, 20, 30, ...} (d) {2, 3, 5, 7, 11, 13} 3. *a* and *c* are both subsets of *F* 5. *b* and *c* are both subsets of *N* 7. (a) $U = \{a, e, i, o, u\}$ (b) Ø (c) {0, 1, 2, 3, 4, 5, 6, 7} (d) Ø 9. (a) $|A| = 4$ (b) $|B| = 5$ (c) $|C| = 10$ (d) $|D|$ is infinite (e) $|E| = 1$ (f) $|F| = 50$

Section 4.2 Odd-Numbered Exercise Answers

1. (a) $A \cup B = \{1, 2, 4, 5, 6, 7, 8, 9\}$ (b) $A \cap B = \{2, 5, 8\}$ (c) $A' = \{0, 1, 3, 6\}$ (d) $B' = \{0, 3, 4, 7, 9\}$ (e) $|A \cup B| = 8$ (f) $|A \cap B| = 3$ (g) $|A| = 6$ h. $|A'| = 4$ i. $|U| = 10$ 3. (a) $A' = \{1, 3, 4, 6, 9\}$ (b) $B' = \{0, 3, 4, 5, 8, 9\}$ (c) $A' \cap B' = \{3, 4, 9\}$ (d) $A \cup B = \{0, 1, 2, 5, 6, 7, 8\}$ (e) $(A \cup B)' = \{3, 4, 9\}$ (f) they are equal 5. (a) $D \cap P$ = {*Raging Bull, Casino, Goodfellas*} (b) $P \cup G$ = {*Raging Bull, Casino, Goodfellas, Home Alone, Lethal Weapon 2, Braveheart, Road Warrior, Hamlet*} (c) $D \cap G = \varnothing$ (d) D' = {*Home Alone, Lethal Weapon 2, Braveheart, Road Warrior, Hamlet*} (e) P' = {*Taxi Driver, Cape Fear, Braveheart, Road Warrior, Hamlet*} (f) $P \cap G$ = {*Lethal Weapon 2*} 7. $|B| = |A \cup B| - |A| + |A \cap B| = 15$ 9. $|A \cap B| = |A| + |B| - |A \cup B| = 14$ 11. (a) A computer that would have at least ONE of the following: a Pentium IV processor, 256 MB of RAM, or a DVD-ROM drive (b) $|P \cup R \cup D| = |P| + |R| + |D| - |P \cap R| - |P \cap D| - |D \cap R| + |P \cap R \cap D| = 131$

Section 4.3 Odd-Numbered Exercise Answers

1. 3. 5. $A' = \{2, 3, 4, 5, 8, 9\}$

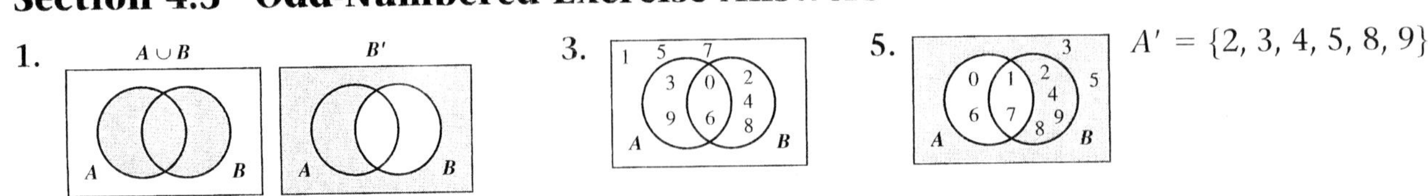

7. 1, 2, 4, 8, 0, 5, 6, 3, 9, 7, A, B

9. 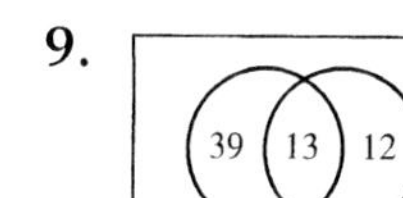

$|A \cup B| = 64$

11. 22, 15, 28, A, B — There are 15 students in both classes.

13. $A \cup C$ — A, B, C

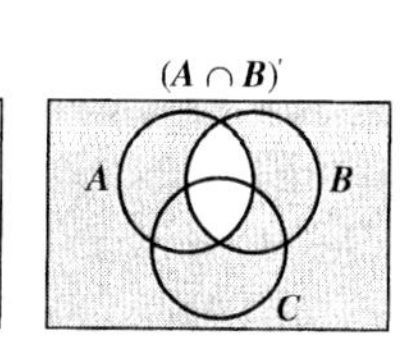

15. A, 12, 5, 26, B, 3, 21, 8, 9, C — $|A \cap B \cap C| = 3$

17. (a) 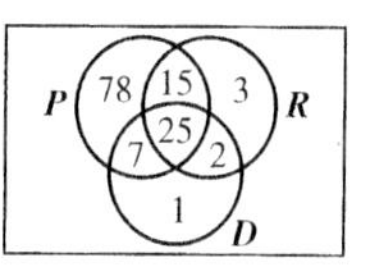

(b) 1 had only a DVD-Writer **(c)** 78 had only a Pentium IV processor

Section 4.4 Odd-Numbered Exercise Answers

1. b, c, e, and g are all propositions **3.** Answers will vary, but typical answers may include: **(a)** $x - 7 \neq 2$ **(b)** 5 is less than or equal to 3 **(c)** It is not 97°F outside **(d)** $12 \leq -4$ **(e)** Math is not fun. **(f)** $9 \geq 2$ **5.**

p	q	~p	~q	~(~p)	~(~q)	~(~(~p))	~(~(~q))
1	1	0	0	1	1	0	0
1	0	0	1	1	0	0	1
0	1	1	0	0	1	1	0
0	0	1	1	0	0	1	1

7. Answers will vary, but typical answers may include: **(a)** Pat always has none of the answers. **(b)** John has no friends.

Section 4.5 Odd-Numbered Exercise Answers

1. (a) Austin can have coffee with cream and sugar. **(b)** Austin can have coffee with cream or sugar. **3.** Think about these situations and how the "AND" and "OR" operators are being treated in each case **5.** Austin can have coffee with sugar or without cream.

7.

p	q	~q	$\sim q \lor p$
1	1	0	1
1	0	1	1
0	1	0	0
0	0	1	1

9.

p	q	~q	$\sim q \land q$
1	1	0	0
1	0	1	0
0	1	0	0
0	0	1	0

11.

A	B	~A	f(A, B)
1	1	0	0
1	0	0	0
0	1	1	1
0	0	1	0

13. (a) 0 **(b)** 0 **(c)** 0 **15.** They are negations of each other, thus demonstrating De Morgan's law. **17.** Abby is playing softball or she is not playing soccer. **19.** Answers may vary. **21.** Answers may vary.

Chapter 4 Glossary Quiz

universal set D element B set A subset C union G empty set E
cardinality F complement J Venn diagram K intersection H $(A \cup B)' = A' \cap B'$ L
logic M unary operator I proposition N

Chapter 4 Review Exercises

1. {Canada, USA, Mexico} **2.** {Greg, Marsha, Jan, Peter, Bobby, Cindy} **3.** {red, orange, yellow, green, blue, indigo, violet} **4.** {11, 13} **5.** {w, i, n, d, o, s} **6.** {5, 10, 15, ...} **7. (a)** Y **(b)** N **(c)** N **(d)** Y **(e)** Y **8. (a)** Y **(b)** N **(c)** N **(d)** Y **(e)** Y **9.** {Ore, Nev, Az} **10.** {0, 1, 2, ..., 9, A, B, ..., F} **11.** {Gilligan, Skipper, the Professor, Mr. Howell, Mrs. Howell, Ginger, Mary Ann} **12.** {Capricorn, Pisces, ..., Sagitarius} **13.** {3, 5, 7, ..., 97, 99} **14.** 4 **15.** 100 **16.** 9 **17.** infinite **18.** 6 **19. (a)** {0 2 3 4 5 6 7 8 *A B C D E*} **(b)** {2} **(c)** {0, 6, *C*} **(d)** {0, 1, 4, 6, 8, 9, *A, C, E, F*} **(e)** {1, 5, 7, *B, D*} **(f)** 8 **(g)** 11 **(h)** 16 **20. (a)** {SIL, B} **(b)** {E, SIL, B, TTMR, S, P, SD, GWH, D, PH, P, A, CL, CA, BR} **(c)** {GWH, D, CA} **(d)** {E, TTMR, S, P, SD, BI, ST, ATPH, CUF, MFF} **(e)** ∅ **(f)** 11 **(g)** 0 **(h)** 20 **(i)** 2 **21.** 43 **22.** 5 **23.** **24.** **25.**

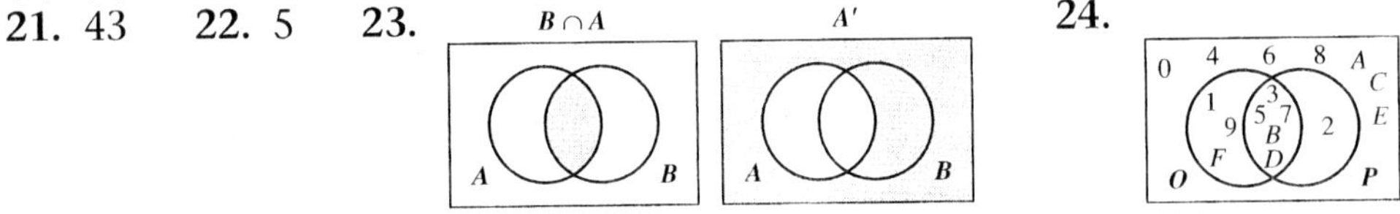

26. **27.** **28.**

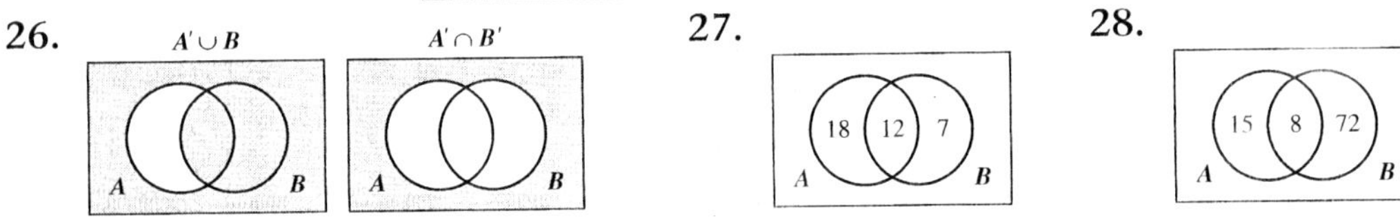

29. 8; **30.** **31.** 5

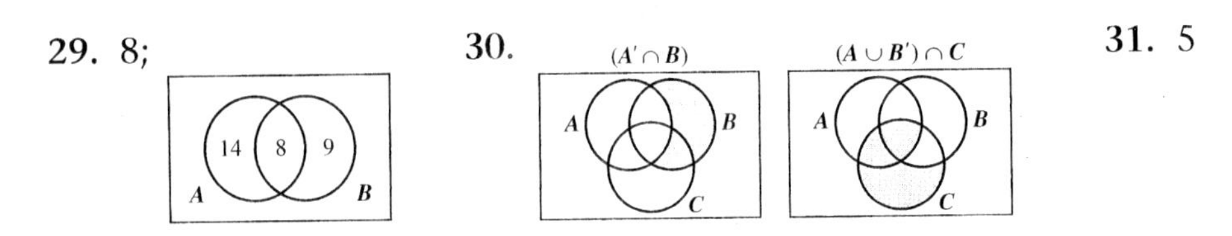

32. (a) **(b)** computer with Nero, 2 CD-drives, and WinMX **(c)** none **(d)** 8

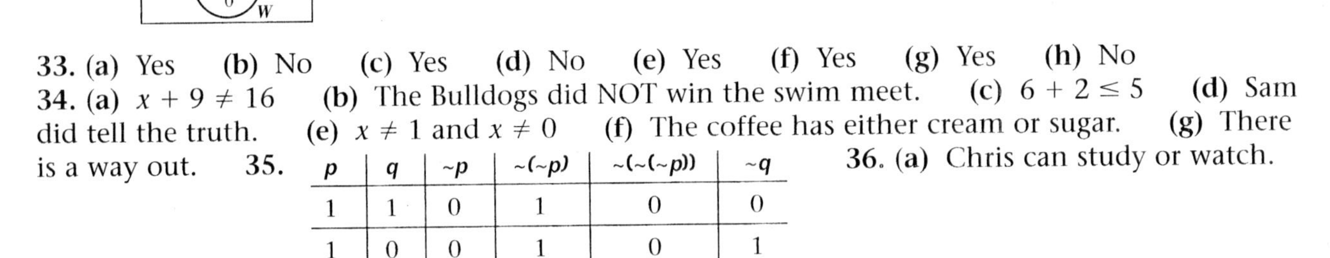

33. (a) Yes **(b)** No **(c)** Yes **(d)** No **(e)** Yes **(f)** Yes **(g)** Yes **(h)** No
34. (a) $x + 9 \neq 16$ **(b)** The Bulldogs did NOT win the swim meet. **(c)** $6 + 2 \leq 5$ **(d)** Sam did tell the truth. **(e)** $x \neq 1$ and $x \neq 0$ **(f)** The coffee has either cream or sugar. **(g)** There is a way out. **35.**

p	q	$\sim p$	$\sim(\sim p)$	$\sim(\sim(\sim p))$	$\sim q$
1	1	0	1	0	0
1	0	0	1	0	1
0	1	1	0	1	0
0	0	1	0	1	1

36. (a) Chris can study or watch. **(b)** Chris can study and watch. **(c)** Chris can study for his math test. **(d)** Chris can study and Chris can't watch football.

37. (a)

p	q	$\sim p$	$\sim p \vee q$
1	1	0	1
1	0	0	0
0	1	1	1
0	0	1	1

(b)

p	q	$\sim q$	$\sim q \vee q$
1	1	0	1
1	0	1	1
0	1	0	1
0	0	1	1

(c)

p	q	$\sim p$	$p \wedge \sim p$
1	1	0	0
1	0	0	0
0	1	1	0
0	0	1	0

38.

A	B	$\sim B$	$f(A, B)$
1	1	0	0
1	0	1	1
0	1	0	0
0	0	1	0

39. (a) 0 **(b)** 0 **(c)** 0

40. Allison is working OR not moving. **41.** many answers **42.** many answers

Chapter 4 Self-Test

1. (a) {Pacific, Atlantic, Indian, Southern, Arctic} **(b)** {23, 29, 31, 37} **(c)** {0, 1, ... 7} **(d)** {s, p, a, n, k, o, i, t} **(e)** {11, 22, 33, ...} **2. (a)** Yes **(b)** No **(c)** Yes **(d)** Yes **(e)** No **3. (a)** {0, 2, 3, 5, 6, 7, 9, b} **(b)** {3, 5, 7, b} **(c)** {0, 2, 4, 6, 8, a} **(d)** {0, 1, 3, 4, 5, 6, 7, 8, 9, a, b} **(e)** {0, 6} **(f)** 6 **(g)** 15 **(h)** 12

4.

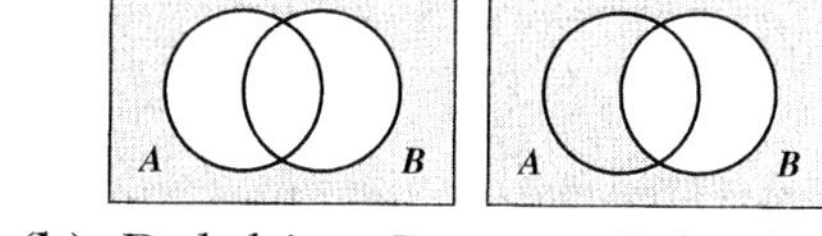

5. (a) Dolphins **(b)** Dolphins, Ravens, Colts, Bengals, Browns, Jags, Broncos, Bills **(c)** Browns, Bill, Broncos, Bengals, Dolphins, Pats, Jets **(d)** Bills, Chargers, Chiefs, Jets, Pats, Raiders, Steelers, Texans, Titans **(e)** 12 **(f)** 1 **6.** **7.** 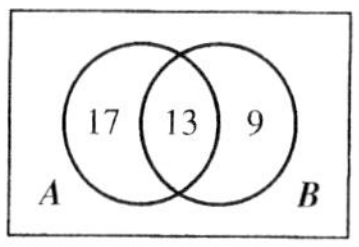 **8.** 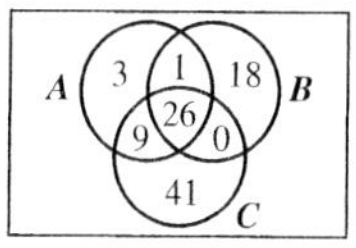 98

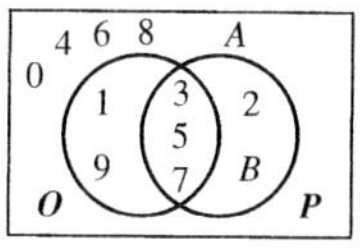

9. (a) $x + 4 \neq 12$ **(b)** yellow and blue, DO NOT make green **(c)** Stefan was NOT ... **(d)** $-3y \not< 15$ **10. (a)**

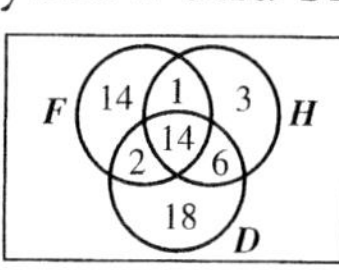

(b) either a flat screen, 100 FB, or DVD and RW drive **(c)** 14 **(d)** 18 **(e)** 6 **11. (a)** 1 **(b)** 1 **(c)** 1

A	B	$\sim A$	$f(A, B)$
1	1	0	1
1	0	0	0
0	1	1	1
0	0	1	1

12. many answers, many answers

13. (a) Yes **(b)** Yes **(c)** No **(d)** yes **(e)** No **14. (a)**

p	q	$\sim p$	$\sim p \wedge p$
1	1	0	0
1	0	0	0
0	1	1	0
0	0	1	0

(b)

p	q	$\sim q$	$q \vee \sim q$
1	1	0	1
1	0	1	1
0	1	0	1
0	0	1	1

(c)

p	q	$\sim p$	$\sim q$	$\sim p \wedge \sim q$
1	1	0	0	0
1	0	0	1	0
0	1	1	0	0
0	0	1	1	1

Chapters 1–4 Cumulative Review Exercises

1.

HEXADECIMAL	DECIMAL	BINARY
$AF7_{16}$	2807	101011110111
$35D_{16}$	861	1101011101
2A.94	42.578125	1010101.100101_2

2. (a) $18EA_{16}$ **(b)** $A1A_{16}$ **(c)** 8817_{16}

3. (a) 7207_8 **(b)** 467 **4. (a)** 00100110_{2^*} **(b)** 11110110_{2^*} **(c)** 10000111_{2^*} **5.** R 2 D 2

6. (a) {1, 2, 4, 5, 6, 7, 8, 9} **(b)** {2, 7} **(c)** {0, 3} **(d)** 6 **(e)** 8 **(f)** {1, 6}

7.

	A	B	C	parity
1	1	1	1	1
2	1	0	1	0
3	0	1	1	1
parity	0	1	1	

8. (a) {C} {K} {CK} {MK} {CMK} {M} {CM} {MY} {CMYK} ∅ {CYK} {Y} {CY} {YK} {CMY} {MYK}

9. (a)

P	Q	P OR Q
1	1	1
1	0	1
0	1	1
0	0	0

(b)

P	Q	$P \wedge Q$	$\sim(P \wedge Q)$
1	1	1	0
1	0	0	1
0	1	0	1
0	0	0	1

(c)

P	Q	$\sim Q$	$P \vee \sim Q$	$(P \vee \sim Q) \vee Q$
1	1	0	1	1
1	0	1	1	1
0	1	0	0	1
0	0	1	1	1

10. (a) 2^{1999} **(b)** 2^{-31} **(c)** $\frac{3^8}{2} = 3280.5$

(d) 50 **11.**

$A' \cap B'$

A B

$A' \cup B$

A B

12.

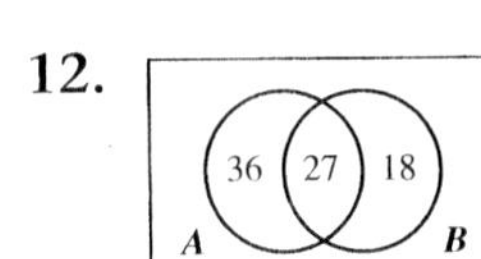

81

13.

A 22 20 33 B
11
1 4
0
C

11

14. many answers **15. (a)** {A, E, I, O, U} **(b)** {0, 1} **(c)** ∅ **(d)** {Jan., Feb., Dec.} **(e)** {4, 6, ..., 48, 50} **16. (a)** 110_2, 6_{16}, 6_8 **(b)** 11011_2, $1B_{16}$, 33_8 **(c)** 1011100_2, 134_8, $5C_{16}$ **(d)** 100100001_2, 441_8, 121_{16}
17. (a) $32y^{22}$ **(b)** x^8 **(c)** 2^{19} **18. (a)** 10.79 **(b)** 2.236 **(c)** 0.5236 **19. (a)** 0.00000060913 **(b)** 502,290,000,000 **20.** 5.89×10^8
21. (a) 10,485,760 **(b)** 5,734,400 **22.** $\bar{x} = 33$, AE = 2, RE = 5.7%
23. (a) 4 **(b)** 8 **24.** 12,710,700 min sleep; 489,435 hrs awake **25. (a)** 1010_2 **(b)** 11_2

Chapter 5

Section 5.1 Odd-Numbered Exercise Answers

1. (a) Complement Property of Addition **(b)** Distributive Property of Multiplication over Addition **(c)** Commutative Property of Addition **(d)** Associative Property of Multiplication
3. (a) Complement Property of Meet **(b)** Distributive Property of Meet over Join **(c)** Commutative Property of Join **5. (a)** $2x + 14y$ **(b)** $x^2 - y^2$ **7.** B **9.** D
11. B **13.** C **15.** C **17.** F

19. Use Boolean Algebra to reduce the expression:

$$x \vee (z \wedge x')$$
$$= (x \vee z) \wedge (x \vee x')$$
$$= (x \vee z) \wedge 1$$
$$= (x \vee z)$$
$$= (1 \vee 1)$$
$$= 1$$

21. Use Boolean Algebra to reduce the expression:

$$xyz \vee xyz'$$
$$= xy(z \vee z')$$
$$= xy(1)$$
$$= xy$$

Section 5.2 Odd-Numbered Exercise Answers

1.

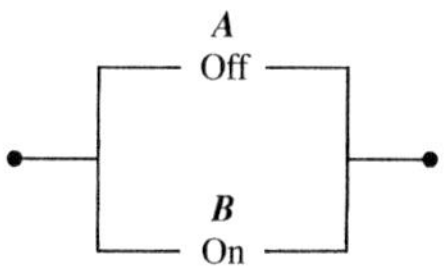

3.

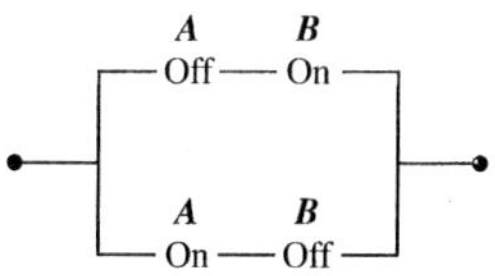

5. $AB'C$ **7.** $BC' \vee ABC'$

Section 5.3 Odd-Numbered Exercise Answers

1. (a) $A'B$ **(b)** $AB' \vee A'B \vee A'B'$ **3.**

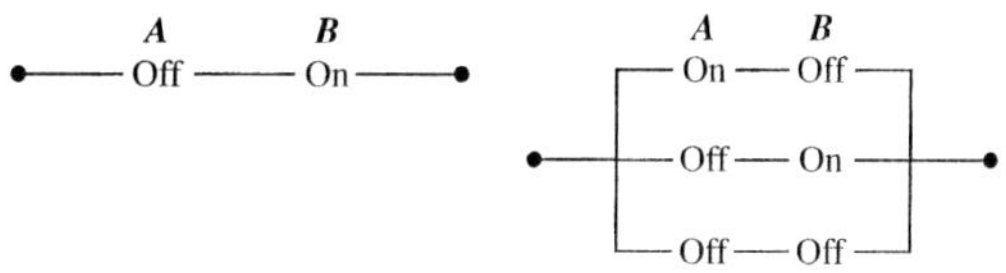

5. (a) $ABC \vee ABC' \vee A'BC \vee A'BC'$

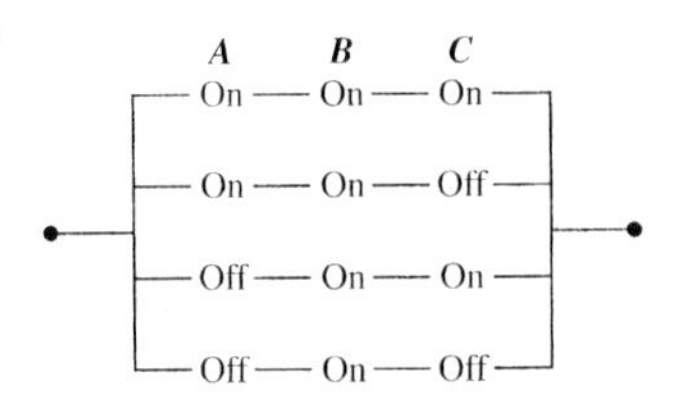

(b) $ABC \vee AB'C \vee A'BC \vee A'BC' \vee A'B'C'$

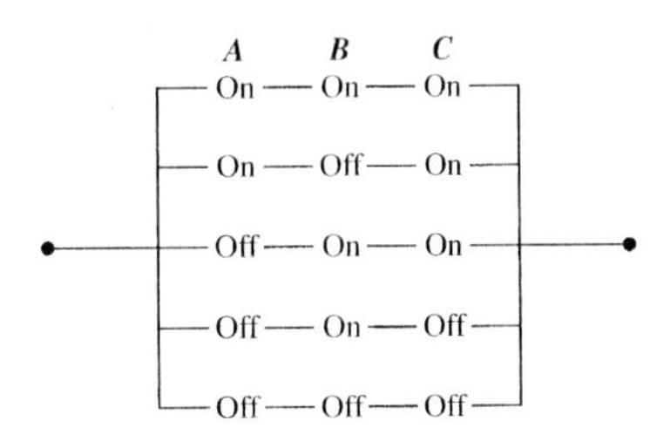

Section 5.4 Odd-Numbered Exercise Answers

1.

3.

5. $A'C \vee A'B$ 7. $ABC \vee (B' \vee C)$ 9.

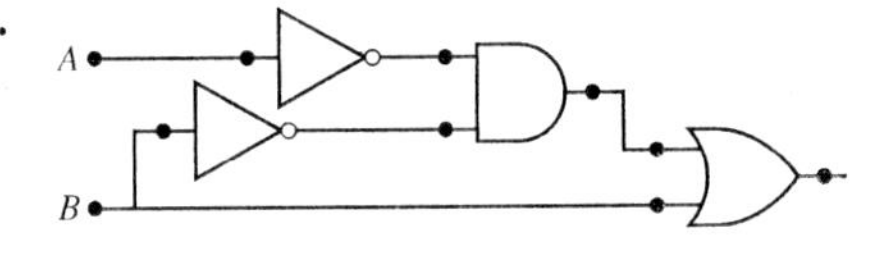

11. $ABC \vee ABC' \vee A'B'C$
$= AB(C \vee C') \vee A'B'C$
$= AB(1) \vee A'B'C$
$= AB \vee A'B'C$

13. $ABC \vee AB' \vee A'BC \vee ABC'$

Section 5.5 Odd-Numbered Exercise Answers

1. $A' \vee B'$

	A	A′
B		√
B′	√	√

3. $AC \vee B'C$

	AB	AB′	A′B′	A′B
C	√	√	√	
C′				

5. $BC \vee B'C'$

	AB	AB′	A′B′	A′B
C	√			√
C′		√	√	

7. $ACD \vee AB'C \vee A'B'D'$ or $ACD \vee B'CD' \vee A'B'D'$

	AB	AB′	A′B′	A′B
CD	√	√		
CD′		√	√	
C′D′			√	
C′D				

9. **(a)** $AB \vee AB' \vee A'B'$

(b)

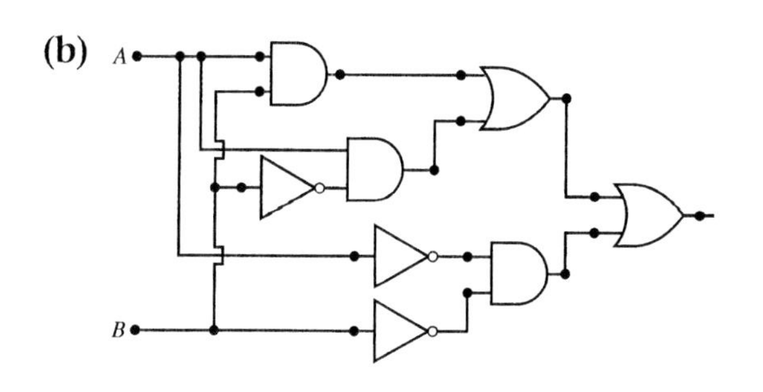

(c) $A \vee B'$

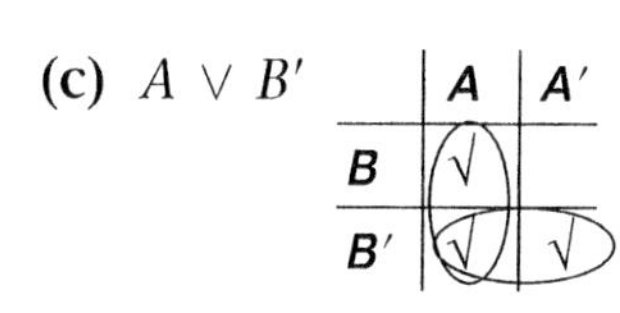

(d)

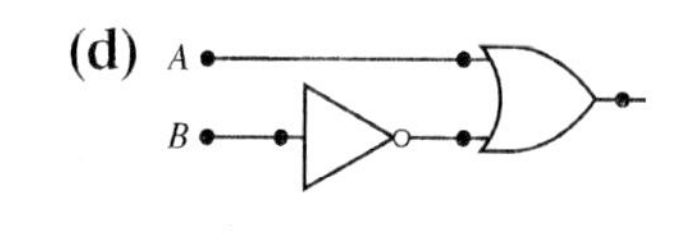

11. $AC' \vee C'E' \vee B'D'E' \vee ABE$ **13.** Circling a group of two allows you to eliminate 1 variable Circling a group of eight allows you to eliminate 3 variables.

Chapter 5 Glossary Quiz

parallel circuit D the commutative property of Boolean algebra A series circuit E the distributive property of Boolean algebra B gated circuits H logic circuit C Karnaugh map I disjunctive normal form (dnf) G switching circuit F

Chapter 5 Review Exercises

1. distributive **2.** comm. **3.** comm. **4.** assoc. **5.** inverse **6.** assoc. **7.** distributive **8.** comm. **9.** $5x + 15y$ **10.** $x^2 + 2xy + y^2$ **11.** $6x + 9y$ **12.** 0 **13.** x **14.** $x' \wedge y'$ **15.** 1 **16.** 1 **17.** 0 **18.** 1 **19.** **20.**

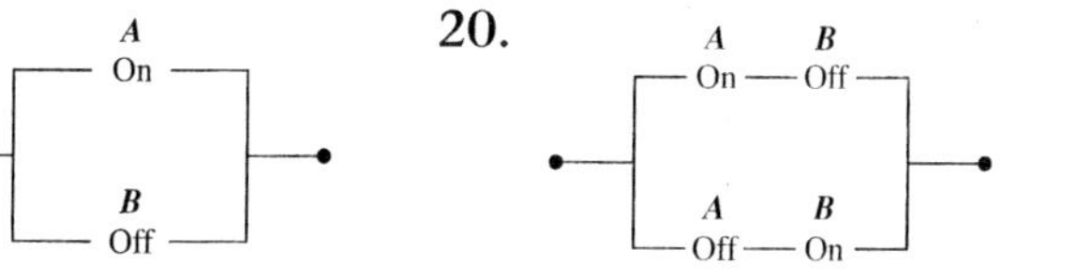

21.

B C
On — Off
A B C
Off — Off — On

22. $AB'C'$ **23.** $A'B'C' \vee A'B'$ **24.** $A'B'$ **25.** $AB \vee AB' \vee A'B'$

26. $AB' \vee A'B'$ or B' **27.**

A B
On — Off
A B
Off — Off

28. $AB'C \vee AB'C' \vee A'B'C$

29. $ABC' \vee A'BC \vee A'BC' \vee A'B'C'$ **30.** $AB'C \vee AB'C' \vee A'BC'$

$f(A, B, C)$
0
0
1
1
0
1
0
0

31.

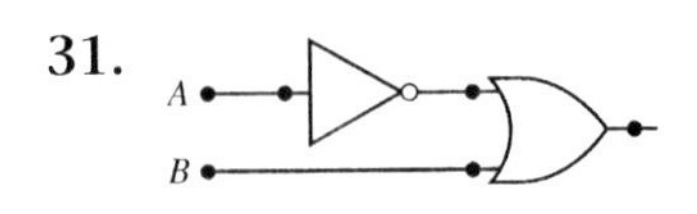

32.

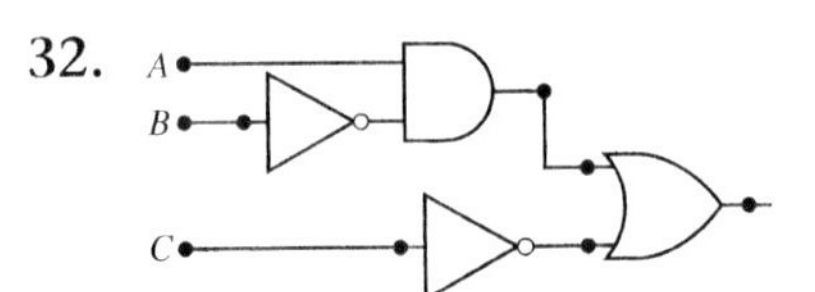

33.

34. $AC' \vee B'$ **35.** $ABC' \vee AB'$

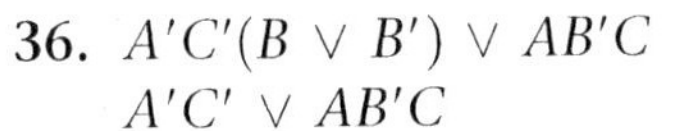

36. $A'C'(B \vee B') \vee AB'C$
$A'C' \vee AB'C$

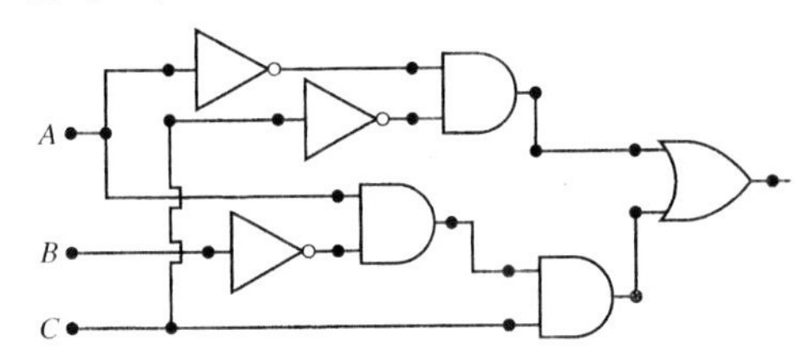

37.

	A	**A′**
B	1	1
B′		1

$A' \vee B$

38.

	AB	**AB′**	**A′B′**	**A′B**
C	1		1	
C′	1		1	

$AB \vee A'B'$

39. $A'B \vee BC \vee AB'C'$

40. $A'D' \vee AB'C \vee ABC'D$

41. $B'C'D' \vee A'BCDE \vee C'D'E' \vee ABC'E'$ **42. (a)** $AB \vee A'B \vee A'B'$

(b)

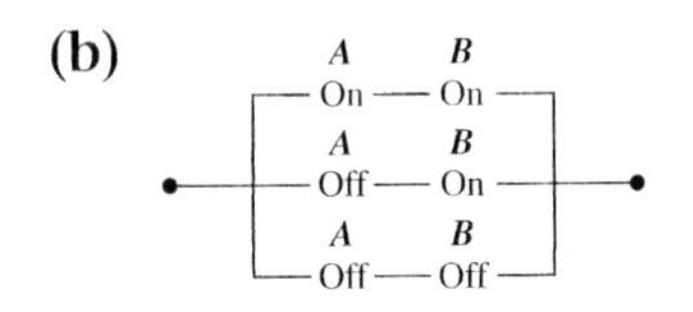

(c)

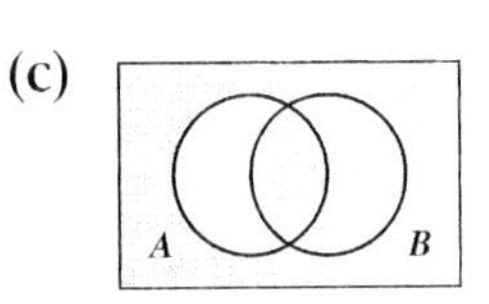

(d)

	A	**A′**
B	1	1
B′		1

$A' \vee B$

(e)

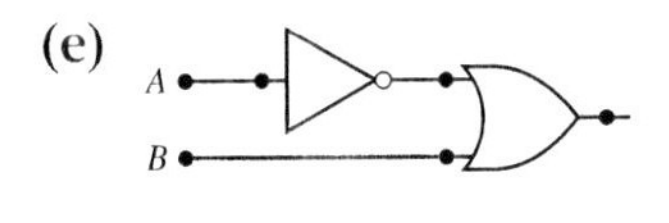

43. (a) $ABC \vee AB'C \vee AB'C' \vee A'B'C \vee A'B'C'$

(b)

(c)

	AB	**AB′**	**A′B′**	**A′B**
C	1	1	1	
C′		1	1	

$B' \vee AC$

(d)

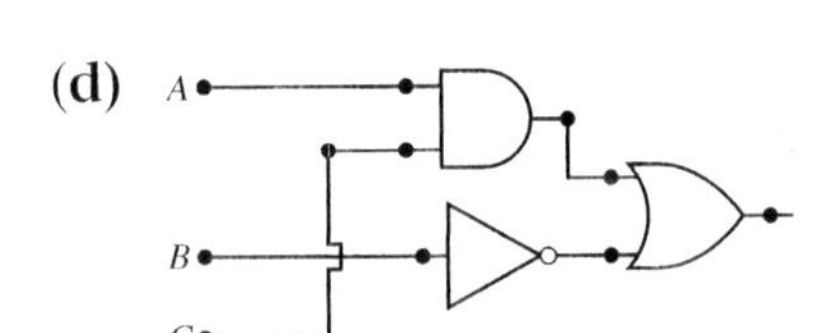

Chapter 5 Self-Test

1. (a) comm. **(b)** assoc. **(c)** dist. **(d)** comm. **2. (a)** assoc. **(b)** inverse **(c)** dis. **(d)** comm. **3. (a)** $28x + 7y$ **(b)** $x^2 - 16$ **(c)** $8x + 3y$ **4. (a)** x **(b)** $x \wedge y$ **(c)** 1 **(d)** 0 **5. (a)** $x \wedge y = 1$ **(b)** $y' = 0$ **6. (a)**

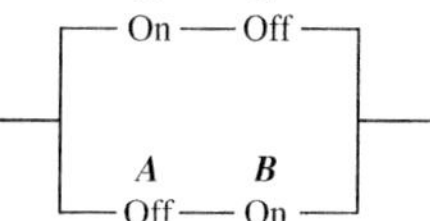

(b)

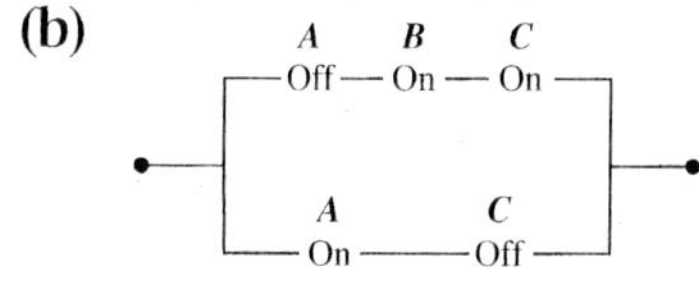

7. (a) $A'B \vee A'B'$

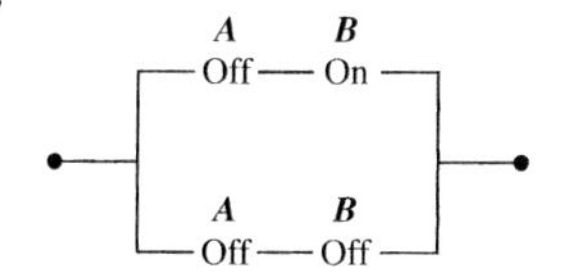

(b) $AB'C' \vee A'BC' \vee A'B'C$

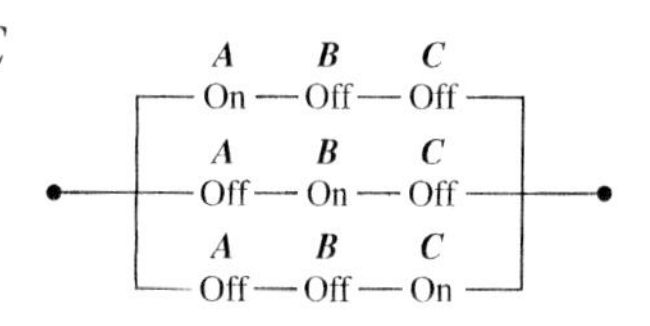

8. $ABC' \vee AB'C' \vee A'BC$

A	*B*	*C*	*f*(A, B, C)
1	1	1	0
1	1	0	1
1	0	1	0
1	0	0	1
0	1	1	1
0	1	0	0
0	0	1	0
0	0	0	0

9. (a)

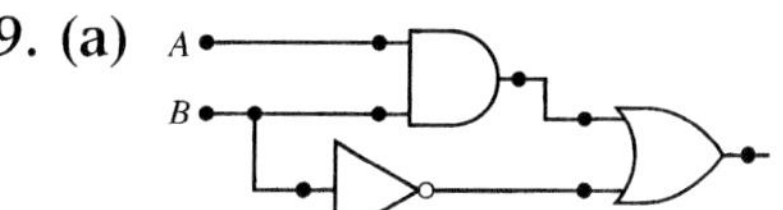

(b)

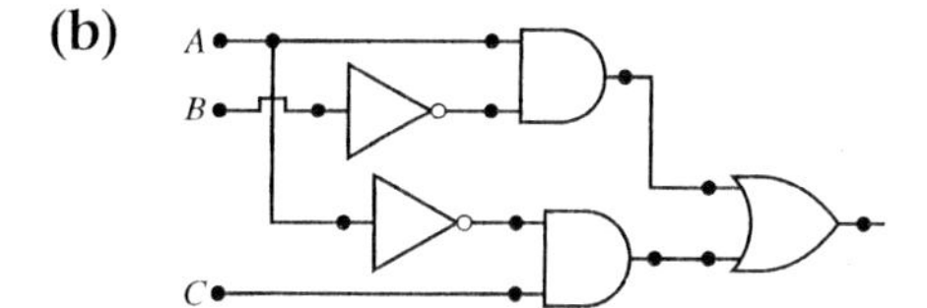

10. $AB' \vee BC'$

11.

	AB	*AB'*	*A'B'*	*A'B*
C		1	1	1
C'				1

$B'C \vee A'B$

12. $C'D' \vee ABD \vee A'B'D'$

13. (a) $ABC' \vee AB'C' \vee A'BC \vee A'BC'$ **(b)**

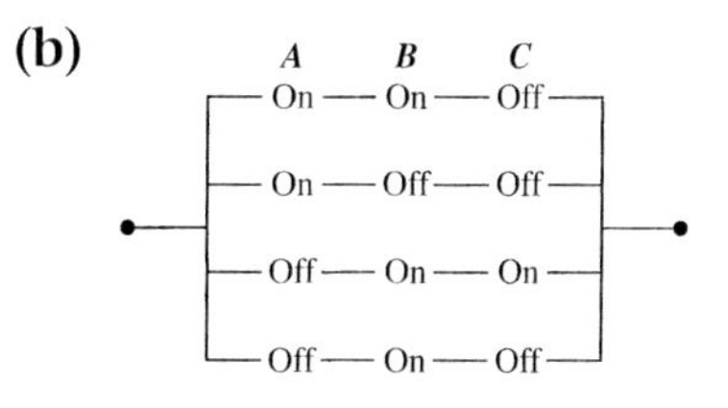

(c)

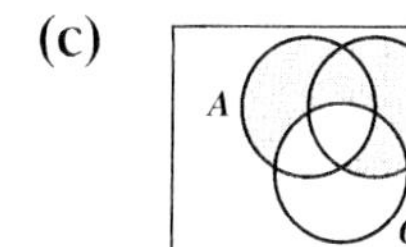

(d)

	AB	*AB'*	*A'B'*	*A'B*
C				1
C'	1	1		1

$A'B \vee AC'$

(e)

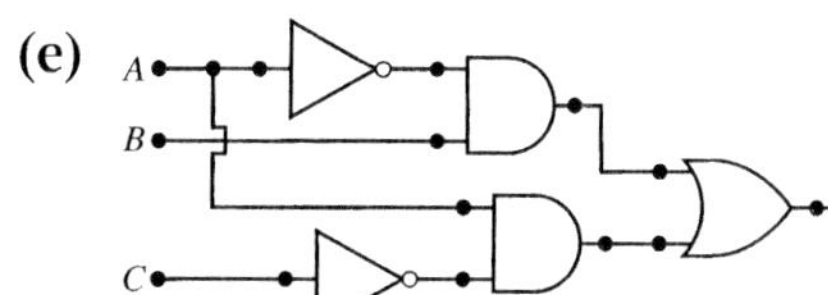

Chapter 6

Section 6.1 Odd-Numbered Exercise Answers

1.

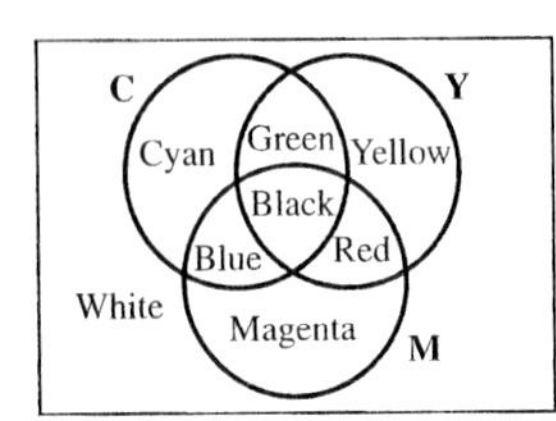

3.

COLOR	CYAN%	MAGENTA%	YELLOW%
BLACK	100	100	100
YELLOW	0	0	100
MAGENTA	0	100	0
CYAN	100	0	0
BLUE	100	100	0
RED	0	100	100
GREEN	100	0	100
WHITE	0	0	0

5.

COLOR	RED	GREEN	BLUE
WHITE	100	100	100
YELLOW	100	100	0
MAGENTA	100	0	100
RED	100	0	0
CYAN	0	100	100
GREEN	0	100	0
BLUE	0	0	100
BLACK	0	0	0

7. The red and blue remain at 0%, but the green would be less than 100%.

Section 6.2 Odd-Numbered Exercise Answers

1.

COLOR	RED	GREEN	BLUE
BLACK	00	00	00
BLUE	00	00	FF
GREEN	00	FF	00
CYAN	00	FF	FF
RED	FF	00	00
MAGENTA	FF	00	FF
YELLOW	FF	FF	00
WHITE	FF	FF	FF

3. 00FFFF **5.** Yellow **7.** 400080 **9.** FF4000 **11.** 0% red, 100% green, and 100% blue is cyan
13. 6% red, 6% green, and 6% blue is less intense gray **15.** $2^{16} = 65{,}536$

Section 6.3 Odd-Numbered Exercise Answers

1. A is (1, 8) and B is (−8, −1) **3.** A (8, 6) B (6, 8)

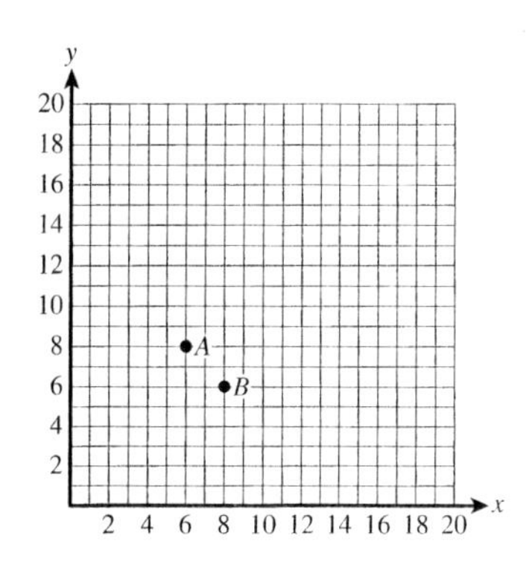

5. A is (1500, 600) **7.** C is approximately (500, 750) **9.** A is approximately (256, 384)
11. A becomes (320, 512)

Section 6.4 Odd-Numbered Exercise Answers

1.	3.	5.	7.	9.
1	10	3	20	127
2	20	8	25	255
3	30	13	30	383
4	40	18	35	511
5	50	23	40	639
6	60	28		767
7	70	33		
	80			
	90			
	100			
	110			
	120			

11. A vertical line would start at the top of the monitor, $\frac{1}{4}$ of the way from the left hand side, and would go to the bottom of the screen. **13.** A point would appear to move horizontally from the middle of the right side to the middle of the left side. **15.** A line would appear starting at (16, 64) (near the left top) and would move to (256, 1024) (part way across on the bottom). **17.** A line would appear starting at (16, 16) (near the left top) and would move to (512, 1008) (part way across near the bottom). **19.** A point would appear to move on a horizontal line near the middle of the screen. It would begin as a very dark green point and become a bright green point by the time it reached the right side.

21.
```
For x =1 to 765
  RGBColor = "x/3, x/3, x/3"
  Plot Pixel (512, x)
Next x

For Yellow, use

For x =1 to 768
  RGBColor = "x/3, x/3, 0"
  Plot Pixel (512, x)
Next x
```

Chapter 6 Glossary Quiz

Cartesian plane C   CMYK A   RGB B

Chapter 6 Review Exercises

1.

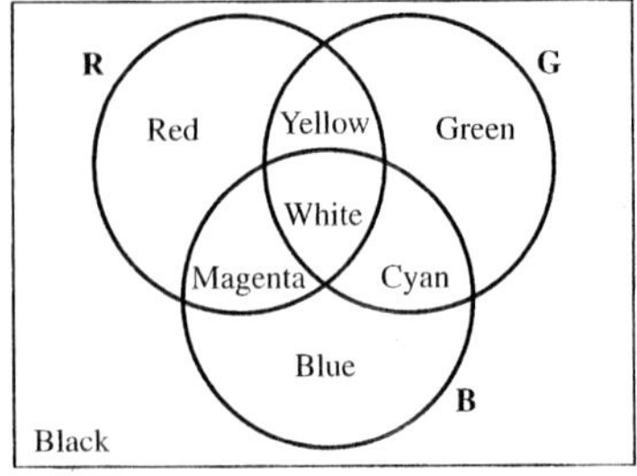

2. **(a)** Blue **(b)** Yellow **(c)** Black **(d)** White

3.

COLOR	RED	GREEN	BLUE
WHITE	100	100	100
YELLOW	100	100	0
RED	100	0	0
BLUE	0	0	100
MAGENTA	100	0	100
BLACK	0	0	0
CYAN	0	100	100
GREEN	0	100	0

4. Green **5.** FFFF00 **6.** 000080

7. R–0%, G–100%, B–0% **8.** R–50%, G–0%, B–100% **9.** *B* (–7, 6), *A* (6, 6)

10.

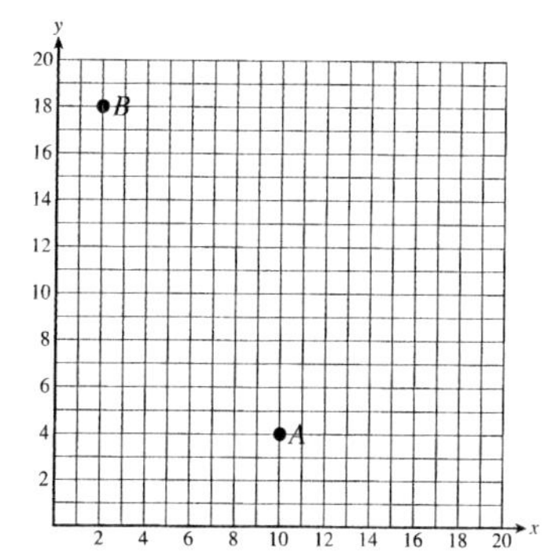

11. (1750, 500) **12.** (4000, 850) **13.** (500, 750)

14. (2000, 1500) **15.** (128, 256) **16.** (512, 576) **17.** (160, 341)

18. 0
1
2
3
4
5

19. 0
5
10
⋮
50

20. 4
5
6
7
8

21. 5
13
21
29
37
45
53
61
69
77
85
93

23. vertical points up the middle of screen

24. diagonal line down and to the right

25. vertical green line in middle of screen

26. horizontal green to red line, makes it a dot instead of a line

Chapter 6 Self-Test

1.

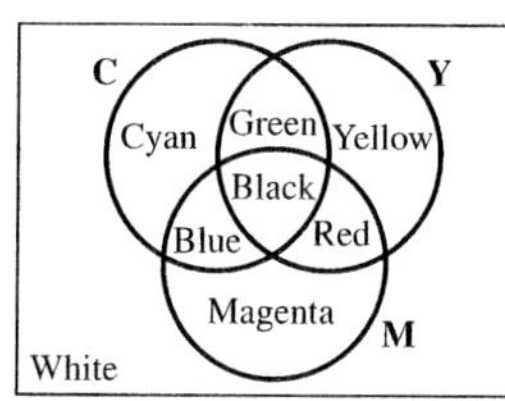

2. **(a)** Yellow **(b)** Blue **(c)** White **(d)** Black

3.

COLOR	RED	GREEN	BLUE
WHITE	100	100	100
BLUE	0	0	100
BLACK	0	0	0
CYAN	0	100	100
YELLOW	100	100	0
MAGENTA	100	0	100
GREEN	0	100	0
RED	100	0	0

4. Yellow **5.** 00FFFF **6.** FF8000

7. R–100%, G–0%, B–100% **8.** R—0%, G—50%, B—100% **9.** *B* (9, 9), *A* (5, –5)

10.

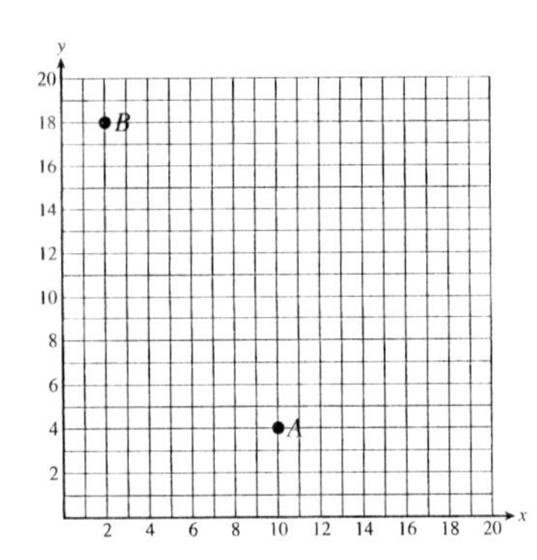

11. (9, 9) **12.** (4, −5) **13.** (−5, 4) **14.** (−8, −6)

15. (512, 192) **16.** (128, 96) **17.** (640, 256)

18. 1, 2, 3, 4, 5, 6, 7

19. 0, 25, 50, ⋮, 250

20. 3, 4, 5, 6, 7, 8, 9

21. 31, 46, 61, 76, 91, 106, 121

22. vertical line in middle **23.** horizontal points in middle
24. diagonal points down and to right **25.** vertical magenta line
26. horizontal green to blue line

Chapter 6 Final Examination

1. (a) 1 **(b)** $8\frac{1}{8}$ **(c)** undef. **(d)** ∅ can't be evaluated **(e)** −3 **(f)** 2.83 **(g)** 4.8×10^4 **(h)** 8 **2. (a)** 15 **(b)** −113 **(c)** −52 **3.** AE: 24; RE: 2.3%

4.

BINARY	DECIMAL	OCTAL	HEXADECIMAL
111101100111_2	3943	7547_8	$F67_{16}$
1001010111_2	599	1127_8	257_{16}
1000001.000001_2	65.015625	101.01_8	41.04_{16}
11.11110010_2	3.945318	3.744_8	$3.F2_{16}$

5. (a) 1965_{16} **(b)** ESE_{16} **(c)** $D40_{16}$

(d) 87_{20} **(e)** 1251_8 **(f)** -2 **6. (a)** {1, 2, 5, 6, 7, 8, 9} **(b)** {0, 1, 3, 4, 9} **(c)** {0, 3, 4} **(d)** 4 **(e)** 5 **7. (a)** OK **(b)** ERROR **(c)** OK **(d)** OK **(e)** ERROR **(f)** ERROR

8.

	AB	AB′	A′B′	A′B
CD	1	1	1	1
CD′				
C′D′	1	1	1	1
C′D				1

$f(A, B, C, D) = CD \vee C'D' \vee A'BC'$

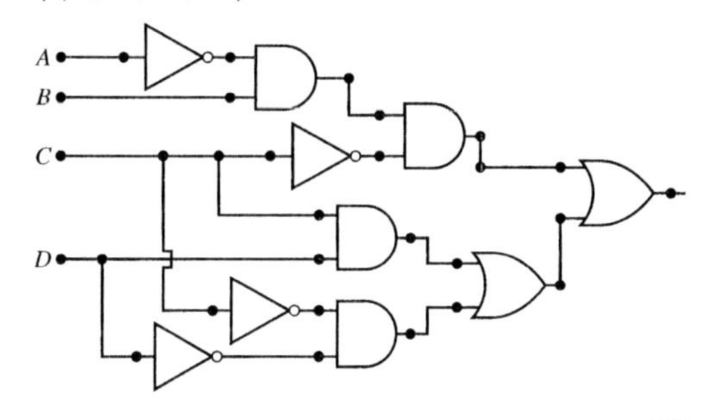

9. $BC'D \vee AB'C'D$ **10. (a)** green-yellow **(b)** 16^4

11.

$(\sim Q \wedge \sim P) \wedge (Q \vee P)$
0
0
0
0

12. 2^{319} **13.**

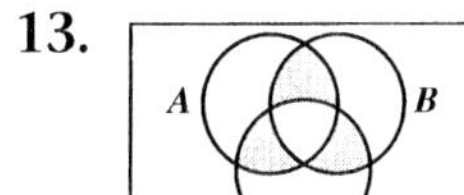

14. $ABC' \vee B'C$

15. (a) 01001011 **(b)** M

16.

1
2
3
4
5

17.

1	−3
2	−1
3	1
4	3
5	5

18. diagonal line up and to the right

19. diagonal points down and to the right **21.** horizontal line from black to white
22. diagonal line from cyan to yellow

Glossary

$(A \cup B)' = A' \cup B'$ One form of De Morgan's law.

absolute error The (absolute value of the) difference between a measurement and the exact value.

accuracy The degree of correctness of a measurement.

algorithm A series of steps that solves a problem.

ASCII An acronym for the American Standard Code for Information Interchange.

base The a in the expression a^n

binary code A system using combinations of two symbols to represent letters or numbers.

binary representation A value written in base two.

bit A binary digit.

byte A collection of eight consecutive bits.

cardinality The number of elements in a set.

Cartesian plane Two scaled axes (the x-axis and the y-axis) that are perpendicular.

the commutative property of Boolean algebra $x \vee y = y \vee x$

complement The set of elements that are in the universal set that are not in a specified set.

CMYK A color scheme (cyan, magenta, yellow, black) used by many inkjet printers—it is called subtractive color, since the pigments absorb certain wavelengths of light to produce color.

decimal representation A value written in base ten.

digit A symbol used to represent a whole number.

disjunctive normal form (dnf) A Boolean expression that consists of a series of terms in which every variable appears in every term.

the distributive property of Boolean algebra

$$x \wedge (y \vee z) = (x \wedge y) \vee (x \wedge z)$$

element An object contained in a particular set.

empty set A set that contains no elements.

exponent A shorthand version of repeated multiplication.

gated circuits A circuit that uses AND gate, OR gates, and the inverter.

gigabyte 2^{30} bytes

hexadecimal number A number written in base sixteen.

intersection The set consisting of all elements that belong in both of two sets.

Karnaugh map A visual representation of a Boolean expression that is written in dnf.

kilobyte 2^{10} bytes

logic The mathematics behind decision.

logic circuit A set of symbols that relate to a Boolean expression.

mantissa The a in $a \times 10^n$, where $1 \leq a < 10$

MB An abbreviation for megabyte.

Mb An abbreviation for megabit.

mean The sum of a set of values divided by the number of data values.

megabyte 2^{20} bytes

micro A prefix meaning one millionth.

milli A prefix meaning one thousandth.

monochrome Having only one color.

nano A prefix meaning one billionth.

octal number A number written in base eight.

overflow error An error that occurs whenever the information is too large for the storage that is allotted.

parallel circuit A circuit related to the Boolean expression $A \vee B$.

parity bit The simplest method of error detection.

pico A prefix meaning one trillionth.

pixel picture elements

proposition A declarative statement that is either true or false.

relative error The ratio of the absolute error over the exact measurement (usually expressed as a percentage).

RGB An additive color (Red, Green, Blue) scheme used by most computer monitors.

scientific notation The form $a \times 10^n$, where $1 \leq a < 10$

series circuit A circuit related to the Boolean expression $A \wedge B$.

set A collection of distinct objects.

significant digits The digits that define a numerical value without regard to the units.

subset A set of elements that are all in a specified set.

switching circuit A Boolean expression represented by a set of switches.

the *exponent*, or *power* The n in the expression a^n.

tribble A group of three bits.

two's complement notation A form used to allow binary representation of negative numbers.

unary operator A sign that changes the value of a number or variable.

union The set consisting of all elements that belong to either of two sets.

units The labels after a number.

universal set The collection of all elements under consideration.

Venn diagram A rectangle that represents the universal set together with inner circles, each representing a specific subset of the universal set.

Index